Mechanics of Residual Soils

Second Edition

Mechanics of Residual Soils

Second Edition

Editors:

G.E. Blight

Department of Civil Engineering, Witwatersrand University, Johannesburg, South Africa

E.C. Leong

Nanyang Technological University, Singapore

CRC Press
Taylor & Francis Group
Boca Raton   London   New York   Leiden

CRC Press is an imprint of the
Taylor & Francis Group, an **informa** business

A BALKEMA BOOK

CRC Press/Balkema is an imprint of the Taylor & Francis Group,
an informa business

© 2012 Taylor & Francis Group, London, UK

Typeset by MPS Limited, Chennai, India
Printed and bound in Great Britain by TJ International Ltd, Padstow, Cornwall

Library of Congress Cataloging-in-Publication Data

Mechanics of residual soils / editors, G.E. Blight, E.C. Leong. — 2nd ed.
    p. cm.
  "A Balkema book."
  Includes bibliographical references and index.
  ISBN 978-0-415-62120-5 (hardback : alk. paper)
  1. Soil mechanics. 2. Residual materials (Geology) 3. Engineering geology.
4. Soils—Tropics. I. Blight, G. E. II. Leong, E. C.
  TA709.5. M4183 2012
  624.1'5136—dc23

                                                          2012006797

Published by: CRC Press/Balkema
               P.O. Box 447, 2300 AK Leiden, The Netherlands
               e-mail: Pub.NL@taylorandfrancis.com
               www.crcpress.com – www.taylorandfrancis.com

ISBN: 978-0-415-62120-5 (Hbk)
ISBN: 978-0-203-11470-4 (eBook)

10 0624131 6

Contents

Preface to second edition

The first edition of "Mechanics of Residual Soils" was published in 1997 by A.A. Balkema, whose successors own the copyright. The book was compiled by members of Technical Committee 25 of the International Society for Soil Mechanics and Foundation Engineering and edited and co-authored by G.E. Blight, then Chairman of Committee 25.

When, 15 years later, the need for a second and updated edition became apparent, it was not considered feasible to assemble the original team of authors who had, by then, moved on to other interests, occupations and locations. As it is generally accepted that subsequent editions of a book may have different author-editor teams, it was decided that the chapters would continue to be presented under the names of their original authors, but the content would be expanded and updated. To ensure a fresh approach to the content of the book the editor of the original book, G.E. Blight, invited E.C. Leong to co-author/edit the second edition, which he has done very successfully.

The main changes and extensions of the content are as follows.

All of the diagrams have been re-drawn, with several errors in the original drafting being discovered and corrected in the process. Chapter 1, on the origin and formation of soils, is now not only illustrated by diagrams, but also by 23 full-colour photographs that are grouped with a further 9 colour photographs near the end of the book. Several new topics have been introduced while original topics, previously only mentioned in passing, have been amplified. Additions and amplifications include:

- a substantial section on pedocretes,
- consideration of the weathering of limestones and dolomites,
- dispersive soils and their adverse effects, and
- an introduction to the mechanics of unsaturated soils.

To eliminate repetition that occurred in the original Chapters 2, 3 and 5, these have been consolidated into a single Chapter 2 on microstructure, mineralogy and classification of residual soils. The re-arranged Chapter 2 appears under the names of the original five authors.

Chapter 3, on soil profile description, remains almost the same, but is now illustrated by a colour reproduction of the Burland colour chart.

Chapter 4, on compaction, has been extended to include aspects of the water-permeability of compacted soils and the design of a compacted soil to meet a required maximum water-permeability. It also considers the mechanics of unsaturated

compacted soils during and after construction, together with an illustrative case history on measurements of pore air pressure in compacted embankments.

The original chapter on permeability (Chapter 7, now Chapter 5) has been extended to consider permeability to both water and air flow, and the measurement of permeability to air flow.

Chapter 6 (originally Chapter 8) on compressibility and settlement, has been extended to include heave in expansive soils, as well as collapse settlement. This chapter now includes a consideration of the mechanics of compression, heave and collapse of unsaturated soils. The authorship continues to honour the late Dick Barksdale who wrote the major part of the original chapter. The material originally appearing as an appendix is now integrated into the main body of the chapter.

Chapter 7, on shear strength, continues virtually unchanged from the original Chapter 9, with the addition of two diagrams and with an expanded explanation of the mechanics of shearing in unsaturated soil.

The final Chapter 8 (formerly Chapter 10) has been expanded to include case histories of settlement and heave of buildings on unsaturated residual soils, settlement of an earth dam embankment constructed of compacted residual soil, amelioration of the effects of heave by in situ flooding, as well as piping failures of hydraulic structures incorporating compacted residual soil. Two additional cases of slope failures in residual soils have also been included.

Overall, the content of Edition 2 has been increased by 50 percent, compared with Edition 1.

Geoff Blight,
Johannesburg,
2012

Eng Choon Leong,
Singapore,
2012

Acknowledgements

We, the editors/authors of the second edition wish to salute the co-authors of the original "Mechanics of Residual Soils": R.D. Barksdale, R.K. Brenner, A.B. Fourie, V.K. Garga, T.Y. Irfan, J.B. Queiroz de Carvalho, J.V. Simmons and L.D. Wesley. Without their pioneering efforts, there would have been neither a first nor a second edition of this book.

We also thank all the geotechnical engineers and engineering geologists who have continued, in the 15 years since the first edition, to investigate the properties of residual soils and to pool their knowledge by means of publications and at conferences.

Geoff Blight has had the good fortune to work and interact with several researchers in the field of residual soils who, at one time or other lived, worked in or visited South Africa. He wishes to acknowledge their contributions to the understanding of residual soils and his growth of knowledge as a result. In alphabetical order, they include:

Gordon Aitchison, George Annandale, Dick Barksdale, Fred Bell, Ian Brackley, Ted Brand, Peter Brenner, Tony Brink, Richard Brummer, John Burland, Dick Chandler, Lou Collins, George Dehlen, George Donaldson, Heinrich Elges, Jim Falla, Andy Fourie, Vinod Garga, Malcolm Jaros, Jere Jennings, Gary Jones, Ken Knight, Peter Lumb, Ken Lyell, Dirk van der Merwe, Jim Mitchell, John Nelson, Frank Netterberg, Philip Paige-Green, Tim Partridge, Mike Pavlakis, Pierre Pellissier, Terry Pidgeon, Harianto Rahardjo, Brian Richards, Eben Rust, Tony van Schalkwyk, Denys Schreiner, Ken Schwartz, George Sowers, Brian Tromp, Kallie Strydom, Ian Watt, Fritz Wagener, Harold Weber, Laurie Wesley, Ant de Wet, Ben Wiid and Tony Williams.

Geoff Blight thanks them, one and all, for their contributions to the mechanics of residual soils.

E.C. Leong thanks Geoff Blight for the opportunity to be part of this second edition. It was a privilege and joy to work with such an experienced and dedicated researcher.

Cathy Snow prepared all of the 242 diagrams and patiently made the many changes, additions and deletions that were part of the process.

We both acknowledge and are extremely grateful for our families' unwavering support.

Geoff particularly thanks his wife of 54 years, Rhona, for patiently typing yet another major manuscript.

Unless otherwise acknowledged, all of the photographs were taken by Geoff Blight.

Author biographies

Geoffrey Blight completed his Bachelor and Master degrees at the University of the Witwatersrand, Johannesburg, and his PhD at Imperial College, London. There he carried out some of the earliest research on the mechanics of unsaturated soils, under the supervision of the legendary Alan Bishop. His early work, published jointly with Bishop in 1960, 1961 and 1963, provided data that is still being used by new generations of researchers on unsaturated soil behaviour. He soon became interested in residual soils, publishing his first work on unsaturated residual soils in 1963. He was a member of the International Society for Soil Mechanics and Foundation Engineering's Technical Committee on the Properties of Tropical and Residual Soils from 1982 to 1997 and served as Chairman from 1994 to 1997. He edited and co-authored the first edition of "Mechanics of Residual Soils", which was produced during his Chairmanship.

He has also authored or co-authored the books: "Assessing loads on silos and other bulk storage structures" (2006), "Geotechnical engineering for mine waste storage facilities" (2010) and "Alkali-aggregate reaction and structural damage to concrete" (2011), all published by CRC Press/Balkema.

Eng-Choon Leong is an Associate Professor at the School of Civil Engineering, Nanyang Technological University, Singapore. He obtained his bachelors and masters degrees at the National University of Singapore and a PhD at the University of Western Australia under the supervision of Mark Randolph. In Nanyang Technological University, he developed research interests in unsaturated soils and residual soils, collaborating with Harianto Rahardjo. His other research interests include soil dynamics, foundation engineering and numerical modeling. He has published widely in international journals and conferences.

He has extensive experience in laboratory and field testing of saturated and unsaturated soils and *in-situ* monitoring of residual soil slopes. He has also developed a number of specialized laboratory testing apparatuses and data acquisition systems for field applications. He is also active in consultancy and national technical committees on standards. He has received several awards for his contributions to accreditation and SPRING, the national standards body of Singapore.

List of abbreviations and mathematical symbols

(In many cases, the same symbol has more than one meaning. The reader will be able to tell from the context which meaning is appropriate. [ ] indicates SI units usually used.)

ROMAN LETTERS

A	Skempton-Bishop pore pressure ratio $\Delta u / \Delta(\sigma_1 - \sigma_3)$	[dimensionless]
A	cross-sectional area	[m^2, mm^2]
A	$(\tan\theta - \sec\theta + 1)$	[dimensionless]
AASHTO	American Association of State Highway and Transportation Officers	
Al_2O_3	alumina	
AMSL	Above Mean Sea Level	
B	Skempton-Bishop pore pressure ratio $\Delta u / \Delta\sigma_3$	[dimensionless]
c	cohesion in total stress terms	[kPa]
c'	cohesion in effective stress terms	
c_v	coefficient of consolidation	[cm^2/s, m^2/y]
C	compressibility	[kPa^{-1}]
C_c, C_r	compression and rebound indices, respectively	[dimensionless]
$CaCO_3$	calcium carbonate	
$Ca(HCO_3)_2$	calcium bicarbonate	
CD	consolidated, drained shear test	
CEC	Cation Exchange Capacity	[mol/L]
CPT	cone penetration test	
CU	consolidated, undrained shear test	
d	depth	[mm, m]
D	depth	[m]
D_c	diffusion coefficient	[s]
D_e	Entrance diameter of sampling tube	[mm]
D_i	Internal diameter	[mm]
D_w	Diameter of wall	[mm]
e	void ratio	[dimensionless]
E_I	elastic modulus	[kPa, MPa, GPa]
E_h	elastic modulus measured in horizontal direction	
Ej	pan evaporation in January	[mm]

ESP	Exchangeable Sodium Percentage	[%]
F	shape factor (Figure 5.13)	[m]
10% FACT	10% Fines Aggregate Crushing Test	[kN]
G, G_s	particle specific gravity, or relative unit weight	[dimensionless]
h	height of suspended water column	[m, mm]
H	Henry's coefficient of solubility of air in water, either in volume terms $[(m^3 kPa)^{-1}]$ or mass terms	[kPa]
H_2CO_3	carbonic acid	
i	flow gradient	[dimensionless]
I_f	influence factor	[dimensionless]
k	coefficient of permeability	[m/s, cm/s, m/y]
k_h, k_v	coefficients of permeability for horizontal or vertical flow, respectively	
K	stress ratio	[dimensionless]
K_A, K_o, K_p	active, at rest, passive principal stress ratios	
LL	liquid limit	[%]
LVDT	Linear Voltage Differential Transformer	
m	mass of air stored in soil	[kg]
m_a	molecular mass of air	[kg/mol]
m_s, m_w	number of moles of solute or water (eqn 1.10)	[dimensionless], [kg/mol]
M	mass of air entering soil per unit time	[m³/s]
$MgCO_3$	magnesium carbonate	
$Mg(HCO_3)_2$	magnesium bicarbonate	
n	porosity	
n_a	air porosity	
n_w	water porosity	
N	Constant used in eqn. 4.3 (page 113)	[dimensionless]
N	Number of blows in Standard Penetration Test (SPT)	
N_c	bearing capacity factor	
$N = 12Ej/Pa$	Weinert's N	[dimensionless]
p, P	pore water suction $= (u_a - u_w)$	[kPa]
$p' = \frac{1}{2}(\sigma'_1 + \sigma'_3)$	mean effective stress	[kPa]
p''	pore water suction $= (u_a - u_w)$	
psd	particle size distribution	
Pa	annual rain precipitation	[mm]
P, Pav	Pressure, average pressure	[kPa] or load [kN]
PI	plasticity index $= LL - PL$ [%]	
PL	plastic limit	[%]
P_L	limit pressure in pressuremeter test	[kPa, MPa]
q	applied stress	[kPa]
q	quantity of flow	[m³/s]
$q' = \frac{1}{2}(\sigma_1 - \sigma_3)$	maximum shear stress	[kPa]
r	radius of liquid meniscus	[mm, μm]

R	universal gas constant	[J/Kmol or kPam^3/Kmol]
R$_s$	stress ratio caused by sampling (eqn. 6.1) page 151	
RH	relative humidity	[dimensionless]
s	pore water suction $= (u_a - u_w)$	[kPa]
s$_e$	air entry suction	
S	degree of saturation	[%, ratio]
S	vane shear strength	[kPa]
SAR	Sodium Adsorption Ratio	[dimensionless]
SEM	Scanning Electron Microscope	
SiO$_2$	silica	
SWCC	Suction Water Content Curve, Soil Water Characteristic Curve (same meaning)	
t, t$_f$	time, time to failure	[s, h, y]
T	basic time lag	[s, h, y]
T	surface tension of a liquid	[N/mm]
T	time factor in consolidation	[dimensionless]
T	Ton	[1 T = 10 kN]
T	Torque	[J]
TDS	Total Dissolved Solids	[mg/L]
TEM	Transmission Electron Microscope	
u	pore pressure	[kPa, MPa]
u$_a$	pore air pressure	
u$_w$	pore water pressure	
U	degree of consolidation	[%]
UU	unconsolidated, undrained shear test	
USCS	Unified Soil Classification System, or Casagrande classification	
v	flow velocity	[m/s, cm/s, m/y]
V	volume	[m^3]
VCL	Virgin Compression Line	
w	gravimetric water content	[%]
XRD	X-Ray Diffraction	

GREEK LETTERS

α	alpha	rheological factor appearing in eqn. 6.13, page 185 or angle defined in Figure 6.32	
γ	gamma	bulk or total unit weight	[kN/m^3]
γ_d		dry unit weight	
γ_{sat}		saturated unit weight	
δ	delta	displacement	[mm, m]
ε	epsilon	linear strain	[dimensionless]
ε_v		volumetric strain or vertical strain in eqn. 6.3 (page 156) & 6.3b (page 203)	
$\varepsilon_1, \varepsilon_2, \varepsilon_3$		principal strains	

θ	theta	angle of contact between water meniscus and soil grain (Figure 1.14a)	
θ		absolute temperature	[K]
$\lambda, \lambda_c, \lambda_d$	lamda	empirical shape factors used in calculating settlements	[dimensionless]
ν	nu	Poisson's ratio	
ρ	rho	bulk or total density or unit mass [kg/m^3]	
ρ_d		dry density	
ρ_{sat}		saturated density	
$\rho, \rho_0, \rho_\infty, \rho_t$		settlements at times zero, infinity, t	[mm, m]
σ	sigma	total direct stress	[kPa, MPa]
σ'		effective direct stress	
$\sigma_1, \sigma_2, \sigma_3$		principal stresses	
σ_g		intergranular stress	
τ, τ_f	tau	shear stress, shear strength	[kPa, MPa]
φ	phi	angle of shearing resistance i.t.o. total stresses [°deg. of arc]	
φ'		angle of shearing resistance i.t.o. effective stresses	
φ^b		Fredlund's angle by which cohesion increases with suction, $\tan \varphi^b = \chi \tan \varphi'$	
χ	chi	Bishop's effective stress parameter	[dimensionless]
χ_s, χ_m		Bishop's parameter appropriate to solute or matrix suction	

OTHER SYMBOLS

∂ partial differential operator

SI AND RELATED UNITS

kg	kilogram	
N	Newton	$[1\,N = 1\,kgm/s^2]$
J	Joule	$[1\,J = 1\,Nm]$
Pa	Pascal	$[1\,Pa = 1\,N/m^2 = 1\,kg/ms^2]$
bar		[1 bar or atmosphere $= 100\,kPa$]
s	second	
h	hour	
d	day	
y	year	

Note: the unit for electrical conductivity, millisiemens is denoted [m s]

Chapter 1

Origin and formation of residual soils

G.E. Blight

1.1 DEFINITIONS RELATING TO RESIDUAL SOILS

The definition of a residual soil varies from country to country, but a reasonably general definition would be:

> "A residual soil is a soil-like material derived from the *in situ* weathering and decomposition of rock or rock fragments which has not been transported from its original location".

In this definition, "rock" refers to continuous rock strata, and "rock fragments" to materials such as grains of aeolian sand and particles of volcanic ash. There may be a continuous gradation from the fresh, sound unweathered rock or rock fragments through weathered soft rock and hard soil or saprolite, which is recognizable as the product of decomposition of the parent rock, to highly weathered material containing secondary deposits of alumina, calcium, iron or silica salts that bears no obvious resemblance to the parent material.

The differences between transported soils, such as fluvial, lacustrine or delta deposits are illustrated by Figure 1.1 (developed from Wesley, 2010). Point A in Figure 1.1a represents the void condition in a transported soil shortly after deposition. V_v and V_s are the void and solid volumes, respectively. The void ratio $e = V_v/V_s$ is high and the effective overburden stress is low (Figure 1.1b). As the overburden increases with time (Figure 1.1b), the soil is compressed and at B, where the overburden is higher, the void ratio has been considerably reduced. At C in Figure 1.1c, the parent rock of the residual soil has a low void content and the voids may consist of small occluded spaces. As weathering of the rock and erosion of the surface progress, the solids volume decreases by leaching of soluble constituents and suffosion of small detached weathered solid particles. (i.e., removal of solids by piping erosion on a micro-scale). Simultaneously, (Figure 1.1d), the overburden is removed and the overburden stress reduces. This allows the weathering rock to expand, cracks to open and accelerates the weathering process. Once the weathered rock (now a residual soil) has developed a system of inter-connected voids (E), it will be considerably more permeable to penetration by air and water. If the climate has a seasonal water deficit (evaporation at the surface exceeds rainfall) soluble calcium, iron or silica salts will be drawn to the surface and will precipitate in the voids as crystalline silica, calcium

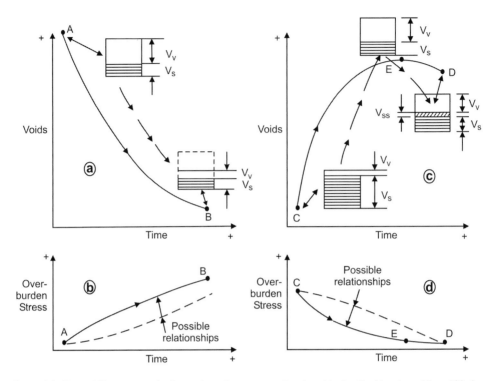

Figure 1.1 Basic differences in the formation of transported and residual soils. (developed from Wesley, 2010).

carbonate, calcium sulphate or iron sesquioxides. This will cause the void volume to decrease along E to D (in Figure 1.1c) by V_{ss} (volume of secondary solids), in many cases causing a cementing of the soil particles and an increase in strength.

Residual soils can have characteristics that are distinctly different from those of transported soils. For example, the conventional concept of a soil grain or a particle size is inapplicable to many residual soils. Particles of residual soil often consist of aggregates of particles or weathered crystals of mineral matter that break down and become progressively finer if the soil is manipulated or compacted (see section 2.1). What appears *in situ* to be a coarse sandy gravel may deteriorate to a fine sandy silt during excavation, mixing and compaction. The permeability of a transported soil can usually be related to its granulometry (e.g., by the well-known Hazen formula). For the reason just explained, this is usually not so with residual soils. The permeability of residual soils is usually governed by its micro- and macro-fabric and jointing and by relict or superimposed features such as slickensiding, termite or other bio-channels. Similar remarks apply to properties such as particle unit weight or specific gravity G_s. The value of G_s, for example will vary with the degree to which the sample being measured has been comminuted, smaller particles, being more solid, usually having higher values of G_s.

1.2   ROCK WEATHERING PROCESSES

Residual soils are formed by the *in situ* weathering of rocks, the three major agencies of weathering being physical, chemical and biological processes. In the weathering process the structure of the parent rock and rock minerals break down, releasing internal energy and forming substances having a lower internal energy which are therefore more stable. Physical processes (e.g., stress release by erosion, differential thermal strain and ice and salt crystallization pressures) comminute the rock, expose fresh surfaces to chemical attack and increase the permeability of the material to the percolation of chemically reactive fluids. Chemical processes, chiefly hydrolysis, cation exchange and oxidation, alter the original rock minerals to form more stable clay minerals (Mitchell, 1976). Biological weathering includes both physical action (e.g., splitting by root wedging) and chemical action (e.g., bacteriological oxidation, chelation and reduction of iron and sulphur compounds (e.g., Pings, 1968).

Most commonly, residual soils form from igneous or metamorphic parent rocks, but residual soils formed from sedimentary rocks are also widespread (e.g., Leong *et al.*, 2002) Chemical processes tend to predominate in the weathering of igneous rocks and metamorphic rocks originating from igneous rocks, whereas physical processes dominate the weathering of sedimentary and metamorphic rocks originating from sedimentary rocks. However, chemical and physical weathering are so closely interrelated that one process never proceeds without some contribution by the other.

Occasionally, residual soils may form by the *in situ* weathering of unconsolidated sediments. The commonest example of this is the loess or collapsing sand formed by the weathering of feldspars in deposits of windblown sand (Knight, 1961; Dudley, 1970; Schwartz & Yates, 1980). Shirasu, found in Japan (Yamanouchi & Haruyama, 1969) is an unconsolidated volcanic sediment, partly weathered *in situ*, whose engineering properties have much in common with loess. Due to the high intensity of weathering in humid tropical climates, fluvial deposits of gravels, sands and silts in stream channels, flood plains and deltas are often found to have weathered *in situ* (e.g., Li & Wong, 2001; Zhang *et al.*, 2007), transforming them from transported to residual soils. Volcanic ash, deposited aerially or transported by mud flows also weathers *in situ*. Ash and other clastic material that has weathered *in situ* is also regarded as a residual soil (e.g., Wesley, 2010).

Hydrolysis is considered to be the most important of the chemical weathering processes (Zaruba & Mencl, 1976). It occurs when a salt combines with water to form an acid and a base. In rock weathering, the salt is usually a silicate and the product of the reaction is a clay mineral. Oxidation is usually preceded by hydrolysis and affects rocks containing iron sulphates, carbonates and silicates. The products of oxidation usually have a larger specific volume than the parent minerals and thus expansion due to oxidation increases the void space and contributes to physical comminution of the rock (e.g., Mason 1949).

The breakdown of one clay mineral to form another can occur through the transfer of ions between percolating solutions and the original mineral. Cations such as sodium and calcium are the most readily exchangeable. Cation exchange does not alter the basic structure of the clay mineral, but the crystal interlayer spacing may change, thus converting, for example, an illite to a montmorillonite, or a sodium bentonite to a

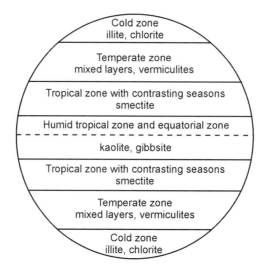

Figure 1.2 Suggested influence of global climate on clay mineral development. (Uehara, 1982).

calcium bentonite. Although the basic structure of the clay may not be changed, the physical properties may suffer extensive alteration.

Bacteria may play the role of catalysts in certain chemical reactions. For example, the oxidation of sulphide minerals may be enormously accelerated by the presence and activity of the bacteria thio-bacillus thio-oxidans and thio-bacillus ferro-oxidans.

Chelation is a process whereby lichens growing on rock surfaces promote the rate of hydrolysis. Jackson & Keller (1970) have shown that the depth of weathering of basalt surfaces is greater and the chemical alteration more extensive under a lichen cover than if lichen is not present.

1.3 THE EFFECTS OF CLIMATE

Climate exerts a considerable influence on the rate and extent of weathering (Weinert, 1964 and 1974; Morin & Ayetey, 1971). Physical weathering is more predominant in dry climates while the extent and rate of chemical weathering is largely controlled by the availability of moisture and by temperature. (Other things being equal, chemical reaction rates approximately double for each 10°C rise in average temperature).

According to Uehara (1982) the clay mineralogy of the soils of the world changes in a predictable way with distance from the equator, as indicated by Figure 1.2. This is a gross over-simplification because as shown by Figure 1.3, climates do not vary uniformly with distance from the equator, but are affected by topography, atmospheric upper air streams and ocean currents, etc. Nevertheless, Figure 1.2 gives a useful concept of the influence of climate on the products of weathering. According to Uehara, high temperatures and year-round rainfall near the equator, favour the formation of low activity kaolin and oxides (Skempton's Activity 0.3 to 0.5 where Skempton's

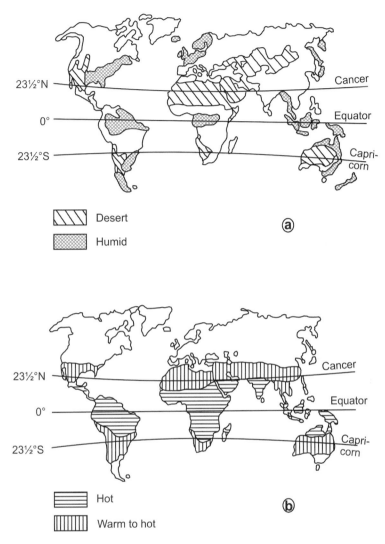

23½°N

0°

23½°S

Cancer

Equator

Capri-
corn

Desert

Humid

(a)

23½°N

0°

23½°S

Cancer

Equator

Capri-
corn

Hot

Warm to hot

(b)

Figure 1.3 Climatic zones of the world showing the distribution of dry and humid, and warm and cold climates.

Activity = Plasticity Index/% particles <2 microns). Rainfall decreases towards the limits of the tropics and high activity smectitic clays predominate (Skempton's Activity 1.5 to 7). The above concept has been presented far more elaborately by Strakhov (1967), whose diagram (Figure 1.4) summarizes the effects of global climate on rock weathering and the formation of the various weathering products – kaolinite, alumina, etc. The influence of temperature and moisture on rock weathering in South Africa has been correlated with Weinert's (1974) climatic index

$$N = \frac{12Ej}{Pa} \qquad (1.1)$$

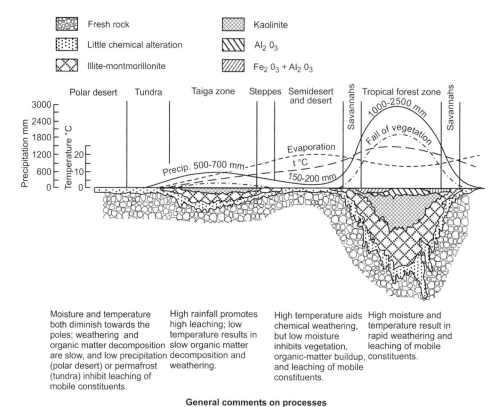

Legend:
- Fresh rock
- Little chemical alteration
- Illite-montmorillonite
- Kaolinite
- Al_2O_3
- $Fe_2O_3 + Al_2O_3$

| Moisture and temperature both diminish towards the poles; weathering and organic matter decomposition are slow, and low precipitation (polar desert) or permafrost (tundra) inhibit leaching of mobile constituents. | High rainfall promotes high leaching; low temperature results in slow organic matter decomposition and weathering. | High temperature aids chemical weathering, but low moisture inhibits vegetation, organic-matter buildup, and leaching of mobile constituents. | High moisture and temperature result in rapid weathering and leaching of mobile constituents. |

General comments on processes

Figure 1.4 Influence of global climate on depth of weathering and weathering products. (Strakhov, 1967).

Where Ej = Potential evaporation during January (i.e., A-pan evaporation), the hottest southern hemisphere month, and Pa is the annual rainfall, or precipitation, all in mm.

A value of N = 5 marks the transition from warm sub-humid conditions in which chemical weathering predominates, to hot semi-arid and arid conditions in which physical weathering is the more important mode. Where N is less than 5, considerable thicknesses of residual soil may occur, and the extent or intensity of weathering is greater. Where N exceeds 5, thicknesses of residual material are usually smaller and the degree of weathering is less. Figure 1.5 shows the sub-division of Southern Africa by Weinert's N = 5 contour. In the area to the east and south, chemical weathering predominates, while to the west and north, physical weathering is predominant. The map also indicates how climate and topography affect rock weathering. The warm Mozambique current flows southward down the east coast, which as a result has a warm humid sub-tropical climate. The cold Benguela current flows northward up the west coast, resulting in a cool desert to semi-desert climate to the west.

Climate has a further possible effect on the properties of tropical residual soils – that of unsaturation. Even in sub-humid tropical or sub-tropical areas, water tables are often deeper than 5 to 10 m and the effects of unsaturation, desiccation and seasonal or

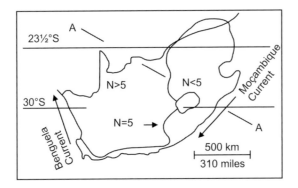

Quaternary
Post African 1 & 2
African
Gondwana & Post Gondwana

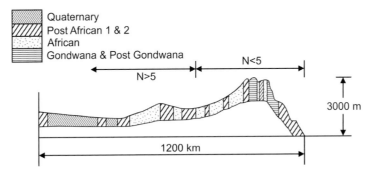

Geomorphic cycle	Cycle initiated x 10^6 years ago	Problem residual soils associated with remnant land surfaces.
Gondwana	190	Remnants cover relatively small surface areas. Erosion very active on steep slopes
Post Gondwana	135	
African	100	Collapsible weathered granites. Cavernous dolomites. Erosion-susceptible soft weathered shales, mudstones and sandstones.
Post African 1	20	Expansive clays. Erosion-susceptible soft weathered shales, mudstones and sandstones.
Post African 2	5	Expansive clays. Cavernous dolomites. Erosion-susceptible weathered shales and mudstones.
Quaternary	2	Collapsing sands. Weathered mudstones.

Figure 1.5 (Above): Subdivision of Southern Africa by Weinert's N = 5 contour. (Below): Transverse section AA showing sub-continental topographical relief and location of ancient erosion surfaces, as well as ages of erosion surfaces.

longer term re-wetting have to be taken into account in geotechnical design. There are many accounts of the effects of unsaturation on the behaviour of soils (e.g., Bishop and Blight, 1963; Blight, 1967; Fredlund & Morgenstern, 1977; Fredlund & Rahardjo, 1993). Various forms of the effective stress relationship for unsaturated soils have been proposed. Blight (2006) lists 10 different forms for the unsaturated soil effective stress equation that were published from 1958 to 2002 by various authors. It now appears accepted, however, that Bishop's equation, proposed in 1959 is the most general and useful form of the effective stress equation:

$$\sigma' = (\sigma - u_a) + \chi(u_a - u_w) \tag{1.2}$$

The effective stress σ' is governed by the stress difference $(\sigma - u_a)$ and the suction $(u_a - u_w)$ where σ is the applied total stress and u_a and u_w are, respectively the pore air and pore water pressures in the soil. χ is a factor (usually <1) that represents the effectiveness of $(u_a - u_w)$ in the unsaturated soil. In most, but not all practical situations, u_a equals the atmospheric pressure and can be put equal to zero. The conventional form of the effective stress equation

$$\sigma' = \sigma - u_w \tag{1.3}$$

can be used with little error for soils that are unsaturated, but reasonably close to saturation. In an unsaturated soil in which $u_a = 0$, u_w will be negative. Hence the pore water stress is added to the total stress to form the effective stress. If the water table is at a depth of 10 m, this adds a maximum of 100 kPa to the effective stress in the soil profile and about 50 kPa to the soil strength. It is this capillary strength that may be lost during periods of prolonged rainfall, or if a water table rises seasonally, of if conditions of unsaturation are changed by a change in land use. Effective stresses in unsaturated soils are discussed in more detail in a note at the end of this chapter.

1.4   THE EFFECTS OF TOPOGRAPHIC RELIEF

For a deep residual soil profile to develop, the rate at which weathering advances into the earth's crust must exceed the rate of removal of the products of weathering by erosion.

Topography controls the rate of weathering partly by determining the amount of available water and the rate at which it moves through the zone of weathering. Precipitation will run off hills and rises and accumulate in valleys and hollows. It also controls the effective age of the profile by controlling the rate of erosion of weathered material from the surface. Thus deeper residual profiles will generally be found in valleys and on gentle slopes rather than on high ground or steep slopes (e.g., Morin & Ayetey, 1971). At least part of the soil accumulated in valleys, and even on slopes, will consist of colluvium eroded from higher elevations and deposited lower down. The engineer must distinguish between these surface layers of transported soil because their behaviour may be quite different to that of the underlying residual soil.

It thus follows that ancient relief may have had a greater role in residual soil formation than does more recent relief. As an example, in southern Africa the history

of land surface development and erosion cycles can be traced back satisfactorily to Gondwana times (190 million years ago) and early Cretaceous or post-Gondwana times (approximately 135 million years ago). Several erosional surfaces that were formed during these periods have been identified in Africa, Australia and South America. Erosional surfaces that were formed later have also been identified, and the most widely occurring of these in Africa is known as the 'African' erosional surface. Initiated during the early Cretaceous era following the break-up of the Gondwanaland supercontinent, the resulting land surface was, on the evidence of surviving remnants, well planed by erosion in most places (Falla, 1985). Elevations of the African surface range from approximately 1 650 m to 1 700 m above sea level in the Johannesburg area of South Africa. The residual soils underlying the surface have distinctive engineering characteristics owing to the great age of the surface (approximately 100 million years). Because the surfaces are so old, profiles of residual soil underlying them are extremely deep (e.g., Figure 1.13).

In the Johannesburg area there are several remnants of the African surface, but erosion has formed several Post-African variants of the original surface. One variant that has suffered widespread lowering and stripping of residual cover is described as the 'Post African 1 and 2 surfaces'. Conversely, surfaces that lie above the general level of the African surface also occur. These have suffered a similar degree of weathering since 'Africa' times. This zone, which is not a planed surface, is characterized by undulating topography, and is described as the 'Above African' surface. These three distinct surfaces or zones are close in geological age, but decrease in age from the Above African to the African to the Post African surface. As will be shown in section 1.6 later, there are distinct differences in the characteristics of the soil profile that underlies each surface. The topographic section in Figure 1.5 shows the elevations and locations of the ancient erosion surfaces that have been identified in Southern Africa, together with a tabulation of the estimated ages of the surfaces and the geotechnical problems associated with soils beneath each surface.

Fitzpatrick & Le Roux (1977) studied soil profiles developed from basic igneous rock on hillsides beneath the African erosion surface. They found that the depth of weathering increased down the slope. Whereas kaolinite and halloysite were the predominant clay minerals at the top of the slope (see Figure 1.6) that at the bottom of the slope was smectite. This is also an example of a case where the hillside and valley soils probably include a certain depth of colluvium eroded from higher elevations and deposited lower down the slope. This is undefined in Figure 1.6, and the boundary between transported and residual soil is very difficult to locate.

van der Merwe (1965) made a study of residual soils derived from a number of basic igneous rock types. In a study of weathered diabase profiles from three different sites he found that the predominant factors, other than climate and rainfall, affecting development of clay minerals were local topography and internal drainage. Samples taken from a site high upon a slope with good runoff showed kaolinite and vermiculite to be the dominant clay minerals. A flatter site indicated a chlorite, vermiculite, montmorillonite and kaolinite weathering sequence. The third site was flat, with impeded drainage. Because of leaching of alkalis and alkali earths and the presence of predominantly reducing conditions, weathering could not proceed to the kaolinite stage and hence montmorillonite was the predominant mineral found. This study indicated that good internal drainage and high rainfall are favourable to the

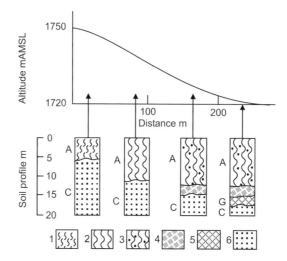

Figure 1.6 Section of black clay toposequence developed under the African surface (surface cover consists of grassy meadow vegetation). 1. Black clay with strong subangular blocky structure (breaking down to crumb structure); 2. Black clay with strong coarse subangular blocky structure (breaking down to medium angular blocky structure with vertical cracks and slickenslides); 3. As for 2, but with small, hard round calcareous nodules; 4. Narrow moist band (approximately 60 mm) of grey friable soft calcium carbonate in yellowish grey clay matrix; 5. Pleyed clay with slickenslides; 6. Yellowish, olive brown friable decomposed dolerite.

development of kaolinite whilst flatter slopes and poor drainage favour the formation of montmorillonite.

Topography also controls the drainage of an area. Local variations in topography influence the amount of moisture retained, the position and configuration of the water table, and thereby the depth of penetration of most chemical weathering processes. The relief of the area thus controls whether a soil develops under conditions of poor or good drainage.

Under conditions of good drainage, topography controls the extent of leaching, an essential mechanism of rock weathering. Highly weathered residual granites, for instance are found to be more porous and therefore more compressible in zones where the annual precipitation and therefore the leaching is greater. This is illustrated by Figure 1.7 which shows some data of Brink & Kantey (1961) relating void ratio to annual rainfall for weathered granites in South Africa. The Figure is, of course, based on current rainfall patterns and therefore represents the effects of the climate in recent times. Obviously, the higher the annual rainfall, the more chemical weathering and leaching will occur, and the higher will be the resultant void ratio.

1.5   GENERAL CHARACTERISTICS OF RESIDUAL SOILS

(Illustrated by colour plates (C) located at the end of the book).

The process of formation of a residual soil profile is obviously extremely complex, difficult to understand and difficult to generalize.

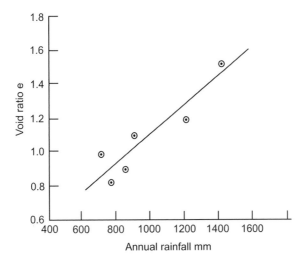

Figure 1.7 Relationship between void ratio and annual rainfall for highly weathered and leached granites in South Africa, under the African erosion surface.

It is evident that apart from a few valid generalizations, it is difficult to relate the properties of a residual soil directly to its parent rock. Each situation requires individual consideration and it is rarely possible to extrapolate from experience in one area to predict conditions in another, even if the underlying hard rock geology in the two areas is similar. For instance, the younger weathered granite soils of the hot, humid Malaysian peninsular may have very different properties to those of the ancient cooler, semi-arid Highveld plateau in South Africa, as have those of the warmer and wetter south western coast of South Africa. In many cases, though, the present state of rock decomposition results from the influence of long past climates, not the climates of the recent past and present.

The chemical changes and sequences of minerals formed during weathering are extremely complex. For example, one suggested weathering sequence leading to the formation of clay minerals is shown in Figure 1.8a (van der Merwe, 1965). The sequence may be arrested at any stage and it is possible that certain stages may be reversed as a result of changes in climate or conditions of drainage.

Figure 1.8b (Gonzalez de Vallejo *et al.*, 1980) shows weathering sequences for volcanic rock observed in Cameroon (West Africa) and Kenya (East Africa). Apart from unknown differences in mineralogy, between the parent rock materials, Cameroon has a lower altitude and hotter, moister climate than does Kenya, which may partly explain the difference in weathering sequence.

As mentioned previously, and indicated in Figure 1.8a, van der Merwe (1965) has shown that on the South African Highveld, reddish kaolinitic soils develop in well drained situations over norite gabbro, whereas blackish motmorillonitic clays develop from identical parent rock in poorly drained situations. This is clearly visible when flying over or even driving through these areas in winter, when the fields have been stripped of their crops and ploughed. It is very noticeable that the ploughed

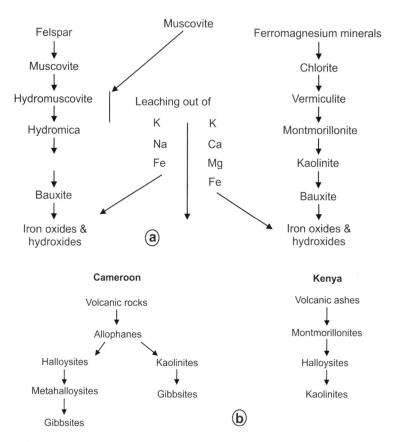

Figure 1.8 (a) Suggested sequence of weathering leading to formation of clay minerals. (van der Merwe, 1965). (b) Sequences in the formation of clay minerals from volcanic rocks in different parts of Africa. (Gonzalez de Vallejo *et al.*, 1981).

surfaces of rises and hillsides are reddish in colour and valleys and hollows are brown to black.

Wesley (1973) concluded that the dark-coloured andosols and red latosols found in Java originate from much the same volcanic parent material but occur in profiles of different ages.

Because weathering proceeds from the surface down and inwards from joint surfaces and other percolation paths, the intensity of weathering generally reduces with increasing depth and reducing intensity of jointing in the material between joint surfaces. In profiles residual from igneous rocks, core stones or boulders of sound parent rock are very often found enclosed within blocks of weathered rock (e.g., Lumb, 1962). This is the typical "onion skin" weathering pattern so often seen in basic igneous rocks such as basalts and dolerites. Typically, a profile of residual soil will consist of three indistinctly divided zones (Vargas & Pichler, 1957; Ruxton & Berry, 1957; Little, 1969) as illustrated in Figure 1.9. The upper zone will consist of highly weathered

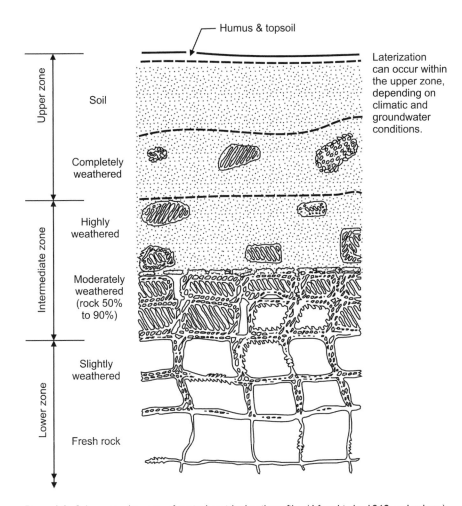

Figure 1.9 Schematic diagram of typical residual soil profile. (After Little, 1969 and others).

and leached soil often reworked by burrowing animals and insects or by cultivation, and intersected by root channels. This zone has always been subjected to at least some transport processes. The intermediate zone also consists of highly weathered material but exhibits some features of the structure of the parent rock and may contain core stones. This zone often contains pedogenic material such as nodules of calcium or iron salts which may give it a mottled or spotted appearance.

Colour Plates C1 and C2 (at the back of the book) illustrate all of the features shown in Figure 1.9. The profile consists of a weathered andesite lava and Plate C1 shows the maximum degree of weathering at the surface with a sparse layer of mainly transported silty, sandy soil, in which the pebbles are amygdales weathered out of the rock, overlying *in situ* weathered andesite containing core stones. The centre of the field of view shows what was originally a large core boulder that has progressively

weathered to a number of smaller onion-skinned core stones, some of which have weathered completely. The sequence shown in Figure 1.9 is continued in Plate C2 (at a location adjacent to Plate C1) which shows (at the top) slightly weathered jointed andesite, with discolouration due to weathering showing on the joints. Just below half-way down the picture is a layer of weathered andesite that represents a time interval between successive lava flows, long enough for the surface of the lava to weather. 300 mm below this is another weathered surface representing a similar time intermission in the volcanic activity. The lava just above the upper weathered layer, and that below the layer is full of amygdales representing gas bubbles (or vesicles) in the molten lava that have now been filled by secondary deposition of light-coloured minerals to form amygdales.

Plate C3 shows a residual profile of a diabase dyke, most of which has weathered to a reddish-brown sandy silt, but which contains a number of large core boulders. The boulders in the centre-right of the photograph form the right hand limit of the dyke, which dips vertically.

Plate C4 shows a weathered andesite lava flow, covered by a thin layer of colluvium and containing core boulders of relatively unweathered andesite. The light-coloured boulders are particularly interesting, as they are of quartzite, and must have been entrained by the flowing lava in much the same way as a glacier entrains boulders from the surface over which it flows.

Plate C5 shows an andesite lava profile, exposed in a rail cutting, that has completely weathered to a great depth (Also see Figure 1.13.). (The lines crossing the photo horizontally are traction wires for electric locomotives.) The boundary between the reddish-coloured residual andesite and the brownish near-surface transported layer is marked by a line of light-coloured pebbles which is usually called the "pebble marker", often occurs near the top surface of a residual profile and usually marks the position of an ancient erosion surface.

Saprolites are materials that have soil-like strength or consistency, but retain modified but recognizable relics of the physical features or fabric of the parent rock. For example a saprolite derived from the weathering of lava may retain the flow structure and amygdales of the parent rock. One derived from a shale often retains the bedding and jointing pattern of its parent rock. Saprolites may occur in profiles of almost any depth.

Plate C6 shows a profile of residual volcanic ash which shows saprolitic structures. The inclined lines of coarse particles represent surfaces covered by coarse particles as the ash layer was built up by successive out-pourings or eruptions. Towards the bottom-left of the photo, a displacement in two parallel coarse layers shows that the ash has been faulted, with the fault running up to the right.

Laterites are usually highly weathered and altered residual soils, low in silica, that contain a sufficient concentration of the sesquioxides of iron and aluminium to have been cemented to some degree. These salts are secondary emplacements resulting from the evaporation of iron- or aluminium-bearing near-surface ground waters. (V_{ss} in Figure 1.1c). Depending on the extent of the emplacement, the material could be described as lateritic or as laterite. Lateritization usually occurs in residual soils, but ancient transported soils may also have been lateritized. Depending on the predominance of either iron or aluminium sesquioxides, these cemented soils are also known as ferricretes or alucretes. Desai (1985) gives definitions of the degree of lateritization

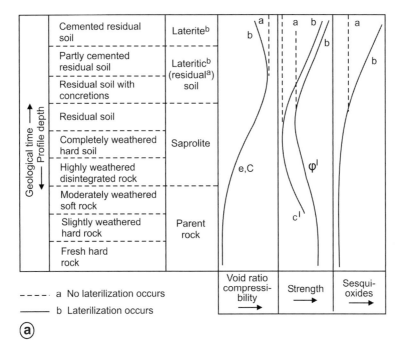

Figure 1.10 (a) Changes occurring in a weathering profile. (Adapted from Tuncer & Lohnes, 1977 and Sueoka, 1988).

in terms of the silica/alumina ratio. Unlateritized soils have SiO_2/Al_2O_3 greater than 2. For lateritic soils SiO_2/Al_2O_3 lies 1.3 and 2 and for true laterites the ratio is less than 1.3. Madu (1980) shows that chemical and physiochemical properties of ferricrete can be related to the strength of the soil, although the correlation is somewhat vague.

The above remarks apply equally to the precipitation of calcium salts in the form of calcium carbonates, in which case the cemented soils are known as calcretes. In other cases silica can be precipitated, in which case a silcrete is formed. The general term "pedocrete" is used to describe any residual or ancient transported soil that has been cemented by secondary deposition of salts.

Figure 1.10a (adapted from Tuncer & Lohnes (1977) and Sueoka (1988) shows in schematic form the progression from fresh rock through saprolite to laterite. Note that the progression from residual soil to laterite, ferricrete, alucrete, calcrete or silcrete is not inevitable, but depends on climatic and topographic conditions being favourable for precipitation of the salts to occur. Figure 1.10b shows an example of this progression (Futai *et al.*, 2004) in a profile of residual gneiss, with the parent rock below 7 m, saprolitic soil between 2 and 7 m and the top residual soil. The Figure does not record any cementing, but Figure 2.4b (for the same profile) shows an increasing content of amorphous iron salts as the soil surface is approached. This profile is described as saturated which is probably why it does not appear to be cemented by the iron salts.

Most of the depth of a residual profile will usually not be subject to salt precipitation. Because of their mode of formation, pedocretes usually occur as fairly near-surface

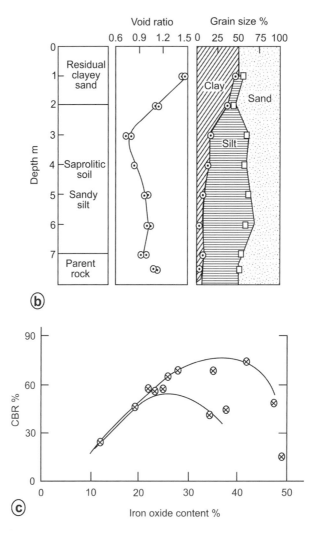

(b)

(c)

Figure 1.10 (b) Weathering profile of gneiss (Futai *et al.*, 2004). (c) Relationship between iron oxide content and CBR (Madu, 1980).

strata of limited depth. Calcretes, silcretes and ferricretes are usually excellent materials for the building of roads and embankments (e.g., Medina, 1989; Hight *et al.*, 1988; Sweere *et al.*, 1988; Gidigasu & Mate-Korley, 1984) and are often sought after for this purpose. They may also, *in situ*, form good shallow founding strata of high strength and low compressibility. Figure 1.10c shows, however, that the progression of properties indicated in Figure 1.10a is not certain and can be subject to wide variations (Madu, 1980).

Plate C7 shows a layer of laterite gravel formed in a profile of weathered shale. The small, spherical pebbles consist of iron sesquioxides, while some resistant slate-like core stones from the residual shale are also present.

Plate C8 shows a layer of calcrete that has formed in a weathered cretaceous mudstone. The increasing strength or hardness of the calcrete layer with depth is shown by the failed attempt to dig a trench through it with a front-end loader (top, right) and the way in which the shelf of calcrete in the deeper excavation has spanned across the overhang where the uncemented residual mudstone has fallen away into the deeper excavation.

1.6   DEPTH AND INTENSITY OF WEATHERING

As stated earlier, the formation of a profile of residual soil depends among other factors, on the rate of erosion of the surface. Obviously, if the rate of removal of material from the surface equals or exceeds the rate of advance of the "weathering front" into the rock, no residual soil will form, and vice versa. It would therefore be expected that in a relatively small area over which climatic conditions are uniform, more deeply weathered profiles will occur beneath older surfaces. Falla (1985) showed that this is indeed so for andesite profiles. His statistical data shows some interesting trends, that can be summarized as follows:

- The total thickness of residual soil, decreases progressively from that beneath the more than 100 million year old "Above Africa" surface (where it averages 59 m) to that beneath the less than 20 million year old "Post African" surface (where it averages only 16 m).
- The predominant soil colour in the older profiles is red, while in the younger profiles it is yellow.

Plate C4 illustrates a lighter coloured younger profile while Plate C5 illustrates a red, older profile.

As mentioned in section 1.1, physical weathering processes tend to predominate in the weathering of sedimentary rocks, and chemical processes predominate in the weathering of igneous rocks. However, every rule has its exception. Plate C9 shows a profile through a diabase sill that has, in weathering, fragmented into a mass of small angular boulders with only limited chemical breakdown, such as that exhibited by Plate C3. The present rainfall at the site of Plate C9 is considerably higher than at the site of Plate C3 and the temperatures are also higher. Hence the only conclusion can be that the sill of Plate C9 was intruded at a later time than the dyke of Plate C3. However, igneous rocks can and do become highly weathered to very great depths. Plate C10, for example is an exposure in a sand quarry that is 40 m deep. Here, a granite has deeply weathered to a kaolinitic clayey sand that is used as a building sand.

Sedimentary rocks can also exhibit either shallow or deep weathering. Plate C11 shows shallow weathering of a mudstone, progressing in consistency from that of soft rock at the base of the exposure to a clayey soil at the top. Plate C12 on the other hand shows an exposure of weathered shale, where the consistency, from the contact with the overlying transported layer near the top of the photo to the bottom, is that of a stiff fissured clay.

As discussed earlier in conjunction with Figure 1.9, weathering is often variable at a particular depth. Plate C13 is a striking visual example. The soil is a weathered

mudstone where leaching along fissures appears to have removed the red colour from the soil adjacent to the fissures, although the consistency of the soil is similar, regardless of colour.

1.7 MORE ABOUT PEDOCRETES

Because pedocretes are so important as construction materials in many parts of the world, this section provides more information about their occurrence, engineering characteristics and properties. The information is mainly drawn from Netterberg (1994). Dr. Frank Netterberg has kindly agreed to this abstraction of knowledge that he has collected over the past 45 years.

Occurrence

Pedocretes are believed to form mainly in seasonally or perennially water-deficient climates such as those defined in Figure 1.5, by a value of Weinert's N greater than 5 (i.e., in areas where 12 times the maximum monthly potential (or pan) evaporation divided by the annual precipitation is greater than 5, $\{12\ E_p(\text{max, monthly})/P(\text{annual}) > 5\}$. Local drainage conditions must also be suitable, as a supply of shallow ground water, charged with potentially pedocrete-forming salts, must be available to be evaporated for the salts to precipitate in the soil. These conditions are provided by seasonally dry water-courses, pans or depressions that may contain water in the rainy season, but dry out in the dry season.

Obviously, the type of salt that precipitates depends on local geology and must be available in the underlying residual soils or weathered rock. This leads to local occurrences of a particular type of pedocrete. For example, referring to the $N > 5$ contour in Figure 1.5, along the semi-arid west coast of South Africa, silcretes occur as mesa cappings, associated with deep weathering beneath the African erosion surface. Inland, silcretes are associated with the deep sand deposits of the arid Kalahari desert. Gypcretes occur mostly in the extremely dry Namib desert.

Calcretes occur in somewhat less arid conditions with the permanent water table supplying calcium- and magnesium-rich water by capillarity for evaporation in the calcrete zone. For example, in Plate C8 the permanent water table has been exposed in the excavation and occurs about 3 m below the base of the calcrete layer. Ferricretes are commonly associated with seasonally fluctuating water tables, with iron-rich ground water rising in the rainy season and being evaporated, depositing iron oxides in the dry season.

Formation and form

The rate of formation of pedocretes is obviously very slow, but is thought to be between 20 and 200 mm thickness per 1000 years. As an indication of the rate of formation, in 1898, during the South African war, an ammunition dump was blown up. Fragments of shell cases and lead slugs in the area of the dump, 90 years later were encased in ferricrete with layer thicknesses of 2 to 5 mm (Blight, 1991). This supports the postulated rate of deposition of 20 mm per 1000 years. Calcretes may attain thicknesses of over 30 m, but are seldom homogeneous over depths exceeding 2 m. Silcrete of

Table 1.1 Some engineering properties of calcrete in various forms.

Form of calcrete	Plasticity index PI%	Casagrande classification	10% FACT kN	Moh hardness	Excavation method
Calcareous soil	Varies widely	Varies widely	N/A	N/A	Doze-shovel
Powder calcrete	NP–22	CL, ML, SC	20–90	2–3	Doze-shovel
Nodular calcrete	NP–25	SC, SM	10–180	1–5	Doze-shovel
Honeycomb calcrete	NP–10	N/A	80–200	2–6	Rip-blast Rip-grid roll
Hardpan calcrete	NP–7	N/A	30–200	2–6	Rip-blast crush

Notes: N/A = not applicable
NP = non plastic
Casagrande classification – see Figure 2.11a
10% FACT = 10% Fines Aggregate Crushing Test
Moh hardness = hardness of calcrete particles on Moh hardness scale

similar thicknesses occurs beneath the African erosion surface in southern Africa and its equivalent in Australia. Calcified alluvial gravels or valley calcretes may exceed thicknesses of 10 m in places.

Pedocretes occur in a number of forms (only those pedocretes of greatest interest as construction materials are listed):

- Calcareous, ferruginized or siliceous gravel, sand, silt or clay or mixtures of these particle sizes that are rich in the specific mineral, but show little evidence of cementation or nodule development.
- Powder calcrete, ferricrete or silcrete consisting of silt- or fine sand-sized particles or aggregates of particles of precipitate interspersed with particles of the host soil.
- Nodular pedocretes consisting of silt- to gravel-sized nodules of hard calcareous, ferruginous or siliceous particles interspersed with the host soil.
- Honeycomb pedocretes in which the nodules have, in part, coalesced with each other to form a coherent skeleton. Ripping is usually required for excavation.
- Hardpan pedocretes are indurated and strongly cemented to a consistency varying from soft to very hard rock. These layers are seldom more than 1m thick and usually overlie less cemented grades of pedocrete (see, e.g., Plate C8).

Engineering properties

Some engineering properties related to the forms of calcrete listed above are summarized in Table 1.1.

Particle specific gravity G

Because pedocrete grains or nodules are often porous, the particle specific gravity G is difficult to define and measure. Netterberg (1994) quotes the values set out in Table 1.2. G for calcretes falls within the range usually found for soils. Values for laterites depend on the content of iron salts and, on average, have values of 3 or more.

Table 1.2 Measured values of particle specific gravity G for pedocretes.

Description	Particle size range mm	G
Calcretes	<0.425	2.47 to 2.80 mean: 2.66
Laterite sands and gravels	Not stated	2.67 to 3.46 mean: 3.06
	<2 mm	2.2 to 4.6

Foundations on pedocretes

As mentioned above, the quality of pedocretes is usually best, close to the surface of a deposit and usually lessens with depth. Deposits are also subject to lateral variability, and foundation exploration should include trenching to explore lateral variations as well as the usual test-pitting and boreholes. Powder and nodular pedocretes may also have collapsible grain structures, while honeycomb and hardpan deposits may soften *in situ* and lose strength when they become wet. Losses of strength of powder calcrete of as much as 90% have been found after wetting.

Small scale karst features are common in calcretes that have weathered *in situ* and have also been found in laterites. Even small sinkholes have been found to occur in weathered calcretes.

Pedocretes in road construction

The widest recorded engineering use of pedocretes has been in road construction. In the whole of southern Africa, silcrete and calcrete have been used in the wearing courses of unpaved roads and for all layers of the road structure, including surface chippings, in surfaced roads. Similar uses are recorded in the south western United States, Australia and Brazil. An interesting variant is the use of powder gypcrete compacted with sea water as the wearing course of unsurfaced roads along the coast of Namibia. These roads are periodically watered with sea-water. Although the rainfall is nearly non-existent, sea fogs ensure that the humidity of the air is high and this atmospheric water, absorbed by the high salt content of the road surface ensures a stable, dust-free road surface. The surface becomes so blackened by carbon from vehicle tyres, that the roads look superficially like conventional black-top (i.e.,asphalt surfaced) roads (see Plate C23).

In general, pedocretes perform better as pavement materials than would be expected from their gradings and plasticity. Shear failures and rutting are rare and if they develop, it is usually in poorly drained areas where storm water collects and stands.

Pedocretes as building blocks and concrete aggregate

Laterite has been used for millennia in the Middle East and Asia to form building blocks. Both laterite and caliche (honeycomb calcrete) blocks are cut from suitable deposits and allowed to dry in the sun where they carbonate and self harden into

stable, relatively erosion-proof blocks. Powder pedocrete can also be used with the addition of Portland cement to form blocks or bricks. Laterite, calcrete and silcrete have been crushed and used as coarse aggregate in concrete for culverts, wing-walls and other road drainage structures and to construct at least one short-span road-bridge. Thirty years later, the structure had given no concrete-related problems.

Pedocretes in embankment and earth dam construction

In semi-arid and arid climates, pedocretes are sought-after materials for embankment and earth dam construction. However, it must be warned that when subjected to continuous seepage of low salt content water, pedocretes may lose strength and slowly dissolve and have also been known to be dispersive. Their use in water retaining structures or structures to retain hydraulically placed mine tailings should only be proceeded with after careful materials testing.

1.8 TRANSPORTED SOILS THAT HAVE WEATHERED *IN SITU*

It was mentioned in section 1.1 that certain transported soils such as volcanic ashes and aeolian sands may weather *in situ* to produce a residual soil. One example has already been described by Plate C6. Plate C14 shows another example. When Mount Pinatuba on Luzon island in the Philippines erupted on 15 June 1991, it poured out vast quantities of ash that blanketed thousands of square kilometres of what had been the most productive rice fields in the world. This was the largest outpouring of volcanic ash recorded since 1941, with the height of the spewing ash plume reaching 55 km. Over the years following the eruption, the recently deposited ash became subject to mud flows that travelled vast distances, burying an even greater area in ash and damming or changing the course of rivers. Plate C14, taken in April 1999, shows the bank of a newly cut meander in a river on Luzon. (From this viewpoint, Pinatuba could just be discerned on the far horizon and was about 30 km distant.) The erosion, cutting through grey ash from the most recent eruption, had exposed buildings of a recently buried village that had been built on volcanic ash from an earlier out-pouring. The lower stratum of ash can be seen to have weathered from grey to a light brown colour. Hence deposits of this sort, including aeolian sands are the result of ongoing processes of alternating deposition and weathering, as also illustrated in Plate C2 by the clearly defined separate and successive lava flows. It is ironic, that whereas the weathered ash is highly fertile, the new ash that now blankets it, is said to be agriculturally sterile. However, the tall growth of grass shown on Plate C14 shows that the sterile phase will probably be short-lived, and it will be possible to farm this area again within a few years.

1.9 THE WEATHERING OF SOLUBLE LIMESTONES AND DOLOMITES

The main agent of weathering of limestones and dolomites is solution of the rock by rain water infiltrating the surface soil cover and percolating down and laterally through

cracks and fissures in the rock. Because rain water dissolves carbon dioxide both from atmospheric air and from soil air as it infiltrates, it forms a weak carbonic acid. This attacks the calcium and magnesium carbonate in the rock to produce calcium and magnesium bicarbonates, as shown by equation 1.4.

$$3CaCO_3 \cdot 2MgCO_3 + 5H_2CO_3 = 3Ca(HCO_3)_2 + 2Mg(HCO_3)_2$$

$$\begin{array}{ccc} \text{dolomite} + \text{carbonic} = \text{calcium} & + \text{magnesium} \\ \text{acid} & \text{bicarbonate} & \text{bicarbonate} \end{array} \tag{1.4}$$

It has been calculated that carbon dioxide-charged water percolating through dolomite can remove as much as 175 kg of dolomite in solution for each ML (i.e., million L or kg) of percolating water (du Toit, 1954), i.e., a 175 ppm by mass solution can be formed. This means that a large spring issuing from a dolomitic source at a rate of 100 ML per day could remove 17.5 ton of solid dolomite per day. Springs or dolomitic "eyes" with flows as large as this are by no means uncommon in dolomitic or limestone terrain.

Figure 1.11 illustrates the process diagrammatically, and Plate C15 shows a series of near-surface rock pinnacles, formed by differential solution along joints in the dolomite, with the residuum removed. The excavation was about 3 m deep. The residuum is often very rich in manganese and karst areas have been mined for manganese by excavating and processing the residuum from between the steep sided rock pillars (e.g., Viljoen & Reimold, 1999).

Two types of geotechnical problems arise from the weathering pattern described by Figure 1.11 or Plate C15, excessive surface settlement and collapse of the surface into sink-holes.

- The residuum between solid pinnacles is extremely compressible (see the loose texture shown in Plate C15), while the pinnacles are unyielding. Foundations that are not designed to span between pinnacles fail, causing destructive damage to the building they support. Plate C16 shows damage caused to a brick building by this cause.
- The second problem is the formation of sink-holes, as a result of the sudden collapse of the residuum into an under-lying cavern. The size of sinkholes in karst terrain can vary from 5 m or less in diameter or width to 50 m or more.

Plate 1.1 is a photograph showing a sink-hole measuring 55 m in diameter and more than 30 m deep that engulfed the main crushing plant of a South African gold mine in 1962. The accident happened suddenly, without any warning signs, at 11 a.m. when the plant was fully operational, and the lives of 29 workers were lost. None of the plant components or bodies of the victims was ever recovered. Plate 1.2 shows a sink-hole, also measuring more than 50 m in diameter that, in 1964 engulfed a house and a family of 5, also without warning, at 2 a.m. There was originally a cluster of 4 houses on the site shown in the photograph. The occupants of the remaining 3 houses were awoken by the noise of the collapse and managed to get to safety before the second house fell into the hole. Shortly after the photo in Plate 1.2 was taken, the remaining houses also disappeared into the more than 30 m deep hole. Again, no

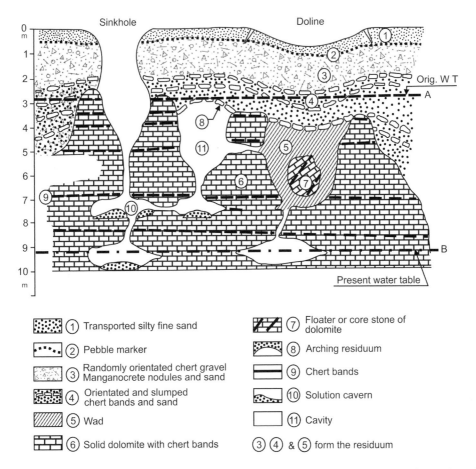

Figure 1.11   The process of weathering in a dolomite and limestone to form a series of pinnacles of solid rock with soft residuum or caverns between them. (After Wagener, 1985).

bodies of victims were recovered. (Both Plates 1.1 and 1.2 were re-photographed from now-defunct newspapers of the time.)

Sink-holes are an ever-present threat to safety in any karst area. They can occur naturally, but often their occurrence is accelerated by two human activities, both related to pumping water from the aquifer that usually lies within the karst (see Figure 1.11). By removing buoyancy from the strata, lowering the water table effectively increases the self-weight of the drained strata by $10 \, kN/m^3$, accelerating both the collapse of soil and rock rubble arches and the settlement of leached residuum between rock pinnacles, both as shown in Figure 1.11. In semi-arid climates, crops are often grown on a large scale by pumping water from a dolomite aquifer and irrigating crops on the surface. The natural condition of soil strata above the water table is one of desiccation. Irrigation reduces the strength of the soil by destroying the capillary stresses engendered by long-term desiccation, and simultaneously, lowers the water table and increases the bulk unit weight, both contributing to increased settlement and sink-hole formation.

Plate 1.1 Sink hole that engulfed entire crushing plant at South African gold mine.

Plate 1.2 Sink hole that swallowed four houses.

Void ratios of leached dolomite or limestone residuum can be extremely high, with measured values ranging from 2.7 to 9.6 in manganese-rich residuum, known as "wad". Particle unit weights or specific gravities (G_s) can also be very high, ranging from 3.2 to 3.5. The net effect is that dry unit weights can range from only 3.3 to 8.6 kN/m^3.

1.10 DISPERSIVE SOILS

Any soil surface will erode if exposed to rapidly flowing water or wind. Some soils erode more readily than others, and are known as erodible or highly erodible soils. However, in certain soils, both transported and residual, the clay minerals will deflocculate and disperse when exposed to water, particularly percolating rain water that has a low dissolved solid content. This is caused by a change of soil chemistry resulting in the repulsive electrical surface charges between colloidal clay particles becoming greater than the attractive van der Waal's forces (Holmgren & Flanagan, 1977).

When this happens, the extremely small dispersed solid particles are carried out of the soil matrix, as if in solution, even in relatively slowly moving seepage water. The result is that internal erosion is initiated and concentrates in zones of exiting seepage. This can cause the formation of erosion pipes that undermine and result in failure and collapse of earth structures, especially water retaining and drainage structures (e.g., Elges, 1985; Paige-Green, 2009).

Since problems with dispersivity and dispersive soils were first recognized in the geotechnical literature (e.g., Aitchison et al., 1963), a number of diagnostic tests have been devised. These include the:

- Pinhole test, in which erodibility is assessed by passing water through an axial hole pierced by means of a hypodermic needle through a triaxial specimen (Sherard et al., 1976).
- Cation exchange capacity (CEC) versus exchangeable sodium percentage (ESP) plot, in which these two measurements are compared (e.g., Gerber and Harmse, 1987).
- Crumb test, in which a crumb of soil is placed in distilled water which is allowed to stand and observed to see if a visible halo of dispersed soil colloids forms in the still water around the crumb (e.g., Reeves et al., 2006).
- Sodium adsorption ratio (SAR) (e.g., Gerber & Harmse, 1987).
- Total dissolved solids (TDS) versus % sodium (% Na) (Sherard et al., 1976).

Bell & Walker (2000) carried out a careful comparison of these five methods and came to the conclusions summarized in Table 1.3. This is probably the best guide to identifying dispersive soils available at present.

As is the case with other properties of residual soils (e.g., compaction, see Chapter 4) both of the physical tests in Table 1.3, i.e., the pinhole and crumb tests, must be performed on soil as close to its natural range of water contents as possible. Even air drying, let alone oven dryness will have a large influence on the result of the test.

Also concerning residual soils, dispersive soils tend to be residual from rocks that are low in calcium and magnesium and high in sodium, hence the concentration on

Table 1.3 Rating criteria for identification of dispersive soils (Bell & Walker, 2000).

Test	Rating of dispersivity							
Pinhole test	Dispersive	5	Moderate	3	Slightly	1	Non-dispersive	0
CEC vs ESP	Highly dispersive	4	Dispersive	3	Marginal	1	Non-dispersive	0
Crumb test	Strong reaction	3	Moderate	2	Slight	1	No reaction	0
SAR	>2	2		1.5 – 2.0		1	<1.5	0
TDS vs% Na	Dispersive	2	Intermediate	1			Non-dispersive	0
Total dispersivity rating	**High >**	**12**	**Moderate 8–11**		**Slight 5–7**		**Non-dispersive >4**	

sodium chemistry in Table 1.3. Soils residual from granites have been found to be dispersive, as have soils residual from mud rocks. Soils containing lithium have also been found to be dispersive. In the field, a good indication that soils are highly erodible, or even dispersive is the presence of "bad-lands" topography, such as the well-known bad-lands of South Dakota in the USA, Basilicata in Italy (Bromhead *et al.*, 1993) or in Turkish Cappadocia (see Plate C17). In all three places, and in others like them, the rapid rate of surface erosion is shown by steep bare slopes of weathered rock, occasionally interrupted by vertical steps and conical or planar caps of more resistant rock.

Physical problems with dispersive soils usually arise from a combination of dispersive tendencies, seepage water (e.g., rain water) with a chemistry that differs from that of the *in situ* pore water, and, most importantly, poor compaction and settlement of the soil in contact with a more rigid interface. Thus piping usually occurs along the side or base of a backfilled trench, along the surface of a concrete retaining wall, or along the side of an outlet conduit to an earth dam. Plate C18 shows a failure by piping between the concrete outlet conduit and the penstock outfall tower of an earth dam constructed of compacted residual shales and mudstones. The soil had dispersive tendencies, but the main reason for the failure, during first filling of the reservoir, was probably arching of the fill over the outlet conduit, resulting in reduced soil pressures against the sides of the conduit, combined with inadequate compaction of the soil against the tower and outlet conduit, the lack of cutoff collars along the conduit and, possibly, the low dissolved solid content of the rainfall-derived seepage water. Of this list, poor compaction and reduced pressures against a rigid interface were probably the most important. (See also section 8.7.)

1.11 RELICT STRUCTURES IN RESIDUAL SOILS THAT MAY AFFECT THEIR ENGINEERING PERFORMANCE

Saprolitic soils, by definition, may contain many of the physical features of the parent rock. These may include relict fault surfaces and shear planes. If these have been subject to repeated movement over aeons of time, these surfaces have often become polished slickensides that, in the saprolitic residual soil, could cause shear failure to occur at low shear stresses. One such surface is shown in Plate C19 that shows an extensive

slickensided surface exposed during removal of residual shale overburden in an open-cast iron ore mine. (The slickenside is the central area which appears lighter-coloured because the polished surface reflects the sunlight.) Plate C20 shows a similar slickenside taken from a planar failure surface in a road cutting in residual weathered shale.

1.12   RAPIDITY OF WEATHERING

Although it is usually assumed that soil-forming processes such as the decomposition of igneous rocks and the formation of cemented soils or pedocretes take place very gradually over time-spans of millions of years, there is also evidence that this is not always the case.

Lava flows on Hawaii island, for example, see Plate C21, show signs of physical and chemical weathering within a few years of deposition and soon support pockets of grass and small shrubs. It can be seen from the photo that the surface of the asphalt road and the painted white lines are still in good condition, yet the lava flow is already supporting vegetation. As mentioned in section 1.7, ammunition buried in the soil during the South African War of 1899–1902 has been found completely encased in lateritic material 90 years later. The very rapid disintegration or slaking of certain mudrocks on exposure to the atmosphere is another example of weathering occurring over a time period of a few years or even months. On the other hand, ancient stone monuments such as the 5000-year-old Stonehenge in England, pre-AD Roman aqueducts in Europe (see Plate C22) and the Nuragi of Sardinia show very little sign of weathering after more than 2000–5000 years of exposure to the elements.

1.13   DETAILED EXAMINATION OF A TYPICAL SOIL
PROFILE RESIDUAL FROM AN IGNEOUS ROCK

This section will deal in detail with the appearance and mineralogy of a typical deeply weathered residual soil profile. The mechanical properties of the same soil will be used to give examples of the compressibility, shear strength and other characteristics of a typical residual soil, later in the book.

Andesite is part of a family of dark coloured rocks in which the predominant dark minerals are hypersthenes, augite and hornblende (Ernst 1969). The composition of andesite in terms of the ratio of alkali (Sodium-Potassium) to total feldspar and the anorthite (Calcium Aluminium Silicate) content of the plagioclase (Na-Ca feldspar) as well as its relationship to other dark coloured rocks is shown in Figure 1.12. The andesite considered in this paper is part of the approximately 2500 million year old Ventersdorp Supergroup of rocks in South Africa (Brink 1979; Falla 1985).

The igneous rocks of the Ventersdorp Supergroup consist mainly of andesitic lavas with interbedded zones of reworked volcanic conglomerates, agglomerates and tuffs. The principal minerals of these rocks are feldspar and augite, occurring as fine laths with occasional phenocrysts, together with amygdales of quartz, epidote and some chlorite, while the volcanic conglomerates consist of clastic accumulations of compacted lava fragments.

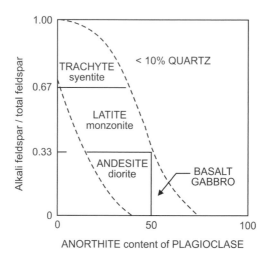

Figure 1.12 Mineralogical composition of andesites, and relationship to other dark-coloured rocks. (Capitals denote extrusive rocks and lower case intrusive rocks. Ernst, 1969).

The constituent minerals in the andesite usually weather to produce a usually well-defined colour sequence in the residual soil. Plate C5 shows the upper 4 to 5 m of an ancient residual andesite profile. The ferro-magnesian minerals (pyroxenes) have altered to chlorite, which imparts a distinct greenish colour to the lower zones of the residual soil. As the chlorite has altered higher in the profile, iron in the lattice structure has oxidized and hydrated to form limonite or yellow ochre, which gives the soil a yellow or yellowish brown colour. In upper zones, above the water table, where the soil may be seasonally desiccated, haematite or red ochre has been produced, giving a characteristic red or reddish brown colour, particularly near the surface. These three colours do not always all appear in the profile, as this depends on local conditions, and the boundaries between colour zones are seldom sharply defined. The tendency for this colour sequence to develop provides a convenient means for recognition and reference to engineering and weathering characteristics.

Iron compounds often become concentrated in the upper zones of a residual soil profile to form ferricrete. As described earlier, this is produced by the seasonal precipitation and gradual accumulation of insoluble iron oxides and hydroxides within the soil mass. Its development depends on the amount of iron available and on repeated annual fluctuations in water table levels. The ferricrete may vary from thick, impermeable, compact, brownish accumulations of nodular "hardpan" to a dark ferruginous staining in the soil. Figure 1.13 shows a typical soil profile residual from andesite in the Johannesburg area of South Africa. This illustrates the upper transported layer of hillwash or colluvium that often occurs at the surface of a transported layer, but no hardpan or ferricrete is present in this profile.

The statistics of the geotechnical parameters for these soils do not show very clear trends, but this is because the whole range of degrees of weathering was present in most of the test holes used for compiling the data. However, the data do show that residual

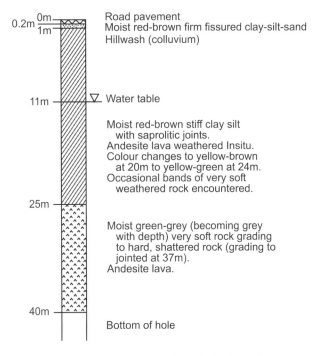

Figure 1.13 Typical profile of residual andesite lava.

andesite soils can vary from clayey soils (clay content approaching 50% and Plasticity Index above 50) to silty soils (clay content less than 10% and Plasticity Index less than 10). Void ratios can be high (greater than 1.5) to moderate (0.6), and compressibilities likewise can vary from C_c of over 0.9 to less than 0.1. More details of geotechnical properties will be given in later chapters.

1.14 AN INTRODUCTION TO THE MECHANICS OF UNSATURATED SOILS

Because residual soils are found so extensively in temperate climates with seasonal rainfall, as well as in semi-arid and arid climates, an introductory note on the mechanics of unsaturated soils follows. Although not the primary subject of the book, an explanation of unsaturated soil mechanics will be found useful by readers who are unfamiliar with the terms and concepts of unsaturated soil mechanics.

The state of unsaturation in a soil may be defined either geometrically, or in terms of the stress in its pore water (u_w). The geometrical definition is simply that a soil having a degree of saturation of less than unity is unsaturated (i.e., $S < 1$). In terms of the stress in its pore water, if a soil at a height h above the water table has a pore water pressure of less than $-\gamma_w h$ (i.e., more negative) then it is unsaturated. If placed in contact with water at a pressure of $-\gamma_w h$, such a soil will absorb moisture and swell.

It should be noted that soils that are unsaturated in terms of the geometrical definition, $S < 1$, may not be unsaturated in terms of the pore water stress definition, and vice versa. Usually, however, the two definitions coincide.

In terms of the geometrical definition of an unsaturated soil, all compacted soils are unsaturated, as a certain amount of air is inevitably trapped in the pores during the compaction process. The soil remains unsaturated until the pore air either dissolves under pressure in the pore water, or dissolves and is carried away in solution by percolating pore water.

Also in terms of the pore water stress definition, any soil that is located above the water table in an area for which the annual evaporation from a free water surface (e.g., a lake or pond) exceeds the annual rainfall, is likely to be unsaturated for at least part of the year. Such areas cover a large proportion of the area of the earth, in fact, all of the unshaded and desert areas shown in Figure 1.3a.

1.14.1 The effective stress equation for an unsaturated soil

The pores of most unsaturated soils are filled partly with air and partly with water. The water exists in capillary lenses around the points of intergranular contact and the air-water interfaces are curved towards the air phase. Hence surface tension forces ensure that the pore air pressure is greater than the pore water pressure. The surface tensions in the air-water menisci also influence the effective stress in an unsaturated soil, as may any salt dissolved in the pore water.

The effective stress equation for unsaturated soils is usually assumed to have the form:

$$\sigma' = (\sigma - u_a) + \chi(u_a - u_w) \tag{1.2}$$

in which u_a and u_w are, respectively, the pore air and water pressures and χ is an empirical parameter representing the proportion of the pore pressure difference $(u_a - u_w)$ that contributes to the effective stress. χ is known as the Bishop parameter (Bishop *et al.*, 1960).

The characteristics of this parameter can be investigated by considering the equilibrium of a typical interparticle contact in an idealized soil made up of identical cubically packed spherical particles (Figure 1.14). If the pore space is saturated, the conventional effective stress equation applies and χ is unity. If water is withdrawn from the soil and air enters the pore space, the configuration of the retained water and the forces acting at an interparticle contact can be modelled as shown in Figure 1.14a. For equilibrium at the contact point, the intergranular stress (actually the net intergranular force P divided by $4r^2$) will be:

$$\sigma_g = \frac{\pi A}{2} \left\{ \frac{pA}{2} - \frac{T}{r} \right\} \tag{1.5}$$

where $A = (\tan\theta - \sec\theta + 1)$ and θ has a maximum value of $45°$ (once θ exceeds $45°$, the air becomes an occluded bubble). If σ_g is assumed to equal σ', and since $p = (u_a - u_w)$, $\sigma_g = \sigma' = \chi p$ and:

$$\chi = \frac{\pi A}{2} \left\{ \frac{A}{2} - \frac{T}{rp} \right\} \tag{1.6}$$

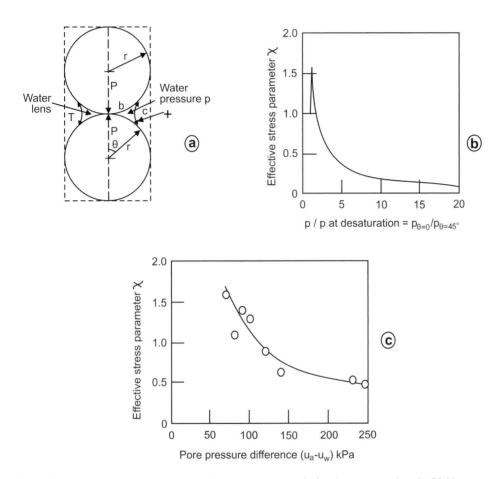

Figure 1.14 (a) Forces at an interparticle contact in an idealized unsaturated soil, (b) Variation of effective stress parameter χ as an idealized soil drains from the saturated to the unsaturated state, (c) Observed variation of the effective stress parameter χ in an unsaturated soil.

Figure 1.14b shows the variation of χ (computed from equation 1.6 with p, from which it will be noted that based on this model, the effective stress in a soil theoretically rises by a factor of 1.57 (or π/2) at the instant of desaturation.

Figure 1.14c shows some values of χ determined experimentally for an unsaturated silt which support the general form of the relationship represented by equation 1.6.

In most, but not all cases, the pore air pressure in an unsaturated soil will be atmospheric (i.e., zero gauge) and hence if $(u_a - u_w)$ exceeds 100 kPa, the pore water will be in a state of absolute tension. This is not a limitation to the theory, as tensile strengths of pure water of up to 20 MPa have been demonstrated experimentally and it is relatively simple to show that ordinary tap water (containing dissolved air) can sustain tensions of about 250 kPa for short periods. In practice, measured tensions in

the pore water of an unsaturated soil range up to about 500 kPa, but may be as high as 1500 kPa or more.

The shear strength of an unsaturated soil in which the pore air pressure is zero can be represented by the equation:

$$\tau = c' + (\sigma - \chi u_w)\tan \varphi' \tag{1.7}$$

As u_w will have subatmospheric values, the term $-\chi u_w$ represents a positive addition to the total stress σ, an addition, moreover, that may be quite independent of σ. When this is combined with the fact that χ may be greater than unity when the degree of saturation of the soil is close to 100%, it will be appreciated that considerable strengths may be developed in unsaturated soils. This is particularly true of clays – many are familiar with the iron-hard consistence of sun-dried clays. Unconfined strengths in desiccated clays of up to 2.8 MPa have been recorded (Blight, 1966).

Another advantage of unsaturated soils is the fact that because the pore space is partly filled with air, changes of effective stress are possible without any drainage of pore fluid. Thus the shear strength of an unsaturated soil increases immediately on the application of a surcharge stress which is not always the case with saturated soils.

An unsaturated soil for which $S = 1$ obeys the effective stress equation for saturated soils, viz:

$$\sigma' = \sigma - u_w \tag{1.3}$$

The parameter χ is obviously related to the degree of saturation S, but is not identical with S. A number of empirical equations relating χ to S have been advanced, of which the most useful appears to be that due to Khalili & Khabbaz (1998) (also see Khalili et al., 2004). This predicts χ as a function of the ratio of the existing suction (s) to the suction at which air would enter the saturated soil (s_e), usually referred to as the air entry suction. Algebraically, the relationship is:

$$\begin{aligned} \chi &= (s/s_e)^{-0.55} \quad \text{if } s \geq s_e \\ \chi &= 1 \qquad\qquad \text{if } s < s_e \end{aligned} \tag{1.8}$$

In Figure 1.15, the bold solid line represents equation 1.8 and the experimental points were measured in tests on a number of soils.

1.14.2 Soil suction

In many situations involving unsaturated soils the pore air pressure is atmospheric or zero (gauge). As mentioned above, the pore pressure difference $(u_a - u_w)$ then becomes $-u_w$ (often written as s or p'') and is known as the soil water suction. The suction of the soil water has two components:

The solute (or osmotic) suction

This arises from the presence of dissolved salts in the pore water that give rise to an osmotic or solute pressure. The solute suction is defined as the negative pressure to

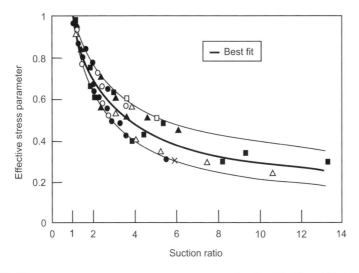

Figure 1.15 Effective stress parameter χ versus suction ratio, after Khalili and Khabbaz (1998).

which a pool of pure water must be subjected in order to be in equilibrium through a semi-permeable membrane (i.e., a membrane permeable only to water molecules) with a pool containing a solution identical with the pore water.

The matrix suction

This is the negative pressure to which a pool of solution identical in composition with the soil water must be subjected in order to be in equilibrium with the soil water through a permeable membrane (i.e., a membrane permeable to both water and solute molecules). The matrix suction is the quantity conventionally measured in a triaxial apparatus where the water in the pore pressure measuring system is separated from the soil water by a water saturated porous stone, not a semi-permeable membrane. (In American English, the word "matric" is used instead of "matrix".)

The total suction

The presence of a solute in water causes a lowering of the vapour pressure in equilibrium with the solution which, according to Raoult's law is directly proportional to the concentration of the solution. The vapour pressure lowering also represents a reduction of the relative humidity of the air in the soil pores and is a manifestation of the solute suction.

The existence of a tension in water also has the effect of lowering the relative humidity in equilibrium with it. Thus the matrix suction is manifested by a lowering of the relative humidity of the air in the soil pores, and vice versa.

The Kelvin equation can be used to relate the relative humidity to an equivalent soil water tension or total suction which is equal to the sum of the solute and matrix suctions. The Kelvin equation is:

$$\frac{R\theta}{m_w} \log_e(RH) = h \tag{1.9}$$

in which

RH = relative humidity,
R   = universal gas constant, 8.31 J/Kmol
θ = absolute temperature, K
m_w = molecular mass of water, 18 (H = 1, O = 16, H_2O = 18)
h   = the height of a suspended column of water, i.e., the negative head of water, that would be in equilibrium with the relative humidity.

At a standard temperature of 20°C (293K), the Kelvin equation can be expressed as:

$$u_w = 311 \log_{10}(RH) \ [MPa] \tag{1.9a}$$

in which (RH) is expressed as a decimal of unity.

According to Raoult's law, the relative humidity in equilibrium with the surface of water containing a dissolved salt is given by:

$$RH = \frac{m_w}{m_w + m_s} \tag{1.10}$$

m_w is the number of moles of water and m_s is the number of moles of dissolved salt (or solute). For example, suppose 1 mole of sodium chloride is dissolved in 1 L of water. The molecular mass of water is 18 g and the number of moles in 1 L is 1000/18 = 55.56. Because NaCl dissociates into N_a^+ and Cl^- there are 2 moles of solute, and;

$$RH = \frac{55.56}{57.56} = 0.965$$

From equation 1.9a:

$$-u_w = 311 \log_{10}(0.965) = 4.812 \ MPa$$

This will be the solute suction and the total suction will be the sum of the matrix and solute suctions.

The relative contributions of the solute and matrix suctions to the effective stress remain to be resolved. Richards (1966) suggested that the effective stress equation be written as:

$$\sigma' = \sigma + \chi_s p_s'' + \chi_m p_m'' \tag{1.11}$$

in which p_s'' is the solute suction,
p_m'' the matrix suction

and χ_s and χ_m are respectively, the χ parameters appropriate to the solute and matrix suctions.

There is some evidence that the values of χ_s and χ_m will differ and also that their values will depend on whether shear or volume change is being considered. There are indications that the solute suction, p''_s, may not directly contribute to shear strength. Casagrande (1965) for example, described attempts to build a causeway across the Great Salt Lake in Utah, USA. He mentioned that the clay forming the lake bottom has a shear strength gradient of only 6 kPa/m depth. This is despite the fact that the clay contains crystalline salt and is therefore subject to a solute suction of the order of 40 MPa. For volume change, however, it may be that χ_s is greater than zero. More probably, solute suction operates indirectly by reducing or increasing the relative humidity in the pores of the soil, thus causing either desiccation or an increase in water content and in this way affects the matrix suction, which in turn dominates the effective stress.

Recent experimental work (Katte & Blight, 2012) has supported the conclusion that solute suction has no direct influence on the shear strength of a soil, but may affect it indirectly by attracting or retaining water in the soil, which then affects the matrix suction, and in turn controls the shear strength.

A practical example of this can be seen in the coastal town of Swakopmund in Namibia. The roads in the town are constructed of beach and dune sand which is first irrigated using sea water and then compacted. The road surfaces are then regularly irrigated with sea water. Although annual rainfall is virtually zero, the cold coastal Benguela current ensures that, each evening, the coast becomes blanketed in fog. The fog particles are absorbed by the salt in the road surface, keeping the surface moist and engendering sufficient matrix suction to bind and strengthen the road surface. The result is excellent low-traffic suburban roads that require a minimum of maintenance. Plate C23 shows one of these roads, that are known locally as "salt roads". The blackening visible in the wheel paths is caused by carbon black abraded from vehicle tyres, giving the roads the appearance of regular "black-top", i.e., asphalt-surfaced roads.

1.14.3 The suction-water content curve (SWCC)

The suction water content curve (or soil water characteristic curve) (SWCC) is to unsaturated soils what the consolidation or e-σ' curve is to saturated soils. Whereas the e-σ' curve is the relationship between void ratio e (or water content w, since e = wG) and consolidation pressure, σ', the SWCC is the relationship between suction (u_w, p'' or s) and water content w. In a saturated soil, the independent or controlled variable is the effective consolidation pressure σ' and hence σ' is plotted horizontally and e vertically. In an unsaturated soil, w is very often the independent variable, and is plotted horizontally, with the suction plotted vertically.

Figure 1.16a is an example of an e-σ' curve plotted in the way usually used for a SWCC. It applies to an undisturbed normally consolidated clay (Taylor, 1948). AB represents the virgin compression line (VCL) while BC is a swell line and CD a recompression line. The reduction in e or w between A and B is irreversible and the area ABC represents the irreversible work done, i.e., the hysteresis (which in this case represents 3.2 kJ/m^3 of clay). The hysteresis between BC and CD, however is very small (0.5 kJ/m^3). Without physically remoulding the clay, and putting in at least work of 3.2 kJ/m^3, it is not possible to reverse line BC to position AB.

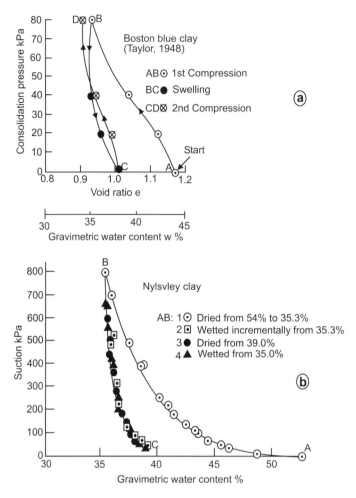

Figure 1.16 Similarities between compression and swelling produced by mechanical stress and suction. (a) Hysteresis during consolidation and swelling in a normally consolidated saturated clay, under mechanical stress. (b) Hysteresis between drying and re-wetting of a normally consolidated clay under suction stress.

Figure 1.16b shows the SWCC for a saturated marsh clay, also normally consolidated. The similarities between the two sets of curves are very obvious, even though the maximum applied suction was 10 times larger than the applied stress in Figure 1.16a (800 kPa versus 80 kPa). The first wet-dry cycle in Figure 1.16b consumed 26 kJ/m³ of hysteretic energy. It should also be noted that path AB was followed by the re-wetting/redrying paths BC/CB and that these paths were followed with indistinguishable hysteresis. In other words, as would be expected, the physical rules applying to mechanical consolidation/swelling apply exactly to drying/re-wetting except that once the clay de-saturates, the effective stress acting in the clay is χu_w and not u_w. In particular, it is not possible to move back from BC to AB without replacing the hysteretic energy that has been consumed.

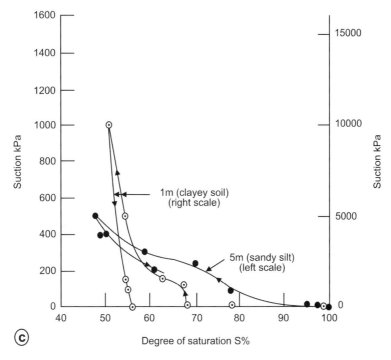

Figure 1.16 (c) Suction/degree of saturation curves for 2 specimens of soil residual from gneiss (after Futai *et al.*, 2004).

Figure 1.16c shows SWCC curves for two specimens residual from a gneiss (Futai *et al.*, 2004). (The profile is shown in Figure 1.10b). The sample from 1 m deep contained about 45% of clay and that from 5 m contained only about 5% of clay. The huge difference in the suction scales used for the two SWCCs shows the effect of differences in particle size distribution on the SWCC. Figure 1.16c has been plotted in terms of degree of saturation S = wG/e. The void ratio e for the 1 m sample was about 1.5, and assuming G = 2.7 for both samples, w at the point of desaturation would have been about 55%. For the 5 m sample, e was 0.95 and w at the point of desaturation would have been 35%. If G had been 3.0, the corresponding water contents at desaturation would have been 50% and 32%. Further information on unsaturated soil mechanics will be found in later chapters.

REFERENCES

Aitchison, G., Ingles, O.G. & Wood, C.C. (1963) Post-construction deflocculation as a contributory factor in the failure of earth dams. In: *1st Aust. N.Z. Conf. on Soil Mech. & Found. Eng., Adelaide*, pp. 275–286.

Bell, F.G. & Walker, J.H. (2000) Dispersive soils: A further examination of the nature of dispersive soils in Natal, South Africa. *Quart. J. Engng. Geol. and Hydrogeol.*, 33, 187–199.

Bishop, A.W., Alpan, I., Blight, G.E. & Donald, I.B. (1960) Factors controlling the strength of partly saturated cohesive soils. *Res. Conf. on Shear Strength of Cohesive Soils, ASCE, Boulder, USA*, pp. 505–532.

Bishop, A.W. & Blight, G.E. (1963) Some aspects of effective stress in saturated and partly saturated soils. *Géotechnique*, 13 (3), 177–197.

Blight, G.E. (1966) Strength characteristics of desiccated clays. *J. Soil Mech. & Found. Eng. Div., ASCE*, 92, (SM6), 19–37.

Blight, G.E. (1967) Effective stress evaluation for unsaturated soils. *J. Soil Mech. & Found. Eng. Div., ASCE*, 93 (SM2), 125–148.

Blight, G.E. (1991) Tropical processes causing rapid geological change. *Geological Soc. Eng. Geol. Spec. Pub. No. 7*, Geological Society, London, UK, pp. 459–471.

Blight, G.E. (2006) Geoenvironmental and management aspects of the behaviour of mining and municipal solid wastes in water-deficient climates. In: Miller, G.A., Zapata, C.E., Houston, S.L. & Fredlund, D.G. (eds) *Unsaturated Soils 2006*, ASCE, Geo. Special Pub. No. 147, ASCE, Reston, USA, pp. 36–80.

Brink, A.B.A. & Kantey, B.A. (1961) Collapsible grain structure in residual granite soils in Southern Africa. In: *5th Int. Conf. Soil Mech. & Found. Eng., Paris*, Vol. 1, pp. 611–614.

Brink, A.B.A. (1979) *Engineering Geology of Southern Africa*. Building Publications, Pretoria, South Africa.

Bromhead, E.N., Del Preto, M., Rendel, H.M., & Cuppola, L. (1993) Overview of geotechnical problems in Basilicata, Italy, resulting from extensive desiccation of clay strata. In: *1st. Int. Symp. on Eng. Characteristics of Arid Soils, London, UK*, pp. 31–40.

Casagrande, A. (1965) Role of calculated risk in earthwork and foundation engineering. *J. Soil Mech. & Found. Eng. Div., ASCE*, 91 (SM4), 1–40.

Desai, M.D. (1985) Geotechnical aspects of residual soils of India. In: Brand, E.W. & Phillipson, H.B. (eds) *Sampling and Testing of Residual Soils*. Hong Kong, Scorpion. pp. 83–98.

Dudley, J.H. (1970) Review of collapsing sands. *J. of Soil Mech. & Found. Eng. Div., ASCE*, 96 (SM3), 925–947.

du Toit, A.L. (1954) *The Geology of South Africa*. London, Oliver and Boyd.

Elges, H.F.W.K. (1985) Dispersive soils. *Civ. Eng. in South Africa*, 27 (7), 347–353.

Ernst, W.G. (1969) *Earth Materials*. Englewood Cliffs, U.S.A., Prentice Hall.

Falla, W.J. (1985) *On the Significance of Climate and Weathering in Predicting Geotechnical Characteristics of Residual Soils Developed on Igneous Rocks*. PhD Thesis, Witwatersrand University, Johannesburg, South Africa.

Fitzpatrick, R.Q. & le Roux, J. (1977) Mineralogy and chemistry of a Transvaal black clay topo-sequence. *J. Soil Sci.*, 28, 165–179.

Fredlund, D. & Morgenstern, N.R. (1977) Stress state variables for unsaturated soils. *J. Geotech. Eng. Div., ASCE*, 103 (GT5), 447–466.

Fredlund, D. & Rahardjo, H. (1993) *Soil Mechanics for Unsaturated Soils*. New York, Wiley.

Futai, M.M., Almeida, M.S.S. & Lacerda, W.A. (2004) Yield strength and critical state conditions of a tropical saturated soil. *J. Geotech. & Geoenv. Eng., ASCE*. 130, (11), 169–1179.

Gerber, A. & Harmse, H.J.vM. (1987) Proposed procedure for identification of dispersive soils by chemical testing. *Civ. Eng. South Africa*, 29, 397–399.

Gidigasu, M.D. & Mate-Korley, E.N. (1984) Tropical gravel paving materials specifications. In: *8th African Reg. Conf. Soil Mech. & Found. Eng. Harare, Zimbabwe*, Vol. 1, pp. 267–273.

Gonzalez de Vallejo, L.I., Jiminez Salas, J.A. & Leguey Jiminez, L. (1980) Engineering geology of the tropical volcanic soils of La Laguna, Tenerif. *Eng. Geol.* 17 (1), 1–17.

Hight, D.W., Toll, D.G. & Grace, H. (1988) Naturally occurring gravels for road construction. In: *2nd Int. Conf. on Geomech. Tropical Soils, Singapore*. Vol. 1. pp. 405–412.

Holmgren, G.G.S. & Flanagan, C.P. (1977) Factors affecting spontaneous dispersion of soil materials as evidenced by the crumb test. *Dispersive Clays, Related Piping, and Erosion in Geotechnical Projects*. ASTM Spec. Geotech. Pub. 623. pp. 218–239.

Jackson, T.A. & Keller, W.D. (1970) A comparative study of the role of lichens and inorganic processes in the chemical weathering of recent Hawaiian laval flows. *Amer. J. Sci.*, 269, 446–466.

Katte, V. & Blight, G.E. (2012) The roles of surface tension and solute suction in the strength of unsaturated soils. In: *Unsat 2012, Int. Conf. Unsatd. Soils, Naples, Italy* (In press).

Khalili, N. & Khabbaz, M.H. (1998) A unique relationship for χ for the determination of shear strength of unsaturated soils. *Géotechnique*, 48 (5), 681–688.

Khalili, N., Geiser, F. & Blight, G.E. (2004) Effective stress in unsaturated soils: review with new evidence. *Int. J. Geomech., ASCE.*, 4 (2), 115–126.

Knight, K. (1961) *The Collapse of Structure of Sandy Sub-soils on Wetting*. PhD Thesis, Witwatersrand University, Johannesburg, South Africa.

Leong, E.C., Rahardjo, H. & Tang, S.K. (2002) Characterization and Engineering Properties of Singapore Residual Soils. In: *Int. Workshop Characterization and Engineering Properties of Natural Soils. Singapore*, Vol. 2, pp. 1279–1304.

Li, W.W. & Wong, K.S. (2001) Geotechnical properties of old alluvium in Singapore. *J. Inst. Engs., Singapore*, 41 (3), 10–20.

Little, A.L. (1969) The engineering classification of residual tropical soils. In: *7th Int. Conf. Soil Mech. & Found. Eng., Mexico City*. pp. 1–10.

Lumb, P. (1962) The properties of decomposed granite. *Géotechnique*, 15 (3), 226–242.

Madu, R.M. (1980) The use of chemical and physiochemical properties of laterites in their identification. In: *7th African Reg. Conf. Soil Mech. & Found. Eng., Accra, Ghana*, pp. 105–116.

Mason, B. (1949) Oxidation and reduction in geochemistry. *J. Geol.*, 57 (1), 66–72.

Medina, J. (1989) Tropical soils in pavement design. In: *12th Int. Conf. Soil Mech. and Found. Eng., Rio de Janeiro, Brazil*, Vol. 1, pp. 543–546.

Mitchell, J.K. (1976) *Fundamentals of Soil Behavior*. New York, Wiley, pp. 49–57.

Morin, W.J. & Ayetey, J. (1971) Formation and properties of red tropical soils. In: *5th African Reg. Conf. Soil Mech. & Found. Eng. Luanda, Angola*, Vol. 1, pp. 45–53.

Netterberg, F. (1994) Engineering geology of pedocretes and other residual soils. In: *7th Congr. Int. Assoc. Eng. Geol., Lisbon, Portugal*, pp. 183–202.

Paige-Green, P. (2009) Dispersive and erodible soils – fundamental differences. In: *A Short Course on Problem Soils*. Johannesburg, South African Inst. Civil Eng.

Pings, W.B. (1968) Bacterial leaching. *Mineral Industries Bulletin, Colorado School of Mines* 2 (3).

Reeves, G.M., Sims, I. & Cripps, J.C. (eds) (2006) *Clay Minerals Used in Construction*. Geological Soc. Spec. Pub. No. 21. London, Geological Society.

Richards, B.G. (1966) Significance of moisture flow and equilibria in unsaturated soils in relation to design of structures built on shallow foundations in Australia. In: *ASTM Symp. Permeability and Capillarity, Atlantic City, USA*, pp. 1.32.

Ruxton, G.P. & Berry, L. (1957) Weathering of granite and associated erosional features in Hong Kong. *Bulletin, Geol. Soc. Amer.*, 68, 1263–1292.

Schwartz, K. & Yates, J.R.C. (1980) Engineering properties of aeolian Kalahari sands. In: *7th African Reg. Conf. Soil Mech. & Found. Eng., Accra, Ghana*, Vol. 1, pp. 67–74.

Sherard, J.L., Dunnigan, L.P., Decker, R.S. & Steele, E.F. (1976) Pinhole test for identifying dispersive soils. *J. Geotech. Eng. Div., ASCE*, 102 (GT1), 69–85.

Strakhov, V. (1967) *The Principles of Lithogenesis*, Vol. 1. Edinburgh, Oliver and Boyd.

Sueoka, T. (1988) Identification and classification of granitic residual soils using chemical weathering index. In: *2nd Int. Conf. Geomech. in Tropical Soils, Singapore*. Vol. 1, pp. 55–62.

Sweere, G.T.H., Galjaard, P.J. & Tjong Tjin J.H. (1988) Engineering properties of laterites in road construction. In: *2nd Int. Conf. Geomech. Tropical Soils, Singapore*. Vol. 1, pp. 421–428.

Taylor, D.W. (1948) *Fundamentals of Soil Mechanics*. New York, Wiley.

Tuncer, E.R. & Lohnes, R.A. (1977) An engineering classification for certain basalt-derived lateritic soils. *Eng. Geol.*, 2 (4), 319–339.

Uehara, G. (1982) Soil science for the tropics. In: *Engineering and Construction in Tropical and Residual Soils. ASCE Geotech. Div. Spec. Conf.*, Honolulu, Hawaii, pp. 13–26.

van der Merwe, D.H. (1965) *The Soils and Their Engineering Properties of an Area Between Pretoria North and Brits, Transvaal*. DSc Thesis, University of Pretoria, South Africa.

Vargas, M. & Pichler, E. (1957) Residual soil and rock slides in Santos, Brazil. In: *4th Int. Conf. for Soil Mech. & Found. Eng., London*, Vol. 2, pp. 394–398.

Viljoen, M.J. & Reimold, W.U. (1999) *An introduction to South Africa's Geological and Mining Heritage*. Johannesburg, Mintek.

Wagener. F.v.M. (1985) Dolomites. *Civ. Eng. South Africa*, 27, 110–120.

Weinert, H.H. (1964) *Basic Igneous Rocks in Road Foundations*. CSIR Research Report No. 218. Pretoria, South Africa, CSIR.

Weinert, H.H. (1974) A climatic index of weathering and its application in road construction. *Géotechnique*, 24 (4), 475–488.

Wesley, L.D. (1973) Some basic engineering properties of halloysite and allophane clays in Java, Indonesia. *Géotechnique*, 23 (4), 471–494.

Wesley, L.D. (2010) *Geotechnical Engineering in Residual Soils*. Hoboken, USA, Wiley.

Yamanouchi, T. & Haruyama, M. (1969) Shear characteristics of such granular soil as shirasu. *Memoirs, Faculty of Eng., Kyushu University*, 29 (1), 63–64.

Zaruba, O. & Mencl, V. (1976) *Engineering Geology*. Amsterdam, Elsevier.

Zhang, G., Whittle, A.J., Nikolinakou, A.M. & Germaine, J.T. (2007) Characterization and engineering properties of the old alluvium in Puerto Rico. In: *2nd Int. Workshop on Characterization & Engineering Properties of Natural Soils, Singapore*. Vol. 4, pp. 2557–2588.

Chapter 2

Microstructure, mineralogy and classification of residual soils

*A.B. Fourie, T.Y. Irfan, J.B. Queiroz de Carvalho,
J.V. Simmons & L.D. Wesley*

2.1 MICROSTRUCTURE AND MINERALOGY RELATED TO WEATHERING

The microstructure and mineralogy of tropical and residual soils are associated with modes of soil formation and occurrence. Microstructure should be considered at two relative scales:

1 ped and within-ped or intra-ped voids,
2 grains and intergranular voids.

Peds are natural aggregations of soil grains which may exist in residual soils on a scale ranging from centimetres to microns. They are the natural units of mechanical interaction of the soil grains, and may break down or form in response to wetting and drying or various methods of mechanical agitation.

Physical breakdown and chemical reactions are weathering processes leading to the formation of tropical and residual soils. Physical breakdown rates are controlled by exposure to water and changing temperature and other sources of energy transmitted to the parent material from the local environment. Chemical reaction processes can be categorized into decomposition, leaching and re-deposition and dehydration.

These processes may proceed simultaneously, cyclically, or sequentially, depending on the climatic conditions and the time of exposure relative to the process reaction rates. Both mineralogy and microstructure are definitively associated with these processes. A wide variety of soils can be produced, depending on the above, and it is necessary to keep a broad perspective when dealing with tropical and residual soils.

2.1.1 Decomposition

Decomposition includes the physical breakdown of the rock fabric and the chemical breakdown of constituent minerals, usually primary rock-forming minerals. Typical products are clay minerals, oxides, hydroxides, and free silica.

Under conditions that include tropical, subtropical, arid and semi-arid climates, transported soils may be modified into materials with residual soil characteristics. Reaction rates vary so that some minerals may have completely decomposed (e.g., feldspars in granite) when neighbouring grains (e.g., quartz) remain virtually unaltered.

Chemical reactions occur more slowly in temperate climates than in tropical climates so that physical breakdown may dominate the soil formation processes in temperate zones while chemical processes dominate in the tropics.

2.1.2 Leaching and re-deposition

Leaching and re-deposition involves leaching and removal of combined silica, alkaline earths, and alkalis. There is a consequent accumulation of oxides and hydroxides of sesquioxides at a level in the profile where evaporation occurs. The leached materials may be redeposited and accumulated in this zone.

2.1.3 Dehydration

Dehydration (either partial or complete) alters the composition and distribution of the sesquioxide-rich materials in a manner which is generally not reversible upon wetting. Dehydration also influences the formative processes of clay minerals. In the case of total dehydration, strongly cemented soils with a unique granular soil structure may be formed.

2.2 MINERALOGY AND OCCURRENCE OF WEATHERING PRODUCTS

Tropical decomposition tends to favour formation of the clay mineral kaolinite. This is by far the most common clay mineral in tropical residual soils. Under suitably moist conditions, halloysites will be formed. Under prolonged decomposition, silica can be removed to the extent that free alumina is present and gibbsite is formed. Usually, part of the silica produced in the soil will be in the form of quartz. Generally, the iron oxides present and/or remaining are sufficient to form goethite and haematite. Iron oxides will form a mineral depending on the in-situ conditions. For example, haematite is only formed in very strong oxidizing conditions, whereas goethite and limonite form where there are conditions of continuous moisture and aeration. Specific minerals are also characteristic of soils from certain parts of the world. For instance, illites are frequently identified in African lateritic soils but not in Brazilian laterites, where normally only kaolinite is present. Montmorillonite or smectite is usually the predominant clay mineral in soils residual from basalts, dolerites, norites and kimberlites, and in many soils residual from sedimentary rocks such as mud-rocks and shales.

Other mineralogical components cited in literature and present in residual soils, may be relatively rare and/or of difficult identification. Such minerals include boehmite, anatase, mixed-layered kaolinite-illite or kaolinite-vermiculite assemblages. Montmorillonite may be present in lateritic soils, for instance, but only as a very-short term transitory mineral in one part of a weathering sequence.

Tropical weathering of volcanic ashes frequently produces an abundance of allophone, a virtually amorphous clay mineral having an unusually high natural moisture content. Allophane may be identified by its characteristically large, irreversible change of plasticity properties upon drying at different temperatures.

Clay minerals tend to be concentrated in the fine fraction of the soil. Iron oxide minerals such as goethite and/or haematite, and also quartz, tend to be concentrated

in the coarse fraction. Gibbsite and boehmite are frequently found as fillings to pores and voids in concreted particle aggregations. Significant quantities of amorphous components have also been identified in tropical lateritic and saprolitic soils (Queiroz de Carvalho 1981, 1985, 1991).

A wide variety of processes may lead to residual soil formation. Highly structured duplex soils tend to form where there is pronounced seasonal wetness and dryness, particularly when the soil profile has matured over a long period of time, as in many parts of Australia and South Africa (e.g., Richards, 1985). The reader is also referred to more detailed descriptions in Chapter 1 of the interactions of landform, climate, parent rock mineralogy, and time.

The classic weathering profile, from mature soil to fresh rock, has been discussed widely, and been subjected to detailed scrutiny by many authors (e.g., Brand, 1988). Within the soil layer, there may be differentiation caused by leaching and dehydration. The soil profile is typically structured into an overlying silt (depleted in clays) and a clay-enriched intermediate horizon, often highly fissured in a blocky pattern. It is necessary to maintain a very disciplined profile description methodology (see Chapter 3) in order adequately to record residual soil profiles.

2.3   DETERMINATION OF MINERALOGICAL COMPOSITION

Soil mineralogy can be assessed in various ways. Very specialized techniques have been developed for particular purposes, but the most common approaches are:

- X-ray diffraction,
- Thermo-gravimetry,
- Optical microscopy including polarization measurements,
- Scanning or transmission electron microscopy combined with some form of spectral element identification.

Mineralogical identification using these techniques requires specialized training and procedures. Often, combinations of techniques are necessary in order to make definite identifications. The processes are not straightforward because the preparation and measurement process usually alters the minerals. Each technique must therefore be fully understood to be of greatest use.

The x-ray diffraction (XRD) technique is by far the most widely used, but is only appropriate for minerals with distinctive crystallography. XRD can be carried out using oriented or randomly selected soil samples placed in a sample holder with or without resin to fix the samples in place. Specialized techniques are necessary for soils containing significant quantities of iron. Glycolation (replacement of interlayer water by ethylene glycol) causes changes to interlayer spacing and is a means of identification of montmorillonites.

Thermogravimetry (TG) identifies minerals based on changes that occur as dehydration takes place through a range of temperatures. It is generally imprecise except for certain simple minerals which have clear and unambiguous thermogravimetric signatures.

Optical microscopy (OM), used to examine thin sections with crossed polars, is a well established assessment process for primary rock minerals. Sample preparation

may involve impregnation with resin, and this may damage the original microstructure to some degree if not done with extreme care. OM can also be applied to some classes of weathered minerals with success. It can be a useful technique for estimating relative abundances of certain minerals, or for assessment of the texture of weathered rocks including fresh and completely weathered grains.

Scanning electron microscopy (SEM) or transmission electron microscopy (TEM) both involve preparation techniques that may displace water and hence damage the original microstructure. Such damage can be minimized with care, and specially designed environmental microscopes can avoid the problem. SEM imaging can be undertaken on lump samples or thin sections. TEM imaging requires dispersed particles or thin sections.

SEM imaging can reveal microstructure details to sub-micron sizes, and is a particularly useful tool for microstructural studies. When combined with some form of microprobe, for example energy dispersive spectroscopy (EDS), measurements can be made of elemental constituents at specific points on the sample surface. From relative abundances of elements, it is then possible to deduce the chemical composition of the material at the point. For many clay minerals, particularly mixed-layered or hydrated forms that are common in residual soils, it is not possible to obtain a conclusive identification unless this process is combined with another technique, for example XRD, which reveals crystal form.

TEM imaging can reveal mineral crystal details at the lattice level, and hence is used to refine knowledge of the clay species. Some form of spectroscopy is normally used to make chemical composition determinations. TEM imaging has been used to visualize the layering within clay minerals, and hence the mineralogical changes which take place during the weathering processes.

2.4   MICROSTRUCTURE OF RESIDUAL SOILS

Soil structure, fabric, and texture are terms referring to the physical arrangement of grains and peds. These arrangements, together with mineralogy, determine engineering behaviour. In conventional soil mechanics, the importance of fabric has been recognized for a long time, and has been well summarized by Rowe (1974). Most aspects of fabric relevant to engineering behaviour are macrostructural (that is, can be observed without magnification).

Microstructure embraces microfabric, composition, and interparticle forces. Investigations of soil microstructure are carried out using optical microscopy, SEM, or TEM. The microstructure of residual soils may reveal cementation and pea contact arrangements, which can lead to a better understanding of the occurrence and engineering performance of such soils.

Collins (1985) extended microfabric studies to residual soils and outlined the models presented in Figures 2.1 and 2.2. These show examples of microfabric organization at three levels:

- elementary level (Figure 2.1),
- assemblage level (Figure 2.1),
- composite level (Figure 2.2).

Elementary level

Elementary particle arrangements (epa) → interaction of elementary particles

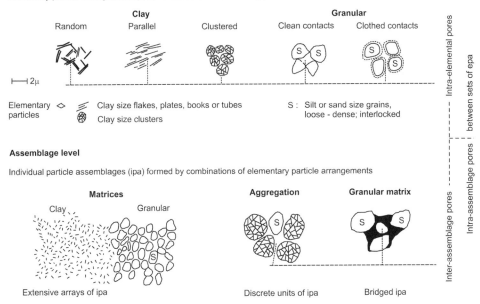

Figure 2.1 Elementary particle arrangements and assemblages.

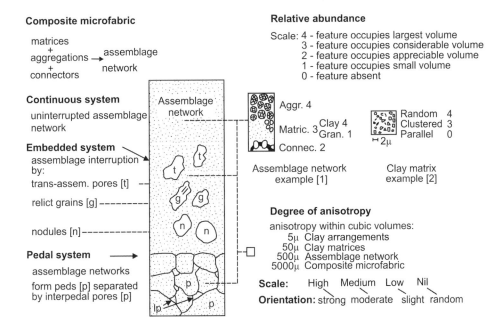

Figure 2.2 Composite level of fabric organization in residual soil.

The **elementary level** is formed by a number of particles of clay, silt or sand size or clay size groups, or even a cluster of clay size particles.

The **assemblage** level comprises a large number of clay or granular particles with definable physical boundaries, and the three types identified with this level are:

1   matrices,
2   aggregations,
3   granular matrices.

The **composite** level has three aspects: the composite microfabric, relative abundance reflecting the heterogeneity, and the degree of anisotropy. Rigorous classification of particular soil microfabrics using a formal scheme is very complex. However, the concepts are important and should be related to observations of particular soils, in order to understand how the soil peds and/or grains are likely to interact, and how fluid flow is likely to occur through the available interconnected pore space.

2.5   MINERALOGY AND MICROSTRUCTURE RELATED TO GEOTECHNICAL PROPERTIES

Mineralogical influences on engineering properties are usually self-evident. For example, soils which contain halloysite as the clay mineral, display high plasticity, sensitivity to drying and also to manipulation, and higher compressibility than kaolinites. The presence of this clay mineral makes the performance of soil particle density (G_s) testing more difficult than usual.

Effects of drying

A distinctive feature of halloysite soils, even more notable with allophane soils, is the irreversible change in soil index properties that accompanies drying from natural moisture content (see also section 2.7). Removal of water induces chemical and mineralogical changes caused by irreversible dehydration processes. Therefore, index testing of known or suspected Group C (see section 2.8.2 and Table 2.1) soil should be performed with the intended use for the soil clearly in mind. If the soil is to be used in its natural state, as it usually is, with minimal disturbance and drying, testing must be applied to soil that is as close as possible to its natural condition or state.

Effects of reworking

Similar experiences can be obtained with Group A soils, as well as transported soils which have sustained subsequent weathering. For example, a non-cemented aged sand composed of volcanic clasts may behave as a sand if subjected to minimal disturbance, but as a high plasticity clay when the clasts are altered and the soil is subjected to reworking (e.g., as in cut-to-fill earthworks).

Reactivity to stabilizing agents

Soils with kaolinite as the clay mineral exhibit normal test behaviour in terms of low to medium plasticity and permeability. On the other hand, when montmorillonite is present (or even halloysite) there will be a high level of reactivity of the soils towards

cement or lime. Usually the soils have a very high plasticity and low permeability. These properties may be modified (for use in road layer work) by addition of stabilizing agents, such as mixtures of Portland cement, slaked lime or ground granulated blast furnace slag (GGBFS).

Effect of wetting

Residual lateritic and saprolitic soils are very often encountered in an unsaturated condition. If the soil is subsequently saturated or its water content greatly increased, the shear strength and compression properties will change because of the destruction of suction forces by saturation.

Effect of secondary minerals

The influence of sesquioxides is related to their mode of occurrence in the soil. If they cement the particles, they tend to cause a reduction in plasticity. However, if there is only very weak cementation, remoulding of the soil can lead to an increase in the interactions between clay particles, and therefore to an increase in plasticity. Also, the presence of goethite/haematite as discrete particles in the matrix soil, can be enhanced by the process of stabilization, forming cementitious compounds and thereby increasing the strength of the soil. Knight (1958) and Collins (1985) attributed increased stiffness in a saprolite to the bridging or bracing between quartz grains by clay minerals and/or sesquioxides. This may also result in an increased cohesional strength for the unsaturated soil.

Baynes and Dearman (1978) concluded that changes of the geotechnical properties of weathered granites were due to microfracturing and increased intergranular porosity, observed in the microstructure of the weathered material studied. Collins (1985) found similar features in weathered granite from Brazil. Queiroz de Carvalho (1991) found that a highly-cemented microstructure was present in soils which were also highly reactive to the process of lime stabilization. The voids in these soils were filled by a form of amorphous aluminous material.

Dispersive behaviour

Dispersive soil behaviour (see section 1.10) is often identified in soils which have been subjected to cyclic wetting and drying under conditions where clay minerals become sodium-dominated. Simmons (1989) found that microstructural aggregation of clays was the most identifiable factor in the occurrence of soils subject to severe dispersive erosion. These clay minerals concentrate in specific zones associated with the evaporation zone of seasonal groundwater changes. The study suggested that microfabric, rather than clay chemistry, had the controlling influence. The erodibility of a soil was found to be greatest where silt-sized aggregations of clay particles were present.

In summary, residual soils must be expected to vary very widely in their structure and mineralogy. Where specific microstructural or mineralogical features are identified, corresponding influences on engineering behaviour may be expected. The most important principle is that residual soils may have much better engineering characteristics than index tests (on the remoulded soil) suggest, particularly where correlations between index tests and the engineering behaviour of transported soils are consulted. Soil properties must always be assessed with regard to the particular purpose, or range

of purposes, expected. Where there is any doubt about the likely behaviour of such soils, field-based assessments and tests for recognition of special behaviour are of much greater value than reliance on empirical correlations.

2.6   EXAMPLES OF THE MINERALOGY OF A RESIDUAL PROFILE

As the first example of the mineralogy of a residual soil, the weathering and resulting mineralogy of the residual andesite soil described in Chapter 1 (section 1.13 and Figure 1.13) will be explored (Blight, 1996).

   Minerals are not equally susceptible to chemical breakdown and mineralogic change as a result of weathering. For each rock there is a sequence of weathering that depends on the minerals present. The first minerals to crystallize out from a magma have the highest internal energy associated with them, and therefore are the most unstable and will break down first during the weathering process. Conversely, the last minerals to crystallize out will have the lowest internal energy and will be the most resistant to weathering. This point is illustrated by Figure 2.3 which shows

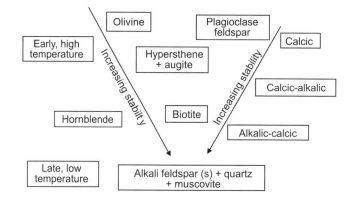

(a) The order of crystallization of minerals in andesite (Bowen, 1928)

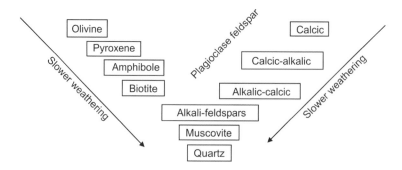

(b) The order of weathering of minerals in andesite (Arnold, 1984)

Figure 2.3 Crystallization and weathering orders of minerals in andesite.

the crystallization and weathering orders of minerals in andesite according to Bowen (1928) and Arnold, (1984). It will be seen that the two schemes are closely related.

Figure 2.4a shows the distribution of minerals observed in a profile of residual andesite from below the ancient "Above Africa" erosional surface near Johannesburg, South Africa (Brummer, 1980). It will be seen that the andesite weathers to form muscovite, chlorite, kaolinite and quartz, although montmorillonite has also been reported to occur (Falla, 1985). The analysis shown in Figure 2.4a does not include the haematite or limonite that gives the soil its characteristic colour. It will be noted that quartz occurs in only part of the profile, although where it does occur, it amounts, at depths of between 5 and 10 m, to almost 50% of the soil. These high quartz contents did not form part of the original mineralogy of the andesite, but resulted from the secondary deposition of quartz in cooling joints in the andesite. It is very common to find quartz veins in residual andesite profiles. Note that one does not usually find a regular progression of minerals with depth from those characteristic of a high degree of weathering to those characteristic of lesser weathering. This is because the profile is composed of a number of successive lava flows, each of which originally had a slightly different mineralogy and each of which was exposed to weathering for a different time before being covered by the next lava flow (see Plates C1 and C2).

The second example is shown in Figure 2.4b (Futai et al., 2004) for a profile of weathered gneiss in Brazil. Here, the soil depth is less and the minerals consist mainly of quartz and kaolinite with a relatively high content of amorphous iron. The micaceous minerals that feature prominently in Figure 2.4a are absent.

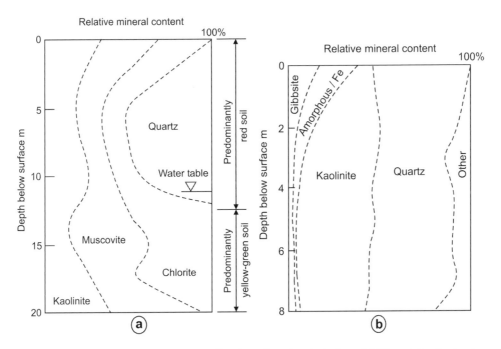

Figure 2.4   (a) Mineral distribution in a profile of weathered andesite lava. (b) Mineral distribution in a profile of weathered gneiss (Futai et al., 2004).

2.7 OVERCOMING DIFFICULTIES IN MEASURING INDEX PARAMETERS FOR RESIDUAL SOILS

It is well-known that oven-drying, and even air-drying, affects the properties of soils, although this effect is usually small for transported soils. Because of their origin, by slow *in situ* decomposition in a largely anaerobic environment, residual soils are particularly prone to changes in properties caused by drying and exposure to air. Drying can cause partial or complete dehydration of the clay minerals and can change them and their properties irreversibly. Even air-dying at ambient temperature can cause changes that cannot be reversed by re-wetting, even if the re-wetted soil is allowed to mature for long periods.

Apart from the relatively well-known effect of drying on index properties of residual soils, drying also affects the composition, compressibility and shear strength characteristics of residual soils (e.g., Frost, 1976).

2.7.1 Water content

The conventional test for the determination of water content is based on the loss of water when a soil is dried to a constant mass at a temperature between 105 and 110°C. In many residual soils, some water exists as water of crystallization, within the structure of the minerals present in the elementary particles (see Figure 2.1). Some of this water may be removed by drying at oven temperature, i.e., not only free water is driven off. This is illustrated by Figure 2.5, which shows the effects on measured water content of different drying temperatures compared with standard tests dried at 105°C. As shown, the water contents may apparently increase significantly with drying temperature. Figure 2.6 shows how the apparent value of the water content increases progressively for four residual clays as the temperature of drying increases from 20°C to 40°C (and relative humidities of 30%) to the standard drying temperature of 105°C. The effect is even more pronounced for soils containing halloysite or allophane (Terzaghi, 1958; Frost, 1976; Wesley, 1973). The option of air-drying soils is problematic as this may take an extremely long time in a humid environment. The following procedure is therefore recommended: Two test specimens should be prepared for water content determinations. One specimen should be oven dried at 105°C until successive weighings show no further loss of mass. The water content should then be calculated in the normal way. The second sample should be air-dried (if feasible), or oven-dried at a temperature of no more than 50°C and a maximum relative humidity (RH) of 30% until successive weighings show no further loss of mass. The two water content results should then be compared: a significant difference (4–6% of the water content obtained by oven-drying at 105°C) indicates that "structural" water is present and is driven off at high temperatures. This water forms part of the soil solids, and should therefore be excluded from the calculation of water content. If a difference is detected using the two different drying procedures, all subsequent tests for water content determination (including those associated with Atterberg limit tests, etc.) should be carried out by drying at the lower temperature (i.e., either air-drying, or oven-drying at 50°C and 30% relative humidity). If possible, the lower drying temperature of 50°C should be used.

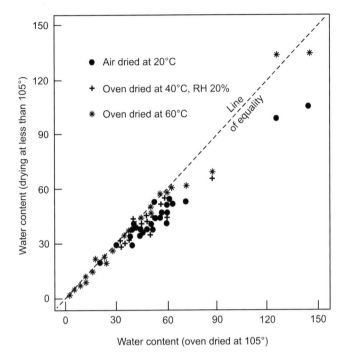

Figure 2.5 Effect of drying temperature on measured water content.

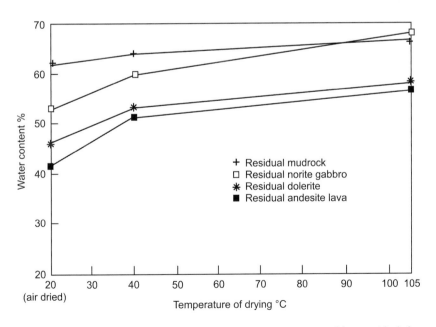

Figure 2.6 Effect of drying temperature on apparent water content of four residual clays.

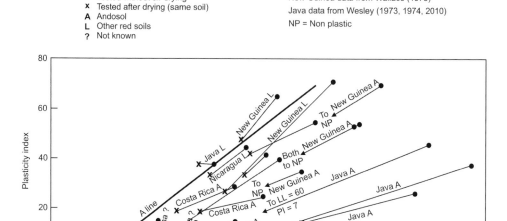

Figure 2.7 Effect of drying on plasticity of volcanic soils (Morin & Todor, 1975, Wesley, 2010).

2.7.2 Atterberg limits

In addition to the problem of water content determination discussed above, two further problems may be experienced when carrying out tests to determine the Atterberg limits, optimum water content and maximum dry density of residual soils. These are the effects of pre-test drying, and of duration and method of mixing.

The effect of pre-test drying

The effect of air-drying specimens prior to carrying out the Atterberg limit tests, rather than testing, starting at natural water content has been observed to result in an apparent decrease in the liquid limit and plasticity index, (Terzaghi, 1958; Rouse *et al*, 1986; Wesley & Matuschka, 1988; Wesley, 2010). This phenomenon is illustrated in Figure 2.7.

According to Townsend (1985), the effect of drying prior to testing may be attributed to:

1   increased cementation due to oxidation of the iron and aluminium sesquioxides, or
2   dehydration of allophane and halloysite, or
3   both 1 and 2 above.

In order to be meaningful, Atterberg limit tests on residual soils should therefore be performed without any form of drying prior to carrying out the test. If some form of drying is unavoidable, for example, if the water content exceeds the plastic limit,

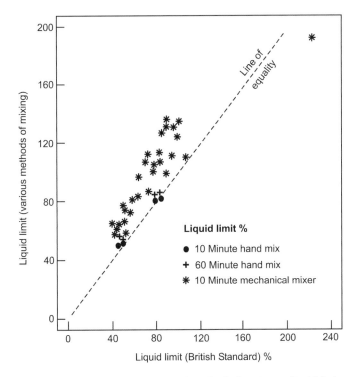

Figure 2.8 Effect of mixing time and method of mixing on liquid limit.

this should be noted on the laboratory report, and details given of the method and duration of drying.

Usually, Atterberg limit tests are carried out on that part of the soil finer than 425 μm. The above recommendation usually precludes separating this fraction. As far as possible, all large soil particles should be removed by working the undried soil with the fingers through a 2.38 mm opening sieve, so as to remove the large particles without breaking them up.

The effect of duration and method of mixing

In general, the longer the duration of mixing (i.e., the greater the energy applied to the soil prior to testing), the larger the resulting liquid limit, and to a lesser extent, the larger the plasticity index. This is because longer mixing results in more extensive break down of cemented bonds between clay clusters and within peds, and thus the formation of greater proportions of fine particles. This effect is illustrated in Figure 2.8. An extreme example of the effect of time of mixing on Atterberg limits is illustrated by Figure 2.9 which shows the increase in number of blows to close the groove in the Liquid Limit cup from 20 after minimal mixing, to 235 after 25 minutes of mixing. The soil in this case was a residual mud-rock.

In order to address this problem, the following procedure is recommended.

Five test specimens should be mixed with water to give a range of water contents suitable for liquid and plastic limit determinations. The minimum amount of air drying

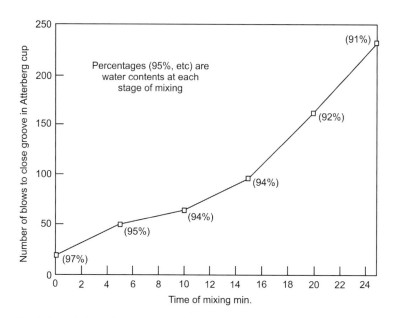

Figure 2.9 Effect of time of mixing on numbers of blows to close groove in liquid limit cup.

should be used, and preferably none at all. This should not be too difficult as in most climates and with most soils the in-situ water content is below the plastic limit. The mixing time should be standardized to 5 minutes, and the mixed specimens should be kept in sealed containers for at least 24 hours (and preferably longer) to allow the water content to equilibrate with the solids before testing.

The liquid limit should be determined using a standard procedure (e.g., British Standard 1377, Part 2) with adequate (5 minutes), but not excessive further mixing. A sub-sample from each of the specimens used in the test should be used for determining the water content, using the procedure established after conducting the evaluation described in section 2.7.1 above. The remainder of each specimen should then be mixed continuously for a further 25 minutes before again determining the liquid limit. A significant difference (i.e., >5% of liquid limit obtained from a test on a specimen mixed for 5 minutes) between the liquid limits from tests using 5 and 30 minute mixing times indicates a disaggregation of the clay sized particles in the soil. If this disaggregation is confirmed by a repeat of the above procedure, the entire programme of testing should:

1   Limit the mixing times to no more than 5 minutes.
2   Make use of fresh soil for each water content point in Atterberg limit tests, as well as in compaction tests.

The soil should be broken down by soaking in water, and not by drying and grinding. The soil should be immersed in water to form a slurry, which is then washed through a 425 μm sieve until the water runs clear. The material passing the sieve should be collected and used for the Atterberg limit tests. The particles retained on the sieve should be dried and weighed. The percentage by dry mass passing the 425 μm sieve should be calculated and recorded.

It is common practice in soil mechanics laboratories to use distilled water for adjusting the water content of soils. This is a practice that should be abandoned, as adding distilled water may affect the soil properties. Where the water table is at a shallow depth, a sample of the ground water should be obtained and stored in a sealed polyethylene or glass container. This water should be used for increasing the water content of the soil, if required.

In cases where the soil contains appreciable quantities of soluble salts, distilled water should definitely not be used as it is likely to cause a change in the properties of the soil. The soluble salt content of the soil is most conveniently assessed by measuring the conductivity of a soil paste at the liquid limit. If the conductivity exceeds 50 ms/m (ms = millisiemens), distilled water should not be used for mixing. In this case an equivalent "soil water" can be prepared by making up a 1.5:1 by mass mixture of distilled water and soil (about a 3:1 by volume mix of water to soil). After allowing the mixture to settle, the clear water is decanted and used for adjusting the water content of the soil.

2.7.3 Particle relative unit weight (or specific gravity) G_s

The particle relative unit weight (G_s), also called the specific gravity, is used to calculate volumetric parameters such as void ratio and porosity. In residual soils, G_s may be unusually high or unusually low depending on whether the "particle" consists of a voidless solid particle of a heavy mineral, e.g., iron sesquioxide or a porous aggregation of elementary particles. It is thus essential that G_s be determined in the laboratory using an accepted standard test procedure, for the purpose of calculating void ratio, dry density, porosity, etc.

The soil to be used in this test should be at its natural water content. Pre-test drying of the soil should be avoided as this tends to reduce the measured G_s as compared with measurements at natural water content. The dry mass of the soil used in the test should be calculated by drying the soil specimen **after** the particle unit weight test has been completed. Depending on the outcome of the evaluation described in section 2.7.1 above, it may be necessary to air-dry the soil, or to dry at a reduced temperature.

2.7.4 Particle size distribution (psd)

As with the above parameters, the particle size distribution of a residual soil may be affected by certain aspects of sample preparation, as described below.

1   **Effect of drying:** The most widely reported effect of drying is to reduce the percentage that is reported as the clay fraction (finer than 2 μm). Drying of the soil prior to testing should be avoided. The soil sample should be split into two sub-samples: one for determining the water content (in order to calculate initial dry mass), and the other for the psd test. The latter sample should be immersed in a solution of dispersant such as dilute alkaline sodium hexametaphosphate, and thereafter washed through the standard nest of sieves. In this case, because dispersant has been added, the washing should also be done using dispersant solution.

2   **Chemical pre-treatment:** This should be avoided wherever possible. Pre-treatment with hydrogen peroxide is only necessary when organic matter is present. If it is

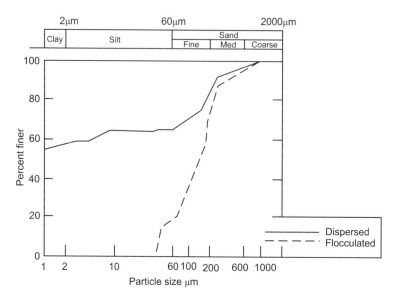

Figure 2.10 Grain size analysis of a lateritic soil without dispersion (broken line) and with dispersion (full line).

considered necessary to eliminate carbonates or sesquioxides, then pre-treatment with hydrochloric acid is used.

3   **Sedimentation:** It is essential to achieve complete dispersion of fine particles prior to carrying out a sedimentation test. The use of alkaline sodium hexametaphosphate is suggested. In some instances a concentration of twice the standard value may be required. If the above dispersant is ineffective, an alternative such as trisodium phosphate should be used. In all cases the dispersant solution should be freshly made before use in the laboratory.

Because of the tendency of residual soils to contain individual particle assemblages (IPAs) or peds (see Figure 2.1), dispersion is an absolute pre-requisite for particle size analysis. This is illustrated by Figure 2.10 (Rodriguez, 2005) showing an extreme example of particle size analyses for a lateritic soil. Without dispersion, the particle size distribution is that of a silty sand with no clay content. Dispersion changes this to that of a sandy clay with an extremely high clay content of 56%.

2.8 CLASSIFICATION OF RESIDUAL SOILS

There are specific characteristics of residual soils that are not adequately covered by methods of soil classification originally designed for transported soils, such as the Unified Soil Classification System (USCS). Among these are the following:

1   The clay mineralogy of some residual soils gives them characteristics that are not compatible with those normally associated with the group to which the soil belongs according to existing systems such as the USCS.

2   The soil mass may display a sequence of materials *in situ* ranging from a true soil to a soft rock depending on degree on weathering, which cannot be adequately described using existing systems based on classification of transported soils.

3   Soil classification systems such as the USCS focus on the properties of the soil in its remoulded state: this is usually misleading with residual soils, whose properties are likely to be strongly influenced by *in situ* fabric and structural characteristics relict from the original rock mass or developed as a consequence of weathering.

The term residual soil needs some clarification as it is not sufficiently well defined by the "translational" definition that opens Chapter 1. In any weathering process that converts rock into soil there will be a gradual transition from rock to soil. In the profile shown in Figure 1.9, the upper and intermediate zones will behave as soils, while the lower zone will behave as soft to hard rock. Hence the term residual soil should be applied only to the upper two zones. The term "saprolite" generally refers to a residual soil with clear structural features inherited or relict from its parent rock, and could be used for features present in the intermediate zone, although any relict features in the upper zone have usually been obscured by weathering or re-working by termites and other agents of disturbance.

2.8.1   The place of the USCS in classifying residual soils

The USCS divides soils, primarily by particle size, into gravels (G), sands (S), inorganic silts (M) and clays (C) and organic silts (OM). It has been shown (Figure 2.1) that because of the origin of residual soils, the concepts of "gravel" and "sand", and even "silt" cannot be applied to residual soils without ambiguity. The USCS, however, does use the Casagrande plasticity chart shown in Figure 2.11a to display the relationship between "soil liquidity" (via the liquid limit LL) and "soil plasticity" (via the plasticity index PI = (LL − PL), where PL = plastic limit). In Figure 2.11a, the (second) letters L, H and I denote low, high and intermediate plasticity, respectively. This chart can be used to display the same relationship for residual silt and clay sized soils, as shown in Figure 2.11b (after Wesley, 1988, 2010). This shows the effect of the predominant clay mineral on the position of the soil on the chart. By inference, the LL and PI of a given residual soil can be used, in conjunction with Figure 2.11b, to establish the likely predominant clay mineral. (Note the enormous extension of the scales in Figure 2.11b from the original Casagrande version of the plasticity chart, by 2.5 times, to accommodate the very high liquid limits and plasticity indices that occur in residual soils of volcanic origin.)

The USCS originally appeared as the "Casagrande soil classification for airfield projects" (Casagrande, 1947; Taylor, 1948). It came complete with a design chart based on the California Bearing Ratio (CBR) where the measured CBR could be replaced by the soil classification. (Before publication, the Casagrande method was used extensively by the US Airforce for designing airfields in newly conquered territories during the Second World War. Considering that it was as likely to be used for residual as transported soils, it would be interesting to know the method's success rate, during its war-time application.)

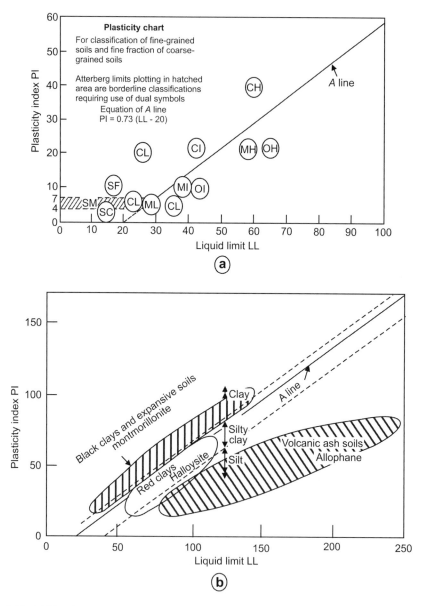

Figure 2.11 (a) The Casagrande plasticity chart. (b) Influence of predominant clay mineral on position on plasticity chart.

2.8.2 Wesley's classification scheme for residual soils

Wesley (1988) proposed a practical system for classifying all residual soils, based on mineralogical composition and soil micro- and macrostructure. Wesley's classification system is intended to provide an orderly division of residual soils into groups which

Table 2.1 Characteristics of residual soil groups (After Wesley, 2010).

Group				
Major Group	Subgroup	Examples	Means of Identification	Comments on Likely Engineering Properties and Behaviour
GROUP A Soils without a strong mineralogical influence	(a) Strong macrostructure influence	Highly weathered from acidic or intermediate igneous or sedimentary rocks	Visual inspection	Large group of soils (including "saprolites"). Behaviour dominated by influence of discontinuities
	(b) Strong microstructure influence	Completely weathered from igneous or sedimentary rocks	Visual inspection, evaluation of sensitivity	Essentially homogeneous. Nature and role of bonding from relict primary or weak secondary bonds important
	(c) Little or no structural influence	Soils formed from very homogeneous rocks	Little or no sensitivity, uniform appearance	Likely to behave like moderately overconsolidated soils
GROUP B Soils strongly influenced by normal clay minerals	Smectite (montmorillonite) group	Dark coloured soils formed in poorly drained conditions	Dark colour (grey to black) and high plasticity	Found in flat and low-lying areas, having low strength, high compressibility, and high swelling and shrinkage characteristics
GROUP C Soils strongly influenced by clay minerals found only in residual soils	(a) Allophane subgroup	Soils weathered from volcanic ash in the wet tropics and temperate climates	Position on plasticity chart, irreversible changes on drying	Very high natural water contents and Atterberg limits. Engineering properties generally good, though high sensitivity may make earthworks difficult
	(b) Halloysite subgroup	Soils derived from volcanic material, especially tropical red clays	Reddish colour, well-drained topography, volcanic origin	Fine-grained soils of low to medium plasticity, and low activity. Engineering properties generally good
	(c) Sesquioxides: gibbsite, goethite, haematite	Laterites, or lateritic clays	Nonplastic or of low-plasticity granular or nodular	Behaviour ranges from low-plasticity silty clay to gravel. End products of very long weathering process

belong together because of common factors in their formation and/or composition that can be expected to give them similar engineering properties. The system is based on a grouping framework designed to enable engineers to place any particular residual soil into a specific category on the basis of common engineering properties.

This scheme, which is set out here in its latest published form (Wesley, 2010), is not intended to provide an all-embracing method for the detailed systematic description or logging of residual soils in the field, or in the laboratory. Neither is it intended as a replacement for any other particular method of classification such as the USCS. The classification is set out in Table 2.1.

2.9 EXAMPLE OF CLASSIFICATION OF A RESIDUAL SOIL

As an example, the soil forming the weathered residual andesite profile illustrated in Figure 1.13 will be classified. The mineralogy of the profile is shown in Figure 2.4a:

1   The clay minerals comprising the soil consist (Figure 2.4a) of kaolinite, muscovite and chlorite, together with quartz. The relative proportions of these minerals vary with depth, but cannot be said to have a strong influence on the properties of the soil. Thus the soil falls within Group A.
2   The saprolitic joints in the soil exercise an important influence on the strength of the soil, so it falls into sub-group (a). (Joints in the parent andesite probably resulted mainly from cooling stresses.)

Hence the soils classifies as: A (a).

REFERENCES

Arnold M. (1984) The genesis, mineralogy and identification of expansive soils. *5th Int. Conf. on Expansive Soils, Adelaide, Australia.* pp. 32–36.

Baynes, F.J. & Dearman, W.R. (1978) The microfabric of a chemically weathered granite. *Bull. Int. Assoc. Eng. Geol.,* 18, 91–100.

Blight, G.E. (1996) Properties of a soil residual from andesite lava. *4th Int. Conf. on Tropical Soils, Kuala Lumpur, Malaysia.* pp. 575–580.

Bowen, N.L. (1928) *The Evolution of the Igneous Rocks.* Dover, New York. USA.

Brand, E.W. (1988) Evolution of a classification scheme for weathered rock. In: *Geomechanics in Tropical Soils.* Rotterdam, A.A. Balkema, 2, 515–518.

Brummer, R.K. (1980) *The Engineering Properties of Deep Highly Weathered Residual Soil Profiles.* MSc(Eng) dissertation, Witwatersrand University, Johannesburg, South Africa.

Casagrande, A. (1947) Classification and identification of soils. *J. Geotech. Eng. Div., ASCE,* June, 295–310.

Collins, K. (1985) Towards characterization of tropical soil microstructure. *1st Conf. Geomech. in Tropical Lateritic and Saprolitic Soils, Brasilia, Brazil.* Vol. 1, pp. 85–96.

Falla, W.J. (1985) *On the Significance of Climate and Weathering in Predicting Geotechnical Characteristics of Residual Soils Developed on Igneous Rocks.* PhD Thesis, Witwatersrand University, Johannesburg, South Africa.

Frost, R.J. (1976) Importance of correct pretesting preparation of some tropical soils. *1st S.E. Asian Conf. Soil Eng.,* Bangkok, Thailand. pp. 210–216.

Futai, M.M., Almeida, M.S.S. & Lacerda, W.A. (2004) Yield strength and critical state conditions of a tropical saturated soil. *J. Geotech. & Geoenv. Eng., ASCE*, 130 (11), 1161–1179.

Knight, K. (1958) Problems of foundations on collapsing soils. *Trans. South African Inst. of Civ. Engrs.*, 8 (10), 304–307.

Morin, W.J. & Todor, P.C. (1975) *Laterite and Lateritic Soils and Other Problem Soils of the Tropics*. Baltimore, USA, US Aid, Lyon Associates.

Queiroz de Carvalho, J.B. (1981) Study of the microstructure of lateritic soils using scanning electron microscope. *Int. Sem. Lateritization Processes, Sao Paulo, Brazil*. Vol. 1, pp. 563–568.

Queiroz de Carvalho, J.B. (1985) Effects of deferation and removal of amorphous silica and alumina on selected properties of Brazilian lateritic soils. *Int. Sem. Laterite, Tokyo, Japan*. Vol. 2, pp. 559–567.

Queiroz de Carvalho, J.B. (1991) Microstructure of concretionary lateritic soils. *9th Panamerican Conf. Soil Mech. & Found. Eng.* Vol. 1, pp. 117–128.

Richards, B.G. (1985) Residual soils of Australia. In: Brand E.W. & Phillipson H.B. (eds), *Sampling and Testing of Residual Soils*. Hong Kong, Scorpion. pp. 23–30.

Rodriguez, T.T. (2005). Colluvium classification: a geotechnical approach, PhD. Thesis; COPPE/UFRJ, Brazil. Quoted in: Lacerda, W.A. (2006) *Landslide Initiation in Saprolite and Colluvium in Southern Brazil: Field and Laboratory Observations, Geomorphology*. Netherlands, Amsterdam, Elsevier.

Rouse, W.C., Reading, A.J. & Walsh, R.P.D. (1986) Volcanic soil properties in Dominica, West Indies. *Eng. Geol.* 23, 1–28.

Rowe, P.W. (1974) The importance of soil fabric and its relevance in engineering practice. *Géotechnique*, 24 (3), 265–310.

Simmons, J.V. (1989) *Preliminary Studies of Dispersive Soil Fabrics from the Burdekin River Irrigation Area using the Scanning Electron Microscope*. Res. Bull. CS38, Department of Civil and Systems Engineering, James Cook University of North Queensland, Townsville, Australia.

Taylor, D.W. (1948) *Fundamentals of Soil Mechanics*. New York, Wiley.

Terzaghi, K. (1958) Design and performance of Sasumua dam. *Proc. Instn. Civ. Engrs., UK.*, 9, 369–394.

Townsend, F.C. (1985) Geotechnical characteristics of residual soils. *J. Geotech. Eng., ASCE*, 111 (1), 77–94.

Wesley, L.D. (1973) Some basic engineering properties of halloysite and allophane clays in Java, Indonesia. *Géotechnique*, 2 (4), 471–494.

Wesley, L.D. (1988) Engineering classification of residual soils. *2nd Int. Conf. Geomech. Tropical Soils, Singapore*. Vol. 1. pp. 77–83.

Wesley, L.D. (2010) *Geotechnical Engineering in Residual Soils*. Hoboken, USA, Wiley.

Wesley, L.D. and Matuschka, T. (1988) Geotechnical engineering in volcanic ash soils. *2nd Int. Conf. Geomech. Tropical Soils, Singapore*. Vol. 1, pp. 333–342.

Chapter 3

Describing the engineering properties of residual soils as observed *in situ*

G.E. Blight & J.V. Simmons

Sites identified for construction of engineered facilities should be explored in detail by qualified and experienced professional geotechnical engineers or by qualified professional engineering geologists who have specialized in work of this nature. The final decision on the suitability and feasibility of a site cannot be taken until subsurface conditions at the site have been fully explored, described and recorded.

The following information should be collected in respect of every potential site for engineering development.

3.1 SOIL ENGINEERING SURVEY

Every civil engineering project should be preceded by a detailed soil engineering survey in order to establish the different types of soil present at the site and the areas and boundaries within which they occur.

- Typical soil profiles, depths to bed rock, the position of the ground water table, preliminary estimates of the permeability or impermeability of the soils and bed rock and their *in situ* shear strength should all be established.
- Any indications of the presence of dykes, faults, sills and other geological features which were not revealed on available geological maps should be carefully recorded.
- All potential problems with soils in the area should be noted (e.g., the presence of expansive or collapsing soils, karst, dispersive soils, etc.).
- Geophysical methods are attractive for site investigation and, in particular, as a means of extending the basic information provided by test pits or boreholes: In karst areas gravimetric surveys and thermal imaging are useful in locating possible sinkhole or subsidence areas which correspond to gravimetric lows or thermal contrasts. Microseismic methods are useful for locating the depth to bed rock. Resistivity surveys may be used to obtain information on water table depths.

3.2 SOIL ENGINEERING DATA

Soil engineering information is most readily obtained by means of large diameter (about 1 metre) boreholes which can be drilled rapidly and economically by means of a pile hole auger. Holes dug by a back actor shovel may also be used. However, safety considerations limit the depth of an unsupported hole to 1.5 m.

Whenever a person enters a test hole for the purpose of recording the soil profile, all recognized and/or required safety precautions must be taken, including supporting the sides of the hole to prevent possible collapse.

Particular care should be taken when entering a hole that:

- has penetrated the water table and into which water is seeping,
- has stood open for more than eight hours since excavation,
- is possibly filled with poisonous or inert gas.

With regard to the latter point, holes dug in the vicinity of an existing mine waste storage or sanitary landfill should be regarded with particular suspicion, as they may fill with methane and carbon dioxide in the case of a landfill, or with carbon dioxide or other gases in the case of a mine waste storage. A grass or bush fire passing over the top of a hole may also cause it to fill with carbon dioxide and carbon monoxide, both of which are heavier than air. It is a good practice to drop a wad of burning rags or paper down the hole to observe whether or not it is extinguished by carbon dioxide. Stand well clear of the hole and wear goggles in case the gas explodes and throws up a shower of dust and soil particles.

Safety demands that no-one ever descends into a hole if he is alone, that the observer wear a hard hat and that he be linked to the surface by a rope fastened around his body by means of a harness designed to keep his head uppermost.

If large diameter drilled exploration holes, pits or trenches are not feasible, it may be necessary to explore the site by means of small diameter (75 mm, 100 mm or 150 mm) boreholes. The actual siting, distribution and number of holes will depend on the characteristics of a particular site and should be decided by the geotechnical engineer or engineering geologist on the basis of available knowledge of the local geology and from features visible at the surface.

In addition to the geological and geotechnical information which is collected during the course of the site survey, the following information should also be obtained:

- any evidence of local seismicity and the magnitude of possible seismic events in the area (available from the local or national geological or seismic survey department),
- mean and extreme rainfall distributions for the area as well as mean and extreme monthly pan evaporation (available from the local or national weather office),
- maximum rainfall intensities for each month for 1 hour, 24 hours, 24 hours in 50 years and 24 hours in 100 years storm events,
- mean monthly wind direction and speed and hourly wind direction and speed with the maximum one minute gust speed in each hour (a series of monthly wind roses is preferable, if available).

3.3 DETAILED INFORMATION ON STRENGTH AND PERMEABILITY

The following detailed information is required to enable the design of possible basement or foundation excavations, basement walls, retaining walls, drainage systems or seepage cut-offs, load-bearing foundations, slab-on-grade floors, etc.

- The permeability of the foundation soil or of particular strata in the foundation soil: The permeability is best measured by means of the *in situ* tests which are

described in Chapter 5. Supplementary measurements of permeability may be made in the laboratory. Depending on the purpose of the exploration, a sufficient number of permeability or infiltrometer tests should be performed around the perimeter and covering the site to ensure that representative permeability values are available for design purposes.

- The shear strength of the foundation soil: In order to design the slopes, retaining walls and foundations, it is necessary to have information on the shear strength of the soil, and in particular the shear strength of any unusually weak strata. The shear strength may be measured using a number of *in situ* methods of testing of which the most suitable appear to be the vane shear test and the pressuremeter which are described in sections 7.3.3 and 7.3.4. Alternatively, consolidated undrained or consolidated drained triaxial tests or shear box tests on undisturbed specimens taken from the field may be performed in order to establish the shear strength parameters of the soil in terms of effective stresses. These tests are described in section 7.2. Whichever methods are used, sufficient tests should be made to establish the shear strength properties of all the soil types that occur on the site. (Note: *In situ* test methods are most suitable in conditions of high water table where access to the soil from the surface is difficult. Where water levels are deep (greater than 5 metres) and in dry climates where covering the surface of the soil with a building is likely to have a pronounced local effect on moisture conditions in the soil, (i.e., wetting of the soil), laboratory methods of measuring shear strength of undisturbed specimens taken from the foundation soil may be more suitable as the effect of changing the ground water regime can be simulated in these tests.

- In many cases it may be necessary to consider both short term and long term conditions for a development, depending on the purpose and type of development being planned, the characteristics of the foundation soil and other factors. If both long term and short term stability are to be considered it may be necessary to have information not only on the *in situ* strength of the foundation soil in terms of total stresses but also to measure the strength parameters in terms of effective stresses by means of triaxial shear or shear box tests in the laboratory (see section 7.2).

- The compressibility of the soil strata is always very relevant and with certain soils, e.g., loose, potentially collapsing sands, or potentially expansive clays, it may be necessary to estimate the settlement or heave that the foundation strata will undergo. If compressibility or expansion is thought to be a potential problem, the compressibility or expansibility of the soil strata or any particular stratum may be measured by means of laboratory or *in situ* tests (see section 6.4).

3.4 PROFILE DESCRIPTION

Because all soil sampling involves some degree of disturbance, it is necessary to distinguish facts related to the soil, from facts related to, or affected by, the method of inspection or sampling. The engineering properties of the material in its natural field state may be difficult to assess, except by including observations of the behaviour of excavation equipment, drills, or probes during the investigation (e.g., by the rate of advance of a drill, or rate of penetration of a driven probe).

Often, for legal purposes or because clients believe that they are getting more cost-effective information, only factual reporting is required. In practice, however, some degree of interpretation is essential. Whatever the case, reporting should always clearly distinguish between factual information and interpreted information. For example, classification information is always based on interpretation. It is good practice to indicate interpretation clearly on logging sheets, either by use of parentheses or by clearly marked sections for "interpretation".

3.4.1 Profile description procedures

Procedures for profile description and soil sampling have been developed by a wide variety of organizations, and are set out in various procedure manuals or codes of practice. Most of these procedures were developed many years ago and still represent acceptable recording procedures (e.g., Hvorslev, 1948; Jennings, Brink & Williams, 1973; Brink, 1979; Cook & Newill, 1988). Particular procedures may be adapted or extended for other purposes. There is a temptation to include detail for its own sake rather than for specific purposes. This detail may, however, become valuable later on, for reasons not earlier anticipated. For example, the results of the original survey and profile may be used for a different purpose, if the original project is abandoned and a different one is later embarked on.

The following procedure list is not exhaustive, but can serve as a basis for most purposes. It may be supplemented or reduced according to requirements. The appropriate sections of locally applied codes of practice may also be substituted. The final choice for use of any such lists lies with the person responsible for the fieldwork.

3.4.2 Site records

The following should be recorded:

- general description of site, general location, vegetation types and distribution across site, access routes (if remote from existing towns or villages),
- dates of site investigation,
- weather during investigation,
- precise location details (co-ordinates, marks, beacons, reference features),
- all field activities (diary, logging forms, equipment used).

Descriptions should be recorded for the following aspects of the soil:

- Moisture condition (M),
- Colour (C) (Best described based on standard colour charts e.g., the well-known Munsell soil colour charts or the Burland chart (1958) – see Plate C24. Multicolours, e.g., mottling, spotting, striping may also occur),
- Consistency (C),
- Soil (e.g., clay, silt, sand, gravel, sandy silt, silty clay, etc.) (S),
- Structure or fabric (zoning, fissuring, slickensiding, cementing, quartz veins, nodules etc.) (S),
- Origin (transported or residual). If transported, likely transport agent, e.g., wind-blown, delta deposit, fluvial. If residual, parent rock, e.g., quartzite, granite (O).

The mnemonic MCCSSO is useful to remember when recording data for a soil profile. Field notes must include entries under each of M, C, C, S, S and O. Field logs will include information from fieldwork alone. Therefore, descriptions of plasticity, moisture condition, etc. will be qualitative field descriptions. Subsequent laboratory testing may cause modification of the plasticity description, and allow inclusion of numerical moisture content values.

The soil profile must be observed in a freshly drilled or dug trial hole, inspection pit or on freshly taken borehole cores It will usually consist of layers which can be discerned by means of changes in moisture condition, colour, and consistency, the presence or absence of joints or fissures, or by changes in the grain size or grain size distribution.

A convenient procedure for down-hole observation is to fasten the zero end of a measuring tape at the surface on the northern side of the hole so that the observer can orientate himself as he observes the depth, thickness, direction of dip, etc. of the various layers, even when he is unsighted because his head is below the soil surface. This is very necessary, as it is difficult to judge orientation when no land-marks are visible.

As the observer descends the hole by means of a chain ladder or bosun's chair the moisture condition, colour, consistency, structure, soil type and origin of each layer of the soil profile (MCCSSO) should be observed and noted as well as the depth and thickness of each.

The observer can either note his observations in a field note book or a small voice recorder, or (a better, safer practice) can call up his observations to an assistant on the surface who records the observations, and prompts the observer to provide information under each of the categories MCCSSO. The assistant will also quickly become aware and act if the observer encounters any difficulties, e.g., a partial collapse of the hole, or a partial or complete loss of consciousness, or confusion as a result of encountering poisonous or asphyxiating gas.

(MCCSSO) moisture condition

This is recorded as dry, slightly moist, moist, very moist or wet. The moisture condition provides a useful indication of water requirements for compaction, should the soil be used in construction as fill. Dry and slightly moist materials will require additional water to attain the optimum moisture content for compaction. Moist materials are near the optimum moisture content while very moist soils will require drying. Wet soils are generally only found below the water table.

(MCCSSO) colour

Colour is used for describing the soil and for identifying continuation of the same layer in different holes. Colours are difficult to describe so that the reader will "visualize" the same colour in his mind's eye as the observer describing the colour. It is useful to use a standardized soil colour chart such as the Burland (1958) chart. This is conveniently small in size and suitable for carrying in an overall pocket. Plate C24 shows the Burland chart which has the segments of colour on one side and the colour names on the reverse. The actual size of the pocket chart is 80 mm × 80 mm.

The natural colour as seen in the freshly exposed soil, i.e., the colour "in profile", should be noted, e.g., "light grey mottled yellow", the predominant colour being noted

first. Some observers carry a small bottle of water with them and form a soil paste with water. The colour of the paste is then also recorded.

(MCCSSO) consistency

Consistency is a measure of the strength of the soil and is based on the effort required to dig into the soil with a geological pick, or to mould it in the fingers. Different sets of terms are used to describe the consistency of granular non-cohesive soils and clayey cohesive soils, as in Table 3.1. A pocket shear vane or penetrometer is also useful to measure the strength. These instruments do not work satisfactorily in stoney or gravelly soils.

(MCCSSO) structure

Intact indicates an absence of fissures or joints.

Fissured indicates the presence of closed joints. The joint surfaces are frequently stained with iron and manganese oxides. If observed, this should be reported.

Slickensided indicates the presence of closed fissures which are polished or glossy and sometimes striated. To observe a slickenside, take a fist-sized clod of soil and carefully break it apart to reveal any continuous glossy or striated surfaces (See Plate C20).

Table 3.1 Categories of soil consistency.

Granular Soils (usually clay free)		Cohesive Soils (usually clayey)		Shear Strength kPa
Very loose	Very easily penetrated by geological pick	Very soft	Easily moulded by fingers. Pick head can easily be pushed in	≤ 10
Loose	Small resistance to penetration by geological pick	Soft	Moulded by fingers with some effort. Sharp end of pick can be pushed in 30–40 mm	$>10 \leq 20$
Medium dense	Considerable resistance to penetration by geological pick	Firm	Very difficult to mould with fingers. Sharp end of pick can be pushed in 10 mm	$<20 \leq 50$
Dense	Very high resistance to penetration by geological pick	Stiff	Cannot be moulded by fingers. Slight indentation by pick	$>50 \leq 100$
Very dense	High resistance to repeated blows of geological pick	Very stiff	Slight indentation produced by blow of pick point	$>100 \leq 200$
Extremely dense	Repeated blows of geological pick make no impression	Hard	Point of pick scratches, but does not groove surface	≥ 200

Shattered indicates the presence of open fissures. Shattered soils are usually in a dry state. Fissures will usually narrow or close completely when the moisture content of the soil increases.

Stratified, laminated, foliated Many residual soils show the relict or saprolitic structure of their parent rock. Observation of this structure may assist identification of the parent rock.

(MCCS<u>S</u>O) soil type

The soil type is based on the predominant grain size.

Boulders are defined as particles of rock larger than 100 mm size. The rock types and range of sizes should be recorded as should the presence of a matrix in the voids, if present. The matrix should also be described. The shape of the boulders should be described as this often aids the interpretation of the origin. Terms are:

well-rounded (nearly spherical)
rounded (tending to spherical)
sub-rounded (corners rounded off)
sub-angular (corners removed)
angular (corners sharp or irregular).

Gravel consists of particles of rock between 100 mm and 2.5 mm size. The method of description follows that for boulders.

Sand consists of particles between 2.5 mm and 0.06 mm (100 mesh sieve) size. Most of the particles sizes are visible to the naked eye.

Silt consists of particles which are between 0.06 mm and 0.002 mm (2 µm) size. Individual particles cannot be distinguished by the naked eye, but the material feels gritty when rubbed against the teeth with the tongue.

Clay consists of particles smaller than 0.002 mm (2 µm) size. Clays feel greasy or soapy when wet.

Most natural soils have a combination of one or more particle size ranges. When describing such a soil, the adjective refers to the lesser size range, and the noun to the predominant size, e.g., a silty clay is a clay containing some silt. A silt-clay has approximately equal proportions of silt and clay and a clayey silt is a silt containing some clay.

(MCCS<u>SO</u>) origin

An attempt should be made to determine the origin of the soil in each layer of the soil profile. On many sites a pebble marker, or layer of gravel indicates the boundary between transported and residual soils. This generally represents a stratum of free drainage which, if drainage is required, may be retained and usefully employed for providing a free flow of water. It also indicates a level below which soil behaviour may be predicted from experience with similar decomposed rock types.

Transported soils

Possible origins of transported soils are as shown in Table 3.2, generally ranked from highest to lowest topographic elevation of occurrence.

Table 3.2 Possible origins of transported soils.

Transported soil type	Transport and deposition by	Soil types* (Usually mixed)
Talus	gravity (from cliff face or upper steep slope)	unsorted angular gravel and boulders
Hill wash or colluvium	sheet wash (erosion from hillside)	boulders gravel sand clayey sand silt clayey silt clay
Alluvium or gulley wash	flow in: river stream gulley	gravel sand silt clay
Lacustrine deposit	stream depositing in pan (pond with no outlet) pond lake	sand silt clay
Estuarine deposit	rivers and tides	sand silt clay
Aeolian deposit	wind	fine sand
Littoral (beach) deposit	waves	medium and coarse sand

*If any of these deposits are sufficiently ancient to have decomposed *in situ*, they become residual soils (see Plate C14).

Residual soils

A knowledge of the local geology and reference to geological maps, provides a guide to the origin of residual soils on any site. Residual soils may be recognized by the preservation of the primary (or saprolitic) rock structures inherited or relict from the parent rock, e.g., bedding planes, amygdales or characteristic rock jointing structures.

It is sometimes possible to identify primary and secondary minerals characterizing the mineralogical composition of their parent rock, e.g., residual soils derived from granite will contain quartz grains, mica flakes and kaolinite derived from the decomposition of feldspars. Those formed from rocks such as andesite, dolerite or diabase, will contain no quartz (other than secondary deposits of vein quartz) and will consist entirely of clay and silt. Amygdales in residual soil will identify its derivation from a volcanic lava such as andesite or basalt. Residual soils developed from sedimentary parent rocks are usually easy to identify from their inherited or relict structure, e.g., bedding structure, particle size distribution across bedding layers, etc.

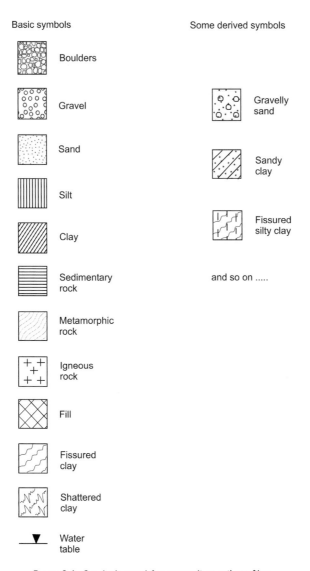

Basic symbols

Boulders

Gravel

Sand

Silt

Clay

Sedimentary rock

Metamorphic rock

Igneous rock

Fill

Fissured clay

Shattered clay

Water table

Some derived symbols

Gravelly sand

Sandy clay

Fissured silty clay

and so on .....

Figure 3.1 Symbols used for recording soil profiles.

3.4.3   Recording of soil profiles

The left hand side of the profile record should consist of a section, with the depth drawn to scale, which shows the various strata of the profile. Soil types and structure are indicated by the standard symbols shown in Figure 3.1. The section should also record the presence of such features as water tables and the pebble marker as well as the full depth of penetration of the hole.

To the right of each stratum in the section, a full description (MCCSSO) should be recorded, as shown in the examples in Figure 3.2.

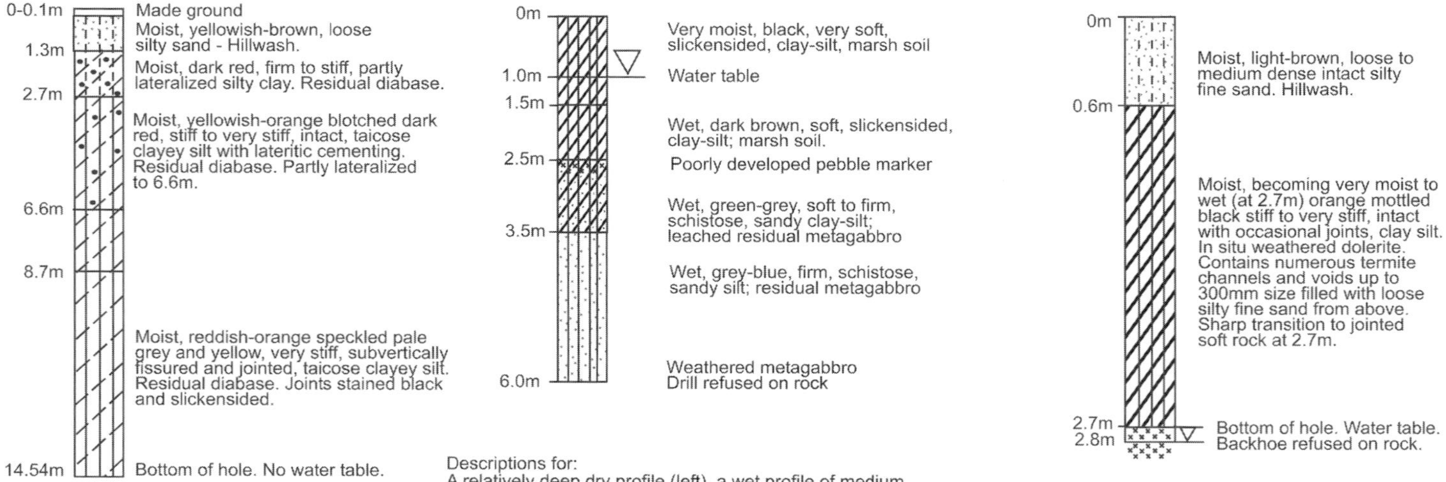
0-0.1m
1.3m
2.7m
6.6m
8.7m
14.54m
Made ground
Moist, yellowish-brown, loose silty sand - Hillwash.
Moist, dark red, firm to stiff, partly lateralized silty clay. Residual diabase.
Moist, yellowish-orange blotched dark red, stiff to very stiff, intact, taicose clayey silt with lateritic cementing. Residual diabase. Partly lateralized to 6.6m.
Moist, reddish-orange speckled pale grey and yellow, very stiff, subvertically fissured and jointed, taicose clayey silt. Residual diabase. Joints stained black and slickensided.
Bottom of hole. No water table.
0m
1.0m
1.5m
2.5m
3.5m
6.0m
Very moist, black, very soft, slickensided, clay-silt, marsh soil
Water table
Wet, dark brown, soft, slickensided, clay-silt; marsh soil.
Poorly developed pebble marker
Wet, green-grey, soft to firm, schistose, sandy clay-silt; leached residual metagabbro
Wet, grey-blue, firm, schistose, sandy silt; residual metagabbro
Weathered metagabbro
Drill refused on rock
0m
0.6m
2.7m
2.8m
Moist, light-brown, loose to medium dense intact silty fine sand. Hillwash.
Moist, becoming very moist to wet (at 2.7m) orange mottled black stiff to very stiff, intact with occasional joints, clay silt. In situ weathered dolerite. Contains numerous termite channels and voids up to 300mm size filled with loose silty fine sand from above. Sharp transition to jointed soft rock at 2.7m.
Bottom of hole. Water table.
Backhoe refused on rock.

Descriptions for:
A relatively deep dry profile (left), a wet profile of medium depth (centre) and a shallow profile (right). Note that the description of each separate soil type is given in terms of the MCCSSO system, recording moisture, colour, consistency, structure, soil and origin. Note also that the 'pebble marker' is a layer of gravel that often marks the transition from transported soils (above) to residual soils (below).

Figure 3.2 Examples of three soil profile records.

Water table

It is important to establish and record the depth of the water table (or phreatic surface) on the virgin site. If there are any indications of the presence of a perched water table, this should be noted and an attempt made to establish its depth as well. The depth of a water table is indicated by the level at which water trickles into the hole. However this level may be that of a perched water table, not the main water table. Also, in soils of low permeability, the flow of water towards the hole may be too slow to show as a trickle. It is therefore always best to install a stand-pipe piezometer in the profile inspection hole before back-filling it. The stand-pipe could be as simple as a length of plastic electrical conduit with the lower end perforated with 3 mm diameter holes over the bottom half metre and the perforated length wrapped in needle-punched felt geofabric. If possible, the hole around the perforated end should be backfilled with clean sand or gravel before back-filling and compacting the rest of the hole with the spoil that came out when digging or drilling it.

It may take several days or even a week for the water level in the piezometer to stabilize. The water level in the stand-pipe can easily be established by using an electrical dip meter, or if not available, pushing a length of small-bore plastic tubing down the standpipe to the bottom. If there is water in the tube, the sound of bubbling will be heard when the tube is blown into by mouth. By blowing and slowly withdrawing the tube, the level of the water can be established as the point at which the bubbling sound stops. The water table depth is then the length of small-bore tube remaining in the standpipe, measured to ground level.

3.5 SIMPLE *IN SITU* TESTS AND SOIL SAMPLING

It is usually convenient to augment the visual and qualitative soil assessment described in section 3.4 by simple *in situ* tests that can be carried out in the test hole at the same time as the visual assessment. The hand held penetrometer and hand vane instruments are particularly useful for this purpose.

Sampling can also be carried out to obtain material for index and compaction testing. These are usually disturbed samples, taken by digging out sufficient soil for the purpose of the tests from the sides of the test holes at appropriate depths. The quantity required will vary from a few kilograms for Atterberg Limit tests and particle size analyses to 20 to 30 kilograms for compaction tests. As the samples are disturbed and the soil will be remoulded at various water contents in order to perform the tests, the samples are usually collected in strong water proofed canvas or plastic bags, carefully labeled and the bags are closed by tying their mouths with thick string or thin rope. Plastic bags (plastic buckets with push-on sealed lids are still better) are preferable as the soil should be kept as close to its *in situ* water content as possible until it is tested. All Atterberg Limit and compaction tests should be started from the *in situ* water content, as air-drying or (worse) oven-drying of a soil can markedly alter its Atterberg limit and compaction characteristics (See section 2.7.2 and 4.5).

It may also be convenient to take undisturbed block samples for laboratory testing to establish shear strength and compressibility parameters. Block samples can only, of course, be taken above the water table. To do this, (see Figure 3.3) it is convenient to trim a pedestal of soil out of the side or bottom of the test pit (depending on available space) to fit an open-ended cylinder (e.g., a 200 mm or 300 mm diameter by 200 or

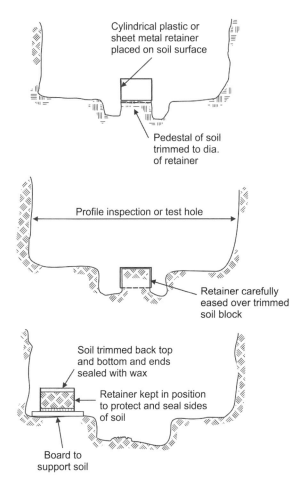

Figure 3.3 Procedure for cutting an undisturbed block sample of soil from a profile inspection or test hole.

300 mm long piece of rigid plastic or thin steel pipe). The cylinder is worked down over the soil pedestal by trimming away excess soil. Once the cylinder fits over the soil pedestal, the pedestal is cut off at its base and excess soil trimmed away. The ends of the cylinder can then be sealed with purpose-made caps, by means of melted paraffin wax, or even by wrapping in several layers of "cling-wrap" plastic wrapping.

3.6 TAKING UNDISTURBED SOIL SAMPLES FOR LABORATORY TESTING

The purpose of taking "undisturbed" samples is to subject the soil in the laboratory to the stress changes and drainage conditions it is envisaged will be applied in the field by the prototype structure, i.e., the soil sample is a model of the soil in the field.

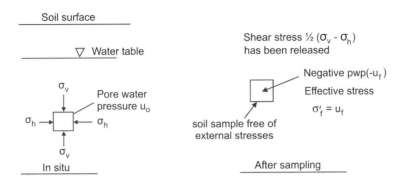

Figure 3.4 Changes in stress on a soil below the water table, caused by sampling.

Therefore it must represent the *in situ* soil as closely as possible in structure, void ratio and water content. Sampling procedures must be suited to the material being sampled.

3.6.1 Sampling of saturated clay soils below the water table

The relationship between void ratio e, water content w and particle relative density G for a saturated soil is:

$$e = wG$$

Hence a saturated soil is incompressible (e is constant) provided the water content does not change. In a clayey soil, removal of a sample from below the water table to the surface should not result in a change of water content as the soil will retain the water by capillarity. The soil will, however, undergo a distortion and a change of pore pressure as a result of the release of the *in situ* direct and shear stresses, as illustrated by Figure 3.4.

3.6.2 Sampling firm to stiff saturated clays above the water table

The usual method is to use a 76 mm internal diameter thin-walled open-drive sampler like that illustrated in Figure 3.5a. The sample is taken by hydraulically pushing the sampler into the soil at the bottom of an augered borehole. Care must be taken not to "overdrive" the sampler, i.e., not to compress the soil by pushing in the sampler further than the length of empty sampler tube. The drill rods are then twisted to shear off the sample at its base. The sampler is withdrawn, and the sample extruded carefully into a plastic film tube, placed in an appropriately marked and identified cardboard tube and waxed to seal and support it during transport to the laboratory and storage before testing. The soil must be extruded from the sample tube in the same direction as it moved into the tube, i.e., first in, first out.

Great care must be taken in packing undisturbed samples for transport to the laboratory. Sample tubes should be laid horizontally in a well-made wooden core box and be packed in a shock-absorbing packing such as plastic foam, polystyrene beads or failing these, sawdust or wood shavings.

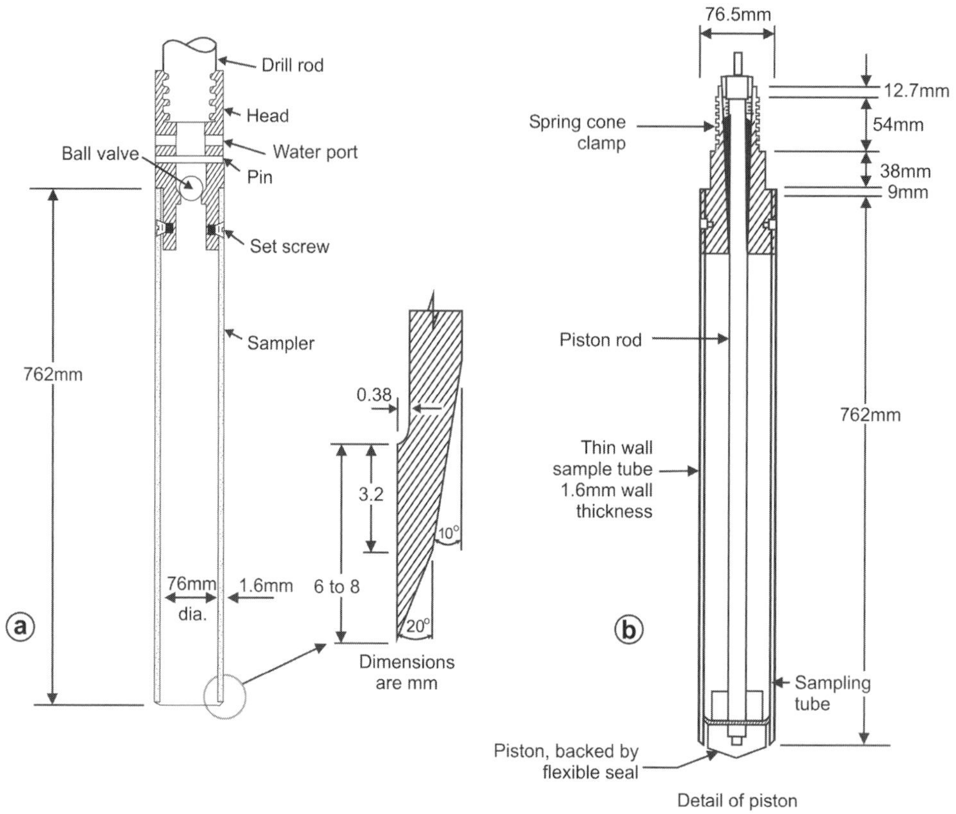

Figure 3.5   (a) Thin-walled tube open-drive sampler, used to sample firm to stiff soils above the water table. (b) Thin-walled stationary piston sampler, suitable for soft soils, and soils below the water table.

3.6.3   Sampling soft saturated clays and silts

When an open-drive sampler is pushed into the soil, the sample is released from the side restriction provided by the surrounding soil, and tends to expand during the initial stages of penetration. However, as the tube is pushed in further, side friction builds up inside the tube and the sample is then slightly compressed, as illustrated in Figure 3.6 by the progression A, B, C. To prevent the disturbance caused by these changes in length, piston samplers, as shown diagrammatically in Figure 3.5(b) are used. The piston is initially flush with the cutting edge. Rods connect the piston to the surface: When the sampler is ready for sampling the piston rests on the surface of the soil to be sampled. The piston is then clamped in position and the sampling tube is pushed past it. Thus, while the sampling tube penetrates into the soil the piston remains stationary at the original level of the top of the sample, ensuring that the length of the sample cannot change during sampling. An automatic locking device or spring cone clamp is incorporated in the sampler head so that the piston will not be forced into the

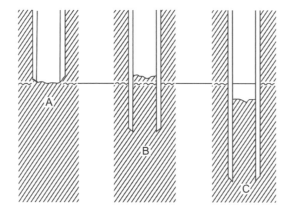

Figure 3.6 Effects of pushing an open-drive sampling tube into an undisturbed soil.

empty sampler tube by hydrostatic pressure before driving and cannot slip down while the sampler is being pulled up after the clamp at the surface has been released. This procedure is known as fixed piston sampling.

Another system sometimes used is that of a free piston. In this case the piston is pushed forward but the rods are not connected to the surface. This means that, when the sampling tube is pushed down, the piston "floats" on the top of the sample. When the sampler is withdrawn the piston automatically locks and the sample is retained in the tube.

Although the piston sampler was developed for sampling soft clays, it has also been used successfully to sample soft saturated silts, below the phreatic surface.

3.6.4 Important soil sampler dimensions and their function

Open-drive and piston sample tubes have a number of characteristics that can be quantified and are of importance in obtaining good undisturbed samples. These are: The amount of disturbance caused to the soil in the sampling operation depends to a large degree on the "area ratio" between the cross-sectional area of the metal annulus of the sampling tube and the cross-sectional area of the sample. For the entrance to the sampling tube shown in Figure 3.7 the area ratio C_a is:

$$C_a = \frac{(D_w^2 - D_e^2)}{D_e^2} \leq 10\%$$

It has been shown that, for large samplers, the disturbance due to displacement of soil by the tube itself is very small provided the area ratio is 10% or less (Hvorslev, 1948). To achieve such a low area ratio the sampling tubes are made from thin-walled, seamless, high strength steel tubing.

Another cause of disturbance is friction on the inside of the tube, which also limits the length of sample which can be taken with a minimum of disturbance. This internal friction can be reduced by providing the sampling tube with a smooth interior surface

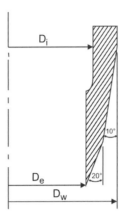

Figure 3.7 Dimensions of the cutting shoe of an open-drive sampler.

having a low coefficient of friction and by making the diameter of the cutting edge D_e slightly smaller than the inside diameter of the sampling tube D_i. This clearance is expressed by the inside clearance ratio,

$$C_i = \frac{(D_i - D_e)}{D_e} = 0.5\% \text{ to } 1.5\%$$

The Recovery Ratio RR is defined by:

$$RR = \frac{\text{Length of sample}}{\text{Distance sampler was driven}}$$

The Recovery Ratio should lie within 96% to 104% for satisfactory samples. Samples with RR's outside the limits of 96% to 104% should be rejected for strength or compressibility tests.

3.6.5 Sampling very stiff and hard soils

In order to sample very stiff and hard soils, a rotary core barrel must be used to recover a satisfactory undisturbed core. To illustrate the progression of sophistication in core barrels, Figure 3.8 shows:

(a) A single tube core barrel, in which the cutting edge consists of a diamond or hard metal studded bit that is slightly larger than the outside diameter of the core barrel. Rotation of the barrel under a downward pressure causes the bit to cut into the rock. Drilling fluid (water or bentonite slurry) pumped down the drill rods both cools the bit and carries the rock cuttings up to the surface of the bore hole. The rock core must be sufficiently hard to resist the erosive effect of the drilling fluid, flowing over the core surface.

(b) Shows a more refined double tube core barrel in which the inner tube is stationary and protects the core against erosion by the drilling fluid.

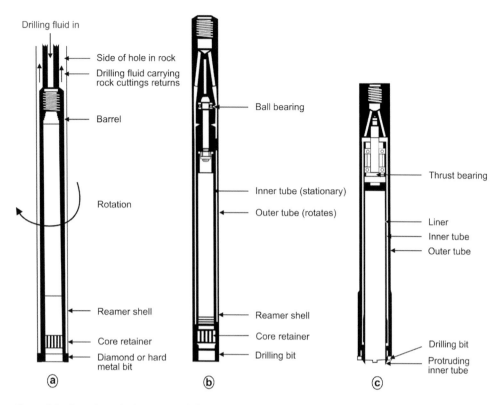

Figure 3.8 Core barrels for rotary drilling and sampling. (a) Single tube core-barrel effective in rock sampling but unsuitable for undistorted sampling of soils and soft rocks. (b) Double tube core barrel, suitable for soft rocks. (c) Triple tube core barrel specially designed to sample soft rocks and hard soil.

(c)   The triple core barrel is specifically designed to sample hard soils and soft rocks. The third tube is a liner that protects the core from erosion and is stationary. The second, or inner tube is also stationary and advances into the soil ahead of the cutting bit. This gives more protection to the advancing base of the core, as the bit cuts into the rock and the cuttings are removed by the drilling fluid.

When drilling cores from soil or rock that is adversely affected by free access to water, combined with stress release caused by the drilling, air or foam can be used as a drilling fluid, instead of water or drilling mud.

REFERENCES

Brink, A.B.A. (1979) *Engineering Geology of Southern Africa.* (5 Vols.) Pretoria, South Africa, Building Publications.

Burland, J.B. (1958) *A Simple Soil Colour Chart for Soil Profiling Purposes.* Final year project, Department of Civil Engineering, University of the Witwatersrand, Johannesburg, South Africa.

Cook, J.R. & Newill, D (1988) The field description and identification of tropical residual soils. Eds.: Publications Committee of 2ICOTS, Pub. A.A. Balkema, Rotterdam. *2nd Int. Conf. on Geomech. in Tropical Soils, Singapore.* Vol. 1, pp. 3–10.

Hvorslev, M.J., (1948) *Subsurface Exploration and Sampling of Soils for Civil Engineering Purposes.* Vicksburg, MS, USA, Waterways Experiment Station.

Jennings, J.E., Brink, A.B.A. & Williams, A.A.B. (1973) *Revised Guide to Soil Profiling for Civil Engineering Purposes in Southern Africa.* Civ. Eng. in South Africa, Jan, 3–12.

Chapter 4

The mechanics of compaction and compacted residual soil

G.E. Blight & J.V. Simmons

4.1 THE COMPACTION PROCESS

Compaction is a process whereby a soil is densified by expending energy on it. *Traffic compaction* results from the work done as vehicles travel over the surface, not with the specific purpose of compacting it, but in order to discharge some other function, e.g., to dump more material. *Roller compaction* arises from the deliberate trafficking of a surface by a vehicle specifically designed to expend energy in compaction, i.e., a compactor or roller. The energy input may arise simply from the weight of the roller moving down as it compresses the fill, or there may be an additional energy input, e.g., resulting from the repeated toppling of the roller as with an impact roller or the gyration of an eccentric weight as with a vibrating roller. Pounding with a falling weight is a fourth method of compaction that is sometimes used. Figure 4.1 illustrates these four ways of expending compactive energy. In addition, rollers may have a variety of surfaces that contact the soil in different ways. Smooth wheeled, footed or pneumatic tyred rollers as well as grid-rollers are all used. This chapter will describe the generic technology of compaction and then proceed to describe specific aspects relating to the compaction of residual soils.

Compaction occurs because the solid particles of fill are forced closer together, thus expelling air and reducing the void volume. It is not common for compaction to result in the expulsion of water from a fill. Some breakage of particles may occur during compaction and the resulting fines then partly fill the reduced void space.

For a given energy input and method of compaction, the density achieved depends on the water content of the fill. In general, an optimum water content will exist at which a given energy input will result in a maximum dry density. If the energy or method of input is changed, both the optimum water content and the maximum dry density will change. As shown by Figure 4.2a, as the energy input is increased, the maximum dry density increases and the optimum water content at which it is achieved decreases.

The largest effect on dry density occurs during the first few passes of a roller. Thereafter as indicated by Figure 4.2b, the effect reduces exponentially. It is seldom worthwhile applying more than five or six passes of a roller. If the required effect is not obtained with this number of passes, either the water content of the fill is too high or too low, the energy input per roller pass is too low or the type of roller is not optimal for the soil type (see also sections 4.6 and 6.8).

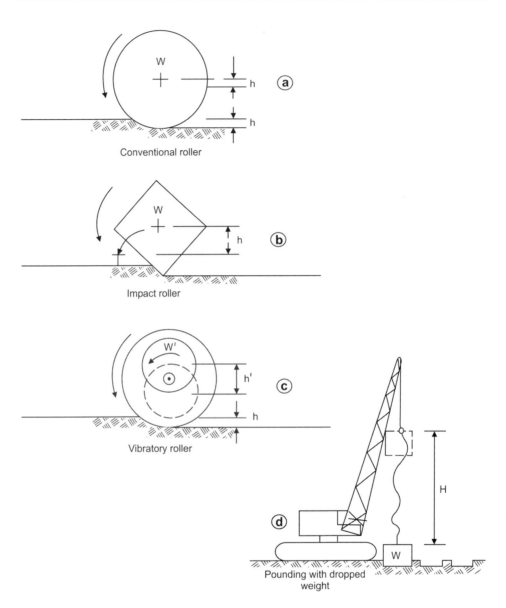

Figure 4.1   Four ways of expending compactive energy: (a&b) Weight W of roller moves down distance h, expending energy Wh on soil. (c) Additional energy W'h' is expended by eccentric weight W'. (d) Pounding soil surface with a falling weight. Input energy is WH.

Compaction results from the imposition by the roller of compressive and shear stresses on the fill material. These stresses are largest immediately under the roller and reduce or disperse with increasing depth below the surface. The type of stress dispersion that occurs is illustrated by Figure 4.2c. Rollers will usually not produce any appreciable effect at a depth greater than about 1m below the surface. Because of this, if a

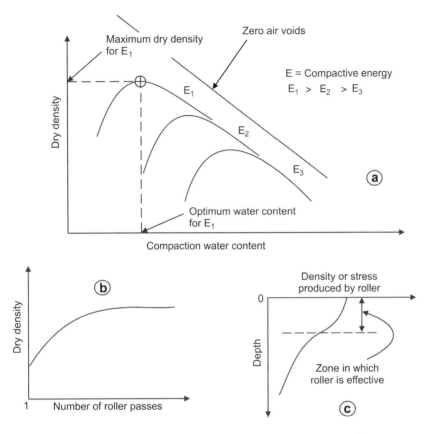

Figure 4.2 Relationships between: (a) dry density, compaction water content and compactive energy. (b) roller passes and dry density. (c) depth and dry density.

reasonably uniform state of compaction is required throughout a fill, the material must be spread in layers not exceeding 300 mm in loose thickness, which will compact to a thickness of about 200 mm. The multi-sided impact roller illustrated in Figure 4.1b and pounding (Figure 4.1c) are exceptions to this, and significant compaction may be (but is not always) attained down to depths of 1.5 m.

Smooth-wheeled and pneumatic tyres are not used to any extent for general earthworks such as earth dam construction, being more suited to the extra-heavy compaction required by road layer works. Impact rollers are usually used to densify deep loose sandy soils, and might be required to densify the foundation layer for a road or railway embankment, if this consists of a loose sandy or silty stratum. Grid rollers are useful if the soil contains aggregations of particles such as often occur in a partly-weathered residual soil and in coal waste, and which require to be broken down by compaction. Footed rollers with feet long enough to completely penetrate the loose layer being compacted are favoured for compacting relatively thin impervious clay layers for the construction of impervious soil liners, as shown in Figure 4.3 (USEPA, 1989).

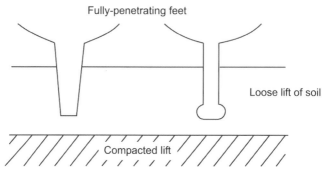

Fully-penetrating feet compact the loose top layer from
the bottom up, improving bond between successive layers.

Fully-penetrating feet

Loose lift of soil

Compacted lift

Partly-penetrating feet act almost like a smooth wheeled
roller until the soil becomes strong enough to carry the
higher stresses imposed by the feet. Bond between layers
is not as good as that achieved with fully-penetrating feet.

Partly-penetrating feet

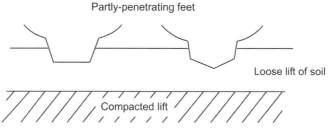

Loose lift of soil

Compacted lift

Figure 4.3 Four kinds of footed rollers on compaction equipment.

Because smooth wheeled rollers give a very smooth surface finish, they are usually used to finish off the top surface of a road base or the top surface of a clay liner that will be covered by a geomembrane. The smooth surface gives good riding quality and promotes the close contact between soil and geomembrane that is essential for the functioning of a composite soil/geomembrane liner.

Laboratory compaction tests can be used to establish approximate values for the optimum water content and the maximum dry density for "light" or "heavy" compaction energies. Laboratory values will generally not agree with the fill characteristics for roller compaction as the method of applying the energy in the laboratory test (usually by repeated blows of a drop hammer applied to thin layers of soil confined in a small rigid-walled mould) differs completely from that used in the field. Nevertheless, because design options usually have to be assessed by means of laboratory tests, it is also important to study laboratory compaction.

4.2 CONSEQUENCES OF UNSATISFACTORY COMPACTION

It can be very difficult and expensive to rectify inadequate field compaction. Because of the time that may be involved in obtaining control test results, substandard work

may be buried before being identified and is then very difficult to rectify. It is therefore important to understand some of the serious consequences of inadequate compaction:

- rutting, cracking, slacks in the surface and excessive overall settlement of highway or railway embankments causing loss of ride quality or dangerous unevenness,
- slumping of embankment slopes, and/or loss of freeboard in water storages, possibly causing overtopping, breaching and disastrous flooding,
- leakage and/or piping erosion in water or waste storages, leading to overall embankment failure, followed by flooding or loss of water, (see section 8.7 and Plate C18).
- inadequate performance may result in utility failure, e.g., incorrect or even reversed camber in a road surface, and also to excessive capital and/or maintenance costs for rectification.

4.3 THE MECHANISMS OF COMPACTION

Essentially, compaction is a process whereby air is expelled from the pores of the soil, reducing the air-filled void volume and, in the process, forcing the solid particles closer together. It is the process of forcing more solid particles into a given volume that increases the dry density and hence increases the soil strength, reduces the compressibility and also reduces the permeability to air and water flow. However, to produce these effects, both shear and compressive stresses have to be applied to the soil, and that is what requires the expenditure of energy.

At low water contents the resistance of the soil to compression and deformation is relatively high. The air-filled void spaces are interconnected and air can freely leave the soil. With a given expenditure of energy (or "compactive effort") only a relatively low compacted dry density can be achieved. As the water content is increased, the resistance of the soil to compaction decreases and higher dry densities result, but as the air-filled voids decrease, the resistance to the escape of air increases, until the air-filled voids become occluded, or sealed off from the atmosphere by surrounding water-filled pore space, This point corresponds approximately with the maximum dry density and optimum water content for the particular compactive effort being used. From the optimum point onwards, as water is added to the soil, it occupies increasing space in the voids, while the air content remains almost constant. The result is that the dry density of the soil decreases progressively with increasing water content. Figure 4.2a shows the "zero air voids" line, a line along which, theoretically, the compacted soil would contain no air.

A set of relationships between water content, suction, strength and dry density is illustrated in Figure 4.4. It must be noted at the outset that the relationships apply after compaction. The relationships shown in Figure 4.4 apply to laboratory compaction, using standard Proctor compactive effort, of a clay derived from the weathering of shales at the site of the Mangla dam in Pakistan. They show compaction water content versus dry density γ_d, shear strength τ and soil water suction $p'' = (u_a - u_w)$. The degree of saturation S for each dry density is also given. It will be seen that as the water content increases, S increases from 62% at $w = 15\%$ to about 90% at optimum water content, and then increases further to reach a constant 94 to 95% as

the water content increases. The suction $p'' = (u_a - u_w)$ (see section 1.14 and equation 1.2) decreases continually as w is increased. The shear strength initially increases as the suction decreases, because the parameter χ in equation 1.2 is increasing, but then reaches a peak value and decreases continually as w is further increased. Note that the maximum (unconsolidated, undrained) shear strength almost coincides with the maximum dry density.

The rapid increase of S with increasing w up to the optimum point shows that the air content of the soil $(1 - S)$ is decreasing from a maximum of 38% at w = 15% to 9 or 10% at optimum, to 5 or 6% for w exceeding 18%.

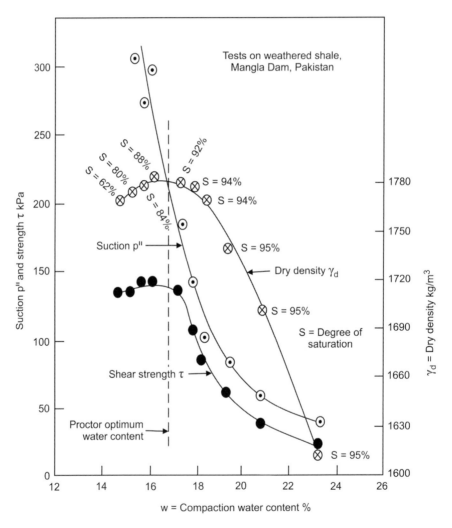

Figure 4.4 Relationships between compaction water content w, dry density γ_d, suction p' and strength τ for a clay residual from shale.

4.4 LABORATORY COMPACTION

Dry density versus compaction water content curves, such as those shown in Figures 4.2a and 4.4 are usually the result of compaction tests carried out in a laboratory. The equipment for laboratory compaction testing consists of a series of compaction moulds, (a typical one of which is illustrated in Figure 4.5) and a compaction hammer with a standard mass and dimensions, that is designed to fall a standard distance in free fall. Thus the energy applied per hammer blow is the weight of the hammer multiplied by the height of fall. For the standard Proctor compaction hammer shown in Figure 4.5, the energy per blow is $25\,N \times 0.3\,m = 7.5\,Nm = 7.5\,J$, and the energy per m^3 of soil compacted in 3 layers with 27 blows per layer (i.e., standard Proctor compaction) is $7.5\,J \times 27\,blows \times 3\,layers \times 1000 = 607.5\,kJ/m^3$. Although the hand-held compaction hammer is a standard, most soil mechanics laboratories use mechanized hammers or compaction machines that distribute the selected number of hammer blows uniformly over the circular surface of the soil layer in the mould. The hand-held compaction hammer is now relegated to the small field laboratory.

To establish a dry density versus water content curve, a number of specimens of loose soil are prepared at a series of increasing water contents. Each specimen is then compacted into a mould with the specified input energy, and weighed, after trimming the soil surface flush and level with the top of the mould. The compacted soil is then removed from the mould and specimens of soil are taken to measure the water content.

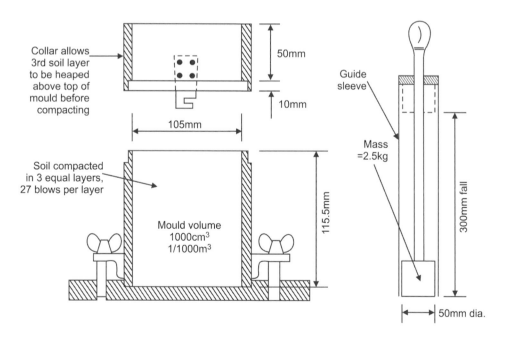

Collar allows 3rd soil layer to be heaped above top of mould before compacting

50mm

10mm

105mm

Guide sleeve

Soil compacted in 3 equal layers, 27 blows per layer

Mould volume 1000cm³ 1/1000m³

115.5mm

Mass =2.5kg

300mm fall

50mm dia.

Figure 4.5 The 105 mm diameter standard Proctor compaction mould and the standard Proctor compaction hammer.

4.5   PRECAUTIONS TO BE TAKEN WITH LABORATORY COMPACTION

Bulk samples of soil taken for compaction testing are not usually sealed against loss of moisture, and often consist of an assemblage of soil clods, which, if allowed to dry, are extremely difficult to break down to achieve a homogeneous material for testing. Drying of a soil sample (which must be avoided) may have at least three consequences.

4.5.1   Moisture mixed into the soil is not uniformly distributed

Most guides to compaction specify that specimens prepared for compaction should be stored in sealed containers overnight to allow the moisture content to equalize throughout the specimen before compacting it. However, as shown by Figure 4.6, this moisture homogenization process takes much longer than a day, and at least a week should be allowed. The water not only has to be uniformly spread over all particles and soil aggregations, but also has to have time to diffuse into the soil aggregations. Even after the 26 days for homogenization shown in Figure 4.6, there is no way of knowing if the properties of the soil tested in the laboratory are the same as the soil layer spread at its *in situ* water content and conditioned for compaction in the field.

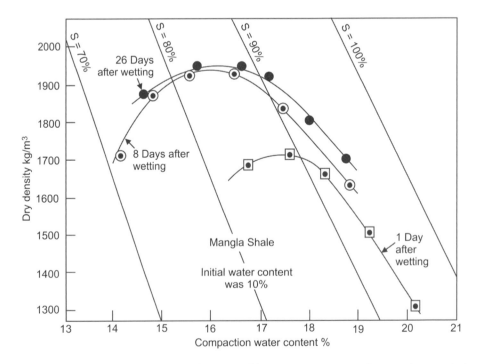

Figure 4.6 Time taken to achieve a stable, repeatable compaction curve after wetting an air-dry clayey soil.

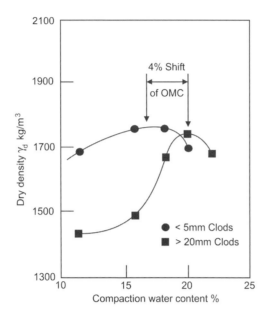

Figure 4.7 Effects of clod size on compaction curves.

4.5.2 Soil aggregations or clods are not broken down

Hard soil aggregations or clods affect dry density in the same way as stones do. The larger and more clods there are in the compaction specimen, the more the true compaction curve will be masked. Figure 4.7 (USEPA 1989) shows an example where larger (>20 mm) initial clods result in a falsely low density for compaction dry of optimum and require a larger optimum water content. A soil with smaller initial clods (<5 mm) shows a higher dry density over the whole water content range, and a 4% lower optimum water content. If the clods had been broken down completely before compaction, an even lower optimum water content could have been expected.

4.5.3 Other treatments that will affect the laboratory compaction curve

Figure 4.8 (Gidigasu and Dogbey, 1980) shows the compaction curve for a soil, taken from site at a water content of 8% and prepared for compaction by adding water, allowing 7 days for dissemination of the added water throughout each sub-sample (the "natural state", A in Figure 4.8).

The same soil was air dried, rewetted and compacted using the same treatment as the "natural" sample, and the resultant compaction curve was very similar (B).

When, however, sub-samples that had received treatment (B) were recompacted, a very different compaction curve resulted. As the soil contained some laterite, which pulverizes under compaction, it is likely that the change in compaction curve resulted from particle break-down caused by the first compaction. Finally, after oven-drying and then rewetting and compacting, a fourth compaction curve resulted. This was

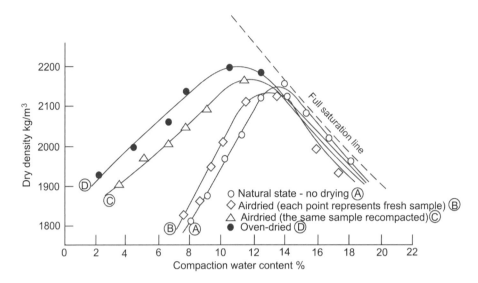

Figure 4.8 Effects of various treatments of a soil, before compaction, on the compaction curve.

probably because the oven drying temperature of 105°C had modified the clay minerals or metallic hydroxides contained by the soil.

It is therefore essential to avoid drying of a soil destined for compaction testing between taking the sample in the borrow pit and conditioning it in the laboratory for compaction testing. If the surface of the soil to be sampled is desiccated, the dried surface layer should be discarded and excluded from the sample. Samples should be stored in sealed plastic buckets and never be recompacted to produce a compaction curve. Use a fresh sample for each point on the curve.

4.6   ROLLER COMPACTION IN THE FIELD

Compaction is undertaken in the field using the best technology available, but depending on available equipment, not always the best possible technology. The compacted product requires compromises between energy and cost expended, and the value of the result obtained. Engineering design must recognize the reality of what can be achieved in the field, as compared with what can be achieved in the laboratory.

The choice of compaction equipment should be made with a view to minimizing earthworks costs while achieving the desired engineering properties for the compacted soil. Earthworks design must include a consideration of what range of equipment is available, and any practical constraints such as weather and site conditions which may influence what can be achieved in the field.

The soil *in situ* in the borrow pit is always variable in composition to some degree, particularly if it is a residual or alluvial soil, as the degree of weathering and yearly layers of deposition will usually be variable. Hence the selection of representative samples for testing can be a major problem. For the same reason good control of quality in compacted fills may be extremely difficult to achieve.

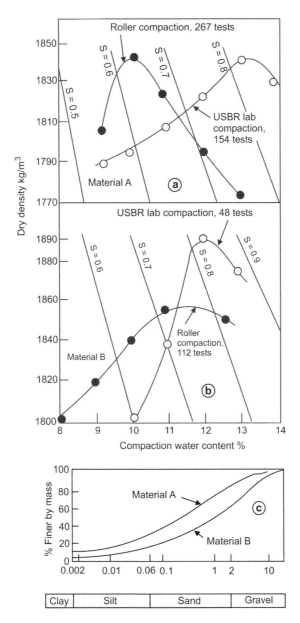

Figure 4.9 (a&b) Comparison of laboratory and roller compaction curves for two soils. (c) Average grading curves for two soils above.

The optimum moisture condition for field compaction is best determined for any compaction equipment by a process of field trials. It is recommended practice to include field trials as an initial component of the construction programme. The trials can be used to optimize equipment selection and operation, and to identify the field optimum moisture condition (see Figure 4.9). Where the soil being compacted is sufficiently

uniform in its physical properties, field trials can also be used to finalize the method of compaction, with construction being controlled by this means.

Another factor to be considered is the process of moisture conditioning. It is relatively easy to add water to a soil surface, but more difficult to ensure that the added water is uniformly distributed within the soil. In high rainfall areas, it may be necessary to dry the soil in order to operate compaction equipment. Air drying may be slow, and must be aided by ploughing or tilling to turn the wetter soil to the surface and expose it to sun and wind.

There are a number of detrimental effects on the engineering properties of the compacted soil, if effective and uniform mixing of water and soil is not achieved. These include:

- dry clods of soil in a wet matrix, resulting in large voids in the compacted mass,
- shearing due to distortion of over-wet fill under the action of the compaction plant, resulting in shear surfaces and loss of shear strength of the compacted mass,
- de-bonding between compacted layers, resulting in loss of shear strength in a horizontal direction. This can be avoided by scarification immediately prior to addition of loose fill layers. De-bonding of layers also has a detrimental effect on permeability,
- poor trafficability of construction plant, and/or ponding of water: this can be avoided by proper attention to grades and levels during construction.

The final product must have the designed engineering properties. It is therefore important that adequate supervision and quality control are used during the fill construction process.

The compaction characteristics of soils may be very dependent on the method of applying the compactive energy. In particular, laboratory compaction curves may bear little resemblance to the compaction curve achievable in the field. This phenomenon is illustrated by Figures 4.9a and b which compare roller and laboratory compaction curves for a residual weathered granite pegmatite clay (Blight, 1962). With soil A, it did not prove possible to achieve the required 100% of laboratory maximum dry density, until it was discovered that the optimum water content for laboratory compaction was 3% wet of that for roller compaction. At that time (1957), the link between permeability and compaction water content was not known. Nevertheless, an area of the fill compacted at roller optimum water content +2% was tested for permeability by excavating a shallow (150 mm deep) pond, filling it with water, covered with a film of engine oil to inhibit evaporation, and observing the rate of seepage. This showed that the permeability was acceptably low, and the compaction requirement was altered on site to "100% USBR laboratory maximum dry density at roller optimum water content +2%". Figure 4.9c shows the grading curves for the two materials. These data also illustrate the variability of material from a single borrow pit. Material A was obtained by blending the upper, more weathered layers of soil, while Material B was a blend of the less weathered underlying soil.

In the case of soil B, roller compaction was not able to achieve laboratory maximum dry density, although the roller and laboratory optimum water contents were almost the same. However, the roller maximum dry density was 98% of the

laboratory maximum, and this was deemed sufficient to achieve the required strength. The *in situ* strength and permeability were checked and found to be satisfactory.

4.7   RELATIONSHIPS BETWEEN SATURATED PERMEABILITY TO WATER FLOW AND OPTIMUM WATER CONTENT

Figure 4.10 (Mitchell *et al.*, 1965) shows compaction curves for three successively increasing compactive efforts, as well as the corresponding variation of saturated

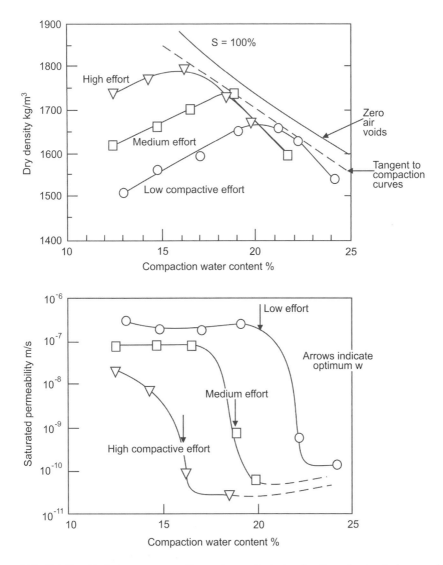

Figure 4.10   Relationship between compaction water content, dry density and saturated permeability to water flow for low, medium and high compactive efforts.

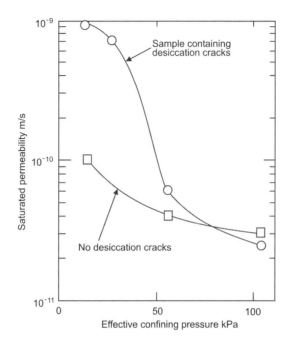

Figure 4.11 Effect of desiccation cracking and confining stress on saturated permeability.

permeability. As the dry unit weight increases with increasing water content, the permeability decreases sharply and reaches a minimum for a water content well in excess of optimum. Thereafter, as the water content is increased further, the permeability increases again, slightly, as indicated in Figure 4.10.

It is important to realize that the permeability achieved at compaction will change if the water content of the soil changes subsequent to compaction. Drying of a compacted soil may have a particularly deleterious effect on permeability if it results in shrinkage cracking. As illustrated in Figure 4.11 (USEPA, 1989) shrinkage cracking may increase the permeability by an order of magnitude, or more. The effect is reduced, but not eliminated, if the cracks are forced to close by applied stress, unless the stress is very large. This is also illustrated by Figure 4.11.

4.8 DESIGNING A COMPACTED CLAY LAYER FOR PERMEABILITY

The method for designing a clay layer, so that it will have less than a specified maximum permeability, is based on a series of tests like that illustrated by Figure 4.10. The procedure is as follows (USEPA, 1989):

Suppose that the specified maximum permeability is 10^{-9} m/s (30 mm/y). On a separate plot of dry density versus permeability, plot all the experimental points that correspond to a permeability of 10^{-9} m/s or less. An example of such a plot appears in Figure 4.12, in which the cross-hatched zone defines the zone of compaction water content and dry density within which the permeability should meet the specified value.

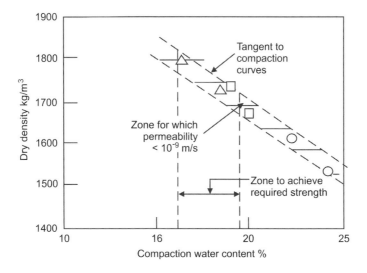

Figure 4.12 Zone for which permeability should be less than 10^{-9} m/s for compaction and permeability data shown in Figure 4.10.

Strength requirements may also limit the compaction water content to a specific range, as indicated in Figure 4.12.

If it is not possible to reach the maximum permitted or designed level of permeability, it is usually possible to reduce the permeability of a natural soil by mixing in a proportion of a clay mineral such as bentonite. Bentonite is a highly expansive clay mineral, that occurs as two main types, sodium bentonite and calcium bentonite, where either sodium or calcium is the predominant exchangeable cation. Sodium bentonite is more expansive and less permeable than calcium bentonite. However, it should be noted that ground water or seepage that passes through the compacted soil may be calcium-alkaline which will convert a sodium bentonite to a less expansive, more permeable calcium bentonite. If the use of bentonite is contemplated, therefore, the chemistry of the seepage water must be ascertained to ensure that the wrong type of bentonite is not used for the specific application.

The bentonite is best mixed into the natural clay as a dry powder, using a mechanical mixer before placing, spreading and compacting the modified soil. If *in situ* mixing is to be used, a "pulvi-mixer" should be used. The quantity required is usually small (5 per cent of sodium bentonite added to a silty sand may reduce the permeability of the compacted soil by a factor of 10^{-4}).

It is particularly important that any layer or sub-layer of a compacted soil mass, that is designed to be of low permeability, should be protected from desiccation by the sun after compaction. The usual method of protection is to spread the next layer of soil to be compacted over a compacted surface, as soon as compaction has been completed. When the soil will be exposed to the effects of desiccation over a long period, e.g., when used as a capping layer, it must be provided with permanent protection against desiccation. For example, the surface could be covered with a 0.5 m thick layer of single-sized gravel.

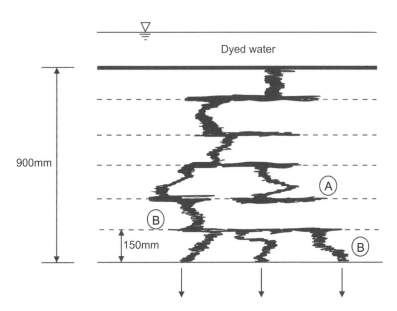

Figure 4.13 Liquid flow between shrinkage cracks and compaction lift interfaces in a soil liner.

4.9 SEEPAGE THROUGH FIELD-COMPACTED LAYERS

The permeability of extensive field-compacted soil layers can be very adversely affected by defects that inevitably occur in a compacted layer. Examples of such defects are shrinkage cracks, poorly mixed or compacted or sandy zones in a layer and poor bond between layers. Figure 4.13, for example, shows the effect of shrinkage or tensile stress cracks, interlinked by zones of poor inter-layer bond in a multi-layer compacted clay liner. The mechanism of leakage was observed by ponding dye on the surface of the liner and, after some time, excavating a hole to observe the seepage path taken by the dye (USEPA, 1989).

It would appear that a multi-layer compacted clay layer should be less pervious than a thin single layer because each layer would tend to interrupt and cut off flow through faults in contiguous layers. As shown by Figure 4.13, this does happen to a limited extent (A in Figure 4.13), but equally, may not happen (B in Figure 4.13).

Figure 4.14 shows measurements of the effect of liner thickness on overall permeability. Theoretically, because permeability is the ratio of flow velocity v to flow gradient i ($k = v/i$), and i would normally be unity in vertical seepage flow under gravity, seepage flow should be independent of liner thickness. In reality, because there is more interruption of defects such as those illustrated by Figure 4.13 in a thick than a thin layer, permeability decreases as liner thickness increases. It will be noted from Figure 4.14 that what was judged to be "good and excellent construction" gave a slightly lesser decrease of k with increasing liner thickness, but the difference between what was considered to be "excellent" and "poor" construction was not very marked. Considering the lower bounding line, a 0.3 m thick liner achieved $k = 10^{-9}$ m/s, whereas a 1.5 m

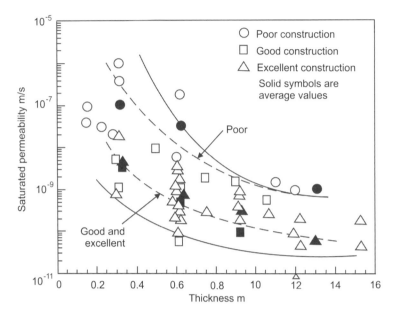

Figure 4.14 In situ measurements of permeability as a function of compacted layer thickness.

thick liner achieved $k = 4 \times 10^{-11}$, a 25-fold improvement for a 5-fold increase in thickness. Thus it has to be concluded that, given similar quality of construction, thicker clay liners will considerably outperform thinner ones, but at the cost of providing the additional compacted clay (USEPA, 1989).

4.10 CONTROL OF COMPACTION IN THE FIELD

In order to set appropriate control criteria, the function of the compacted fill must be understood, and a number of additional constraints must also be considered:

- construction requirements related to weather and climate that will affect borrow pit and compaction water contents (limited time for completion, sunshine, rain, wind, low or high temperatures, etc.),
- moisture conditioning requirements for borrow pit material,
- availability and standards of resources for quality control,
- if available, performance data from field compaction trials either for a method specification or optimum compaction characteristics for equipment used,
- availability and standards of any reference laboratory testing on which specifications have been based.

Table 4.1 is a summary of tests that have been used for compaction control. These can be used as a guide to selection of appropriate testing. The frequency of testing (e.g., number of m^2 of a chosen thickness or m^3 per control test) should be related to a number of additional factors, and must be determined separately.

Table 4.1 Test measurements used for compaction control.

Test	Comments	Standard ?
1. In situ density		
Sand replacement	Preferred, most soils, slow	Yes
Core cutter	Fine grained soils without stones, slow	Special
Nuclear meter	Uniform soil, calibration, rapid	Yes
Balloon densometer	Difficult, unreliable, slow	Yes
2. In situ water content		
Oven drying	Slow	Yes
Microwave drying	Calibration required, rapid	Yes
Nuclear meter	Uniform soil, calibration required, rapid	Yes
3. In situ strength		
In situ CBR	Calibration difficult, special equipment	Yes
Penetrometer	Versatile, fast, calibrate to other characteristics	Yes
Shear vane	Fast, fine grained cohesive soils only, calibrate	Yes
4. Permeability in situ		
Infiltration from surface	Simple ponding test, slow, evaporation difficult to assess or prevent	No
Drill/auger hole (lateral seepage)	Simple, slow, easier to measure seepage	No
Covered double ring infiltrometer	Simple, slow, accurate	No
5. Laboratory tests (on undisturbed samples)		
Constant or falling head permeability	Constant head slow, falling head quicker, representation of field conditions doubtful	No
UU triaxial strength	Fast, select test conditions to suit purpose	Yes
CU triaxial strength	Slow, select test conditions to suit purpose	Yes

Table 4.2 "As compacted" control parameters.

1	*In situ* dry density
2	*In situ* water content
3	*In situ* dry density within a range of water content
4	*In situ* strength
5	*In situ* permeability
6	Laboratory strength properties correlated to *in situ* measurements
7	Recipe specification

Table 4.2 lists five parameters that are commonly used for compaction control.

4.10.1 *In situ* dry density

In principle this is simple and direct. However, accurate field measurement of volume is time consuming and subject to procedural errors. Indirect measurements using nuclear moisture/density meters require careful calibration checks. Nuclear equipment may

not be suitable for many remote sites, and the calibration can be affected by lightning occurring within 1 km from the site, as well as adjacent high voltage power lines.

The greatest disadvantage of density testing is that it offers only limited information on the fill properties that are really required for the function of the fill. The method of field compaction, and the particle sizes involved, will not usually correlate with the standard laboratory methods. At best, only inferences can be made about strength and permeability. Test results are compared to a laboratory value. Considerable material variability may occur in the field. Unless a corresponding laboratory reference test is performed on the same sample for each field density measurement (which is often done), there is a risk of field measurements being incorrectly interpreted.

In situ density has traditionally been the most popular method of measurement, due to its adoption and wide use over a long period of time.

4.10.2   *In situ* water content

Water content can be measured reasonably rapidly, and has the advantage that field variability can be assessed relatively easily by taking many measurements. Provided that the strength and permeability characteristics of the material are understood in relation to water content, water content can be used as an effective control parameter. However, the desired properties of strength and/or permeability are dependent on density as well as water content. The use of water content therefore should be supplemented by measurement of other properties. Alternatively, if established by a field trial, the method of compaction may be specified so as better to achieve the requisite density. Water content could then be used as the sole control parameter. However, this would apply only in unusual cases.

4.10.3   *In situ* dry density within a range of water contents

In section 4.8, a method was described for designing the permeability of a compacted soil. This requires achieving a result within permissible ranges of both water content and permeability, as illustrated in Figure 4.12.

Figure 4.15 shows the statistics for the control of compaction water content and dry density that relate to a large soil compaction project. Each of the three stages of the histograms relates to progressively longer periods of time and, overall, the data show that control adequately met the set targets, and improved as the project progressed.

4.10.4   *In situ* strength

In principle, this is the most effective method of control, if strength is a direct requirement for performance (see, e.g., Figure 4.4). In Figure 4.4 maximum strength coincides with maximum dry density. However, as shown in Figure 4.16 (after Wesley, 2010a and b) this may not happen with every soil. If strength and water content are chosen as control parameters, the acceptable ranges of water contents and strength must both be set. However, it seems unlikely that even when the degree of saturation had dropped to 75% in Figure 4.16, the strength would still have been increasing. Wesley adds a rider to say that the air voids (1 − S) in a compacted soil should not exceed 8%, i.e., S should be no lower than 92%. This would limit the compaction water content for Figure 4.16 to Optimum +3% or above and for Figure 4.4, to Optimum or above. Strength

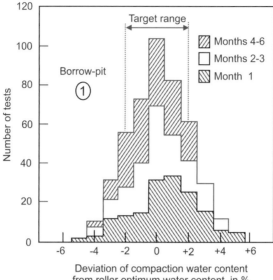

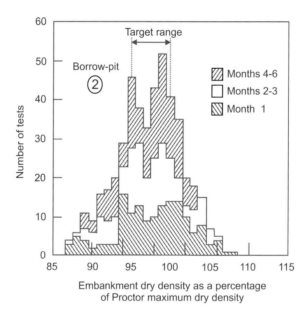

Figure 4.15   Progressive histograms for compaction water content and dry density, showing adequate and improving control as the project progressed.

should be measured with a rapid test which is not subject to significant interpretation problems. A variety of rapid strength measurements can be used, ranging from hand-held or hand-operated vanes or penetrometers to *in situ* CBR tests and larger or heavier penetrometers. Difficulties may occur if large particles present in the soil compromise the performance and interpretation of the measurement technique. Technologically

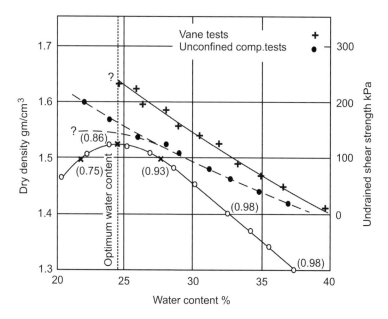

Figure 4.16 Compaction test on clay with corresponding measurements of undrained shear strength (Wesley, 2010b). Figures in ( ) are degree of saturation S.

advanced compaction machinery or monitors, which use accelerometers to assess the response of a fill to vibration, must be calibrated to site conditions. Strength is also greatly dependent on water content (and hence soil water suction) and will change if the water content changes subsequent to compaction. Thus water content must always be used as an essential second control parameter.

4.10.5 *In situ* permeability

This is very effective where permeability is the most important characteristic required for field performance. The greatest disadvantages are that field permeability testing is time consuming, very prone to errors, and it may not be possible to undertake a sufficient number of tests to give a statistically representative picture of field conditions.

Approximate field permeability tests, using simple infiltration methods, are easily performed provided that they can be done in areas which do not affect ongoing construction activities. In this case, permeability testing even if approximate, is a very useful guide.

Another disadvantage of permeability testing is that strength is usually also required. There are no effective correlations between strength and permeability, so that other forms of field tests are required anyway. Permeability cannot be used as a sole control parameter.

4.10.6 Laboratory strength properties correlated to *in situ* measurements

The advantages of laboratory testing relate to control and repeatability. Where this can be combined with correlation to an effective field measurement, very efficient compaction control can be achieved with a high degree of confidence.

The most widespread correlations are of laboratory-measured strength and field water content, for materials which have been adequately tested and whose variability is well understood. Thin-walled tube or core samples can be taken in the field, and tested at a site laboratory. The usual site laboratory strength test is the unconfined compression test (see Figure 4.16) or the unconsolidated undrained (C_{uu} or "Quick") triaxial test.

4.10.7   Recipe specifications

A recipe specification may call for compaction using a specified type and weight of roller, working within a specified range of water contents and applying a specified minimum number of roller passes. In the case of footed rollers, passes may be substituted by a minimum coverage "C" where:

$$C = \frac{A_f}{A_d} \times N \times 100\% \tag{4.1}$$

A_f = area of foot, A_d = area of drum, and N = number of roller passes. A typical range of C would be 150–200%, the aim being to ensure that, other things being equal, the whole area of the layer being compacted is subjected to the higher "under-foot" pressure at least once.

(An attempt was made in the Vietnam war during the early 1960s, to compact the surface of an area of soil by driving a herd of elephants across it. It was discovered that not only is the foot contact pressure of an elephant very low, but also that all of the animals in the second and subsequent rows trod exactly in the footprints of their predecessors, so that coverage was minimized!).

4.11   SPECIAL CONSIDERATIONS FOR WORK IN CLIMATES WITH LARGE RATES OF EVAPORATION

In arid and semi-arid conditions it is usually planned to construct all, or the major part of the earth works during a single dry season. This expedient usually reduces the cost of, for example, diverting a river during construction and minimizes delays due to wet weather, but limits the construction period to a maximum of about six months.

Placing and compaction operations often proceed on the basis of a 24-hour working day. Unless the working area is very large, up to three compacted layers, each 150 mm thick, may be deposited over the entire compaction area each working day. This precludes the usual method of compaction control by checking the embankment dry density and the compaction water content and rewatering and rerolling if either is not up to standard. Even if rapid methods of determining water content and density are used, the construction schedule will not usually allow time for excavation, preparation and recompaction of substandard layers.

Because of this, methods of control must be used which ensure that the soil is at the correct water content before compaction starts, and that the optimum number of roller passes is used to produce an adequate dry density.

Large water losses due to evaporation can take place during the dry season, especially if it is hot, and the *in situ* water content in the borrow pits decreases as the

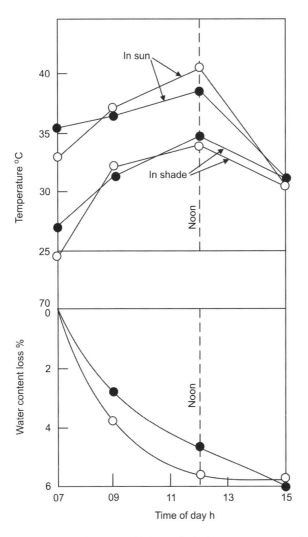

Figure 4.17 Water content losses from two 200 mm thick, loose, uncompacted soil layers in arid conditions. Losses are average through the entire layer thickness.

season progresses. As a result, the water content of the soil usually has to be increased considerably before compaction can be undertaken.

The water content is sometimes increased by a sprinkler system in the borrow area. A few hours may elapse between spreading the soil in place and the start of compaction; in this time a considerable amount of water can be lost from the soil. The delay results from the practical necessity of keeping the slower moving compaction rollers and water tankers out of the way of the faster earth-moving equipment.

Typical water losses from two 200 mm thick layers of loose uncompacted soil on the surface of the embankment of a dam in Central Africa are shown in Figure 4.17.

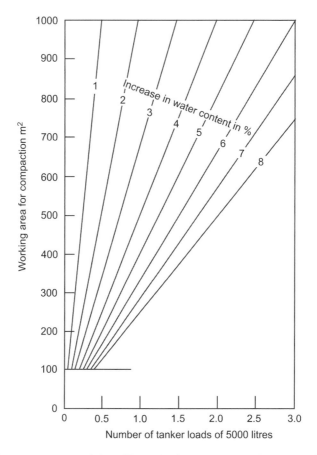

Figure 4.18 Water content control chart. Tanker loads to give required increase of water content.

The shade and sun temperatures on the surface of the soil at the time are also shown. The water contents represent the average through the thickness of the layers. A total loss of water content of 6 percent took place in eight hours, 4 percent being lost within the first three hours. Because of these large potential evaporation losses, it is best, under such conditions, to add and mix in water to the soil on the embankment immediately before compaction. Ideally, the water should be allowed some time to disperse into the soil, but this is not practical when evaporation losses are high.

Using the known capacity of the water tankers, a chart can be constructed relating the area of embankment to be treated and the required increase in water content to the number of full tanker loads to be discharged over a given area. An example of such a chart is shown in Figure 4.18. Evaporation-time curves can be established for work on the day and night shifts. Twice or more in each shift, water content samples can be taken from the incoming soil and the amount of water to be added to bring it to the required compaction water content estimated, using Figure 4.18 in conjunction with the appropriate evaporation-time curve.

4.12 ADDITIONAL POINTS FOR CONSIDERATION

4.12.1 Variability of borrow material

Variability of borrow material may relate both to particle size distribution and to cohesive or frictional attributes. Compaction assessment should include monitoring of borrow materials to ensure that specification requirements can be met.

Field compaction trials are strongly recommended because they enable the work to be controlled as much as possible by factors that have been proven in the field. There is no problem with field trials being undertaken as part of the permanent works, but it is important to realize that trials can be slow and painstaking compared with full production. Both the supervising engineer and the constructor must recognize this and allow for it in their schedules and budgets.

The particular advantages of field trials are related to the practicalities of field conditions and to time constraints.

4.12.2 Compactor performance

Knowledge of compaction performance must be translated from laboratory test results to results which directly reflect the performance of the compaction equipment. For example, field maximum dry density and optimum moisture content must be determined for the combination of soil being compacted and compaction equipment, and related to the required number of roller passes, thickness of placed layers, etc. (See Figure 4.9).

4.12.3 Testing frequency

The necessary testing frequency is very dependent on items 4.12.1 and 4.12.2 above. If materials are variable, or if compactor performance is poor or erratic, a higher testing frequency is required than if these items prove satisfactory.

Selection of lot sizes (i.e., number of m^2 or m^3 per test) and testing frequency should be based on site conditions, and adjusted appropriately. It is generally accepted that sampling of a compacted layer for dry density and compaction water content should take place on a square grid pattern as this provides a method of choosing test locations that is independent of personal bias. It is, however, important to vary the grid points from one layer to the next. The grid size or spacing is usually 10 to 15 m, corresponding to 100 to 44 tests per hectare, or 1 test per 15 to 34 m^3 of a 150 mm thick compacted layer.

Rapid field tests can be identified and selected so as to minimize the delays caused by waiting for laboratory test results.

4.13 COMPACTION OF RESIDUAL SOILS

Some special characteristics of residual soils must be understood clearly, if the compaction process is to be understood and the effort and cost of compaction is to be optimized. The following characteristics, while not a complete list, are associated

with special considerations for efficient compaction of residual soils. Residual soils, especially those of volcanic and igneous origin, often have:

- high *in situ* moisture contents,
- metastable clay minerals,
- soil structures that are lightly cemented,
- weathered soil particles that break down under compactive effort,
- sesquioxide minerals that are affected by wetting and drying.

The starting point for any engineering use of soil is the *in situ*, borrow-pit or "as won" condition. Many tropical and subtropical environments are characterized by frequent or seasonal rainfall. Handling of soils under these conditions may be difficult and the characteristics of the soils themselves may add further complexity to the problem of effective compaction.

Residual soils are widely used as construction materials: as fill for embankment dams and road embankments, as selected layers in highways and airfield construction and as impervious compacted clay liners. Certain residual soils, such as those containing smectite or halloysite clays may be unsuitable for uses requiring strength and volumetric stability, either because of inadequate strength, or excessive change of volume with varying water content, or because of loss of strength on wetting. However, smectitic and halloysitic materials have been used to form successful impervious layers in water-retaining embankments and as seals to waste storages. Examples of such uses are Sasumua dam described by Terzaghi (1958), the Arenal dam described by Rodda *et al.* (1982) as well as numbers of municipal solid waste landfills (e.g., USEPA, 1989).

The parent rock is usually variable in composition, particularly if it is an igneous or a highly faulted sedimentary rock. The degree of weathering will also be variable. Hence the selection of representative samples for testing can be a major problem. For the same reason good control of quality in compacted fills of residual material may be extremely difficult to achieve.

As examples, Figure 4.19 (Blight, 1989) shows the variability with depth of the grading analysis and Atterberg limits in a profile of residual weathered norite gabbro. Note the sharp transition from a clay to a sand at a depth of 4 m. There is usually a similar variation in lateral extent. Plate C25 (at back of book) shows an excavation in a weathered norite clay similar to that described by Figure 4.19. The excavation was made to construct a storm water retention pond. The undulating surface shows the variable contact between the residual clay and the less weathered silty sand. Some of the light-coloured silty sand has been stock-piled at the far side of the borrow-pit. This variability makes selection of consistent material or consistent blending of material very difficult to achieve in practice.

Drying of a residual soil from its *in situ* water content may change both its index and physical properties, including compaction properties (also see Chapter 2). Hence soil samples have to be treated and tested with the greatest care if the results of compaction tests are to be at all meaningful. The influence of sample preparation and laboratory procedure on the compaction characteristics of a lateritic soil (Gidigasu, 1974) is well illustrated by Figure 4.8. Not only was the optimum water content of the soil significantly altered by air- or oven-drying before compaction, but the maximum dry density also changed. The difference between the curves for the air-dried soil in

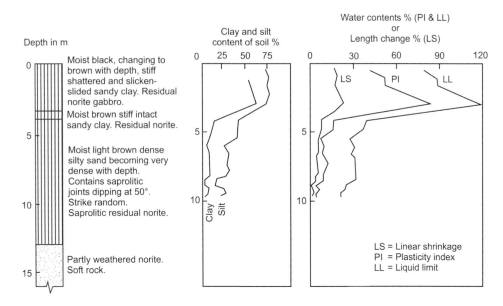

Figure 4.19 Variation in vertical direction of soil composition and index properties for a residual weathered norite gabbro profile.

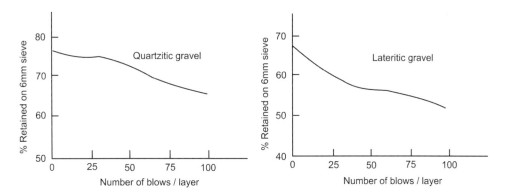

Figure 4.20 Effect of increasing compaction on gravel-size content of two residual granite soils.

which each point represents compaction of a new sample, and that for re-compaction of the same sample should be noted in particular. In compaction tests on residual soils, because of the progressive break-down of friable particles that usually occurs during compaction, each point on the compaction curve should represent compaction of a fresh sample, which is discarded once it has been compacted. The same soil sample should never be re-compacted. The point is further illustrated by Figure 4.20 (Gidigasu and Dogbey (1980)) which illustrates the progressive breakdown of particle size under compaction of a quartzitic gravel and a lateritic gravel, both residual from the weathering of granite.

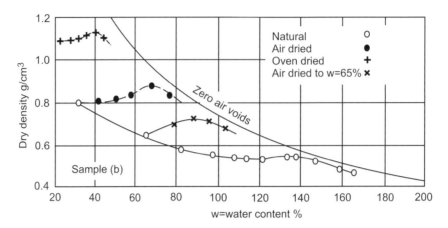

Figure 4.21 Compaction tests on an allophane clay (Wesley, 2010b).

Wesley (2010a) produced the compaction curves, shown in Figure 4.21, for a weathered volcanic ash believed to consist mainly of allophane clay. The *in situ* water content was 165%. As the soil was slowly and uniformly air-dried in the laboratory, compactions were carried out at a series of water contents. At water contents of 65% and 30% (that represented air dryness relative to oven dryness) and 0% (oven dry) compaction curves were carried out on re-wetted samples, with the results shown. Wesley does not mention the time allowed for the soil with added water to cure, i.e., for the water to spread homogeneously through and be absorbed by the soil grains of the re-wetted soil. However, as shown by Figure 4.6, the curves for the re-wetted soil may have approached that for the "natural" soil more closely if the re-wetted samples had been allowed to cure for several days or even a month after re-wetting.

4.14 THE MECHANICS OF UNSATURATED COMPACTED SOILS DURING AND AFTER CONSTRUCTION

The mechanics of unsaturated compacted soils need to be considered fully to understand their behaviour:

- during and at the end of construction while effective stresses in the soil are governed by the compaction water content and applied total stresses, and
- in the long term when effective stresses have changed as a result of drainage and climatic conditions, and the water content has also changed from the compaction water content.

During the construction of a fill the compacted soil is subjected to compression at constant water content under increasing overburden stresses. At modern rates of construction the process of compression is usually essentially undrained. Figure 4.22 shows the three dimensional relationship between volumetric strain and the two components of effective stress $(\sigma - u_a)$ and $(u_a - u_w)$ for the undrained compression of an unsaturated compacted soil (line ABCE).

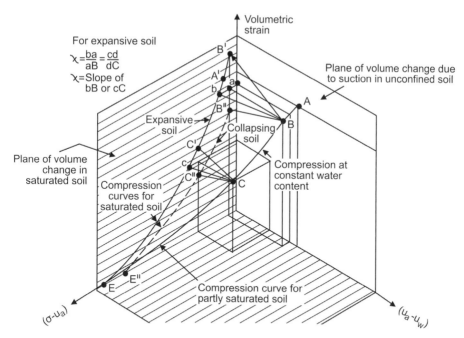

Figure 4.22 Three-dimensional stress-strain diagram for the isotropic compression of partly saturated compacted soils.

As the compressive strain increases, so the suction $(u_a - u_w)$ decreases, eventually disappearing when the soil becomes saturated by compression at point E. Figure 4.23 shows similar data plotted in the $(u_a - u_w)$, compaction water content plane for a compacted silty clay residual from shale. At a given value of $(\sigma - u_a)$ the suction depends on the compaction water content as shown in Figure 4.23, becoming less as the compaction water content is increased. Although the suction decreases continuously as the compaction water content is increased, significant suctions exist over a wide range of compaction water contents.

Figure 4.24 shows that a compacted gravelly clay residual from conglomerate, compacted at 0.5% wet of Proctor optimum water content requires a considerable height of overburden to be placed on it before it becomes saturated, i.e., for $(u_a - u_w)$ to be reduced to zero. In this instance, the suction at zero overburden stress was 250 kPa, hence for the cases illustrated by Figures 4.23 and 4.24, there was a significant potential for swelling if the water content were to be increased subsequent to compaction. However, soils compacted well dry of optimum water content have a potential to collapse when wetted. Figure 4.22 shows the schematic paths that would be followed in the event of either post-compaction expansion (BB′A′) or post-compaction collapse (BB″C″), after wetting.

After construction, the water content of the compacted soil will gradually come to equilibrium with its immediate environment. If the soil forms part of a water-retaining structure, the water content will usually increase as the seepage regime through the soil becomes established. In a semi-arid or arid climate, if no water is retained by the compacted soil structure, and if water tables are deep, as they usually are, except

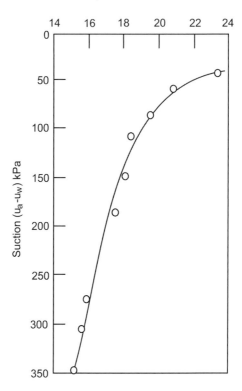

Figure 4.23 Variation of suction with compaction water content under unconfined conditions. Soil is a silty clay residual from weathered shale (Mangla Dam, West Pakistan). Proctor Optimum water content is 16.2%.

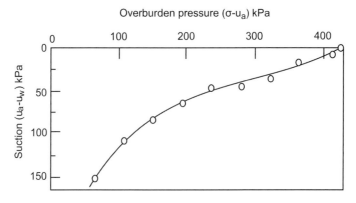

Figure 4.24 Variation of suction with overburden pressure in a compacted gravelly clay residual from a weathered conglomerate. Water content was constant at Proctor Optimum +0.5%. At zero overburden pressure, suction was 250 kPa.

in low-lying areas, the water content of the compacted soil will reduce to below the compaction water content, until it reaches equilibrium with climatic conditions. The compacted soil will, in such circumstances, shrink or settle. In wet climates, the water content will increase as a result of infiltrating rain or a rising water table to above the compaction water content, although the presence of air entrapped in the voids will ensure that the soil never becomes completely saturated. It is in these long term conditions that the potential for post-compaction swell or collapse, illustrated in Figure 4.22 become possibilities.

4.15   PORE AIR PRESSURES CAUSED BY UNDRAINED COMPRESSION

When an unsaturated soil, such as a freshly compacted material, is compressed under undrained conditions, e.g., in a rapidly constructed compacted soil embankment, the pore air pressure will rise according to Boyle's law, and then diminish as the compressed pore air dissolves in the pore water by Henry's law of solubility. The solution of the pore air, being a diffusion process of air into water, takes some time to occur, and as it occurs, causes the pore air pressure to decrease. This, in turn, by increasing $(\sigma - u_a)$, allows the applied stress further to compress the soil. There is, therefore, a complex reaction between Boyle's and Henry's laws, the compressibility of the soil and the effective stress components $(\sigma - u_a)$ and $(u_a - u_w)$.

The pore air pressure in a freshly compacted soil is not atmospheric (i.e., zero gauge), but is positive. Figure 4.25a shows the initial pore pressure measured at the sealed base of an AASHTO compaction mould immediately after compacting soil in the mould. The figure shows how the air pressure gradually dissipates by escape of the air into the atmosphere at the top of the compaction mould combined with solution of the air into the pore water. Figure 4.25b shows the gradual reduction of pore air pressure in a freshly compacted soil specimen measured when enclosed in a latex rubber membrane and immersed under mercury. Although latex rubber is slightly pervious to air, mercury does not dissolve air, and hence the soil was kept within boundaries that were completely impervious. The decline in pore air pressure therefore resulted entirely from solution of pore air in the pore water.

Figure 4.26 shows changes in pore air and water pressure as well as in volume when the confining stress on a saturated compacted soil is suddenly reduced. After an initial simultaneous and instantaneous reduction in both pore air and water pressures, and expansion of the soil, the soil continued to expand for about 2 hours. After the sudden reduction of the confining stress, the pore water pressure decreased instantaneously by 30 kPa and the soil began to de-saturate as the pore air came out of solution in the pore water and u_a and u_w separated. Continuing de-solution of air eventually caused the pore air and water pressures to start increasing, even though the volumetric strain had approached a steady value.

Calculations based on the simultaneous application of Boyle's and Henry's laws to the compression of the pore fluid of a partly saturated soil have been widely used to predict pore air pressures at the end of construction of rolled fill dams. The calculation seems to have been developed by Bruggeman *et al.* (1939) but is usually attributed to Hilf (1948).

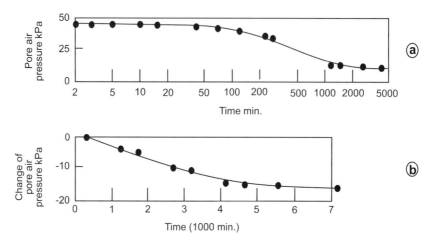

Figure 4.25 (a) Pore air pressure caused by compaction of a clay. (b) Change in pore air pressure in a sample of compacted clay under true undrained conditions. (Sample unstressed).

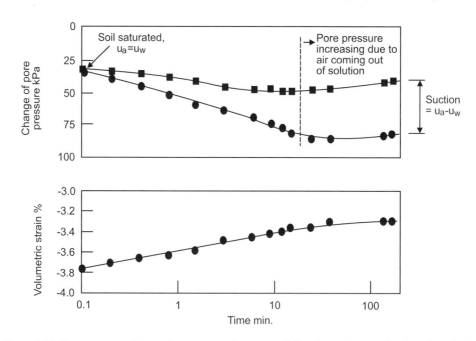

Figure 4.26 Pore pressure changes in a sample of compacted clay after sudden reduction of confining pressure.

The relationship between pore air pressure and volume change has usually been expressed as:

$$\Delta u_a = \frac{u_{ao} \cdot \varepsilon_v}{n_a + Hn_w - \varepsilon_v} \qquad (4.2)$$

which applies until the soil becomes saturated by compression. As the soil becomes saturated n_a tends to zero, and:

$$\Delta u_{sat} = \frac{u_{ao}\,\varepsilon_v(sat)}{Hn_w - \varepsilon_v(sat)} \tag{4.2a}$$

in which:

u_{ao} = initial pore air pressure in compacted soil
ε_v = volumetric strain of the soil
n_a = volume of the air per unit volume of soil (i.e., the air porosity)
n_w = volume of water per unit volume of soil (i.e., the water porosity)
H = Henry's coefficient of solubility of air in water, usually taken as 0.02 per volume per atmosphere pressure
$\varepsilon_v(sat)$ is given by:

$$\varepsilon_v(sat) = \frac{\rho_d e_o (1 - S)}{G_s \rho_w}$$

Equation 4.2 is often (apparently incorrectly) called Hilf's equation. Equation 4.2 can be considerably simplified by considering mass relationships rather than volumetric ones. If this is done (Blight, 2000), an expression for the absolute value of the pore air pressure in terms of the degree of saturation S, the universal gas constant R, the absolute temperature θ in Kelvin and the molecular masses of water, m_w and air, m_a, results:

$$u_a = \frac{NR\theta}{(1 - S)\cdot(m_w/S\rho_w) + R\theta/H} \tag{4.3}$$

where N = {(mass of air)/m_a}/{(mass of water)/m_w} [dimensionless]
ρ_w = mass density of water [kg/m³]
H = Henry's constant in mass terms [kPa]

The pore air pressure at which the soil becomes saturated is then obtained by putting $S = 1$, and is:

$$u_a(S = 1) = NH \tag{4.3a}$$

The relationship between pore air pressure (measured relative to atmospheric pressure) and degree of saturation S is graphed in Figure 4.27.

Once the pore air pressure has been calculated from equation 4.3 or estimated from Figure 4.27, the pore water pressure will equal the pore air pressure u_a, minus the soil suction. As examples,

$u_a = +50\,kPa$, $(u_a - u_w) = +100\,kPa$, then $u_w = 50 - 100 = -50\,kPa$

(below atmospheric)

$u_a = -50\,kPa$, $(u_a - u_w) = +100\,kPa$, then $u_w = -50 - 100 = -150\,kPa$

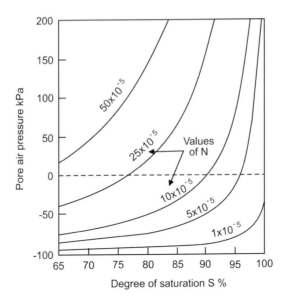

Figure 4.27 Calculated variation of pore air pressure u_a with degree of saturation S in undrained compression of partly saturated soil.

The following numerical values of the various constants appearing in equation 4.3 are required for calculations of this sort.

m_w = molecular mass of water = 0.01802 kg/mol
m_a = molecular mass of air = 0.02897 kg/mol
R = universal gas constant = 8.313 kPam^3/Kmol

Values of Henry's constant in mass terms are tabulated below:

Absolute temperature K	Henry's Constant H kPa/mol of air/mol of water
273	4320
278	4880
283	5490
288	6070
293	6640
298	7200

In calculating the curves shown in Figure 4.27, θ was taken as 293K (20°C). It is noteworthy that, at values of S below about 95%, negative equilibrium air pressures (i.e., pressures below atmospheric pressure) are predicted.

The question also arises as to whether this is possible in a field situation, as opposed to a laboratory test such as the measurements shown in Figure 4.25. Figure 4.28 shows a series of construction pore pressures measured in the compacted clay core of the Bridle Drift rockfill dam on the south eastern coast of South Africa

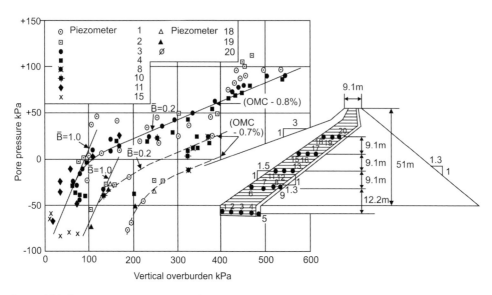

Figure 4.28 Relation between measured pore pressure and overburden stress for Bridle Drift dam, South Africa.

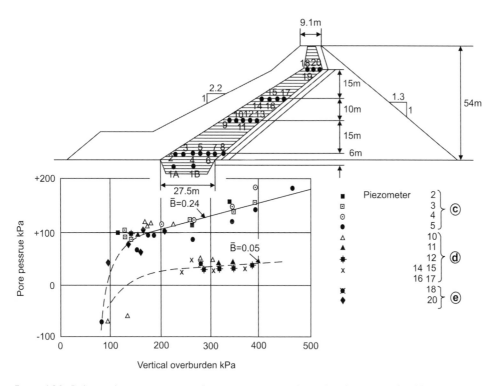

Figure 4.29 Relation between measured pore pressure and overburden stress for Manjirenji dam, Zimbabwe.

(33°S, 27.5°E) (Blight, 1970). The soil used for the clay core was residual from weathered shales. The piezometers were of the twin-tube hydraulic type with fine-pored ceramic tips. However, because the water in the tubes of this type of piezometer usually cavitates at pore pressures below about −70 kPa (0.7 bar), even when situated close to sea level, it is likely that the recorded negative pore pressures actually represent negative pore air pressures and that the real pore water pressures were less (i.e., more negative) than the values shown in Figure 4.28 by the amount of the suction. Hence, this example seems to show that negative pore air pressures can exist in a field situation.

Figure 4.29 shows similar data for construction pore pressures measured in the clay core of the very similar Manjirenji rock-fill dam in Zimbabwe (21.6°S, 31.7°E). The soil used for the core was residual from a weathered granodiorite. The data are not as complete as in the case of Bridle Drift dam and, as the dam is situated at an altitude of 1500 m AMSL, cavitation of the water filling the double hydraulic tubes was even more likely. Nevertheless, Figure 4.29 also shows what are probably measureable negative pore air pressures in a field situation.

4.16 SUMMARY

Most of the compaction technology applied to transported soils is applicable to residual soils. Differences in the properties of transported and residual soils must, however, be taken into consideration. Because residual soils are derived from weathered rocks, variability of the soil will reflect variability of the parent rock. Variability of the source material may relate both to particle size distribution and to cohesive or frictional attributes. Compaction assessment should include monitoring of source materials to ensure that specification requirements can be met.

Field compaction trials are strongly recommended because they enable the work to be controlled as much as possible by factors that have been proven in the field. There is no problem with field trials being undertaken as part of the permanent works, but it is important to realize that trials can be slow and painstaking compared with full production. Both the supervisor and the constructor must recognize this and allow for it in their schedules and budgets.

The particular advantages of field trials are related to the practicalities of field conditions and to time constraints:

- Knowledge of compaction behaviour and performance can be translated from laboratory test results, to results which reflect the performance of the compaction equipment. For example, with the initial guidance of laboratory tests, field maximum dry density and optimum moisture content can be determined for the compaction equipment, working in the actual site conditions, and be related to the required number of roller passes, thickness of placed layers, etc., to achieve the results demanded by the design. Laboratory tests can be used to set required standards such as compaction water contents and dry densities, but these results cannot necessarily be directly applied to field conditions or be met by the performance of field compaction equipment.
- Selection of lot sizes and testing frequency can be based on site conditions.

- Rapid field tests can be identified and selected so as to minimize the delays caused by waiting for laboratory test results.
- Method-based specifications can be proven and finalized. It is important that performance of method-based specifications be checked in the field. Methods of verification should be identified and recorded, and decisions made about the frequency of periodic laboratory or field tests which are necessary as a check on the quality being achieved.
- Compacted soils are often used to construct massive structures such as earth dam or highway embankments, in which cases it may become necessary to predict the generation of construction pore pressures and their effect on end-of-construction pore pressures. As compacted soils are always unsaturated after compaction, the mechanics of the unsaturated soils must be considered.
- The effects of post-construction changes to water contents related to the function of the compacted soil and the local climate must always also be considered.

Because of the experience of the authors, this chapter has tended to concentrate on applications in semi-arid to arid climatic conditions and to soils residual from sedimentary and ancient igneous rocks. The reader is referred to Wesley's (2010a) book on residual soils for more comprehensive information on the behaviour of compacted volcanic soils in wet climates.

REFERENCES

Blight, G.E. (1962) Controlling earth-dam compaction under arid conditions. *ASCE Civil Engineering*, pp. 54–55.
Blight, G.E. (1970) Construction pore pressures in two sloping-core rockfill dams. *10th Congr. on Large Dams, Int. Comm. on Large Dams (ICOLD), Montreal, Canada.* pp. 269–290.
Blight, G.E. (1989) Design assessment of saprolites and laterites. Invited Lecture, Session 6, *12th Int. Conf. on Soil Mech. & Found. Eng., Rio de Janeiro.* Vol. 4, pp. 2477–2484.
Blight, G.E. (2000) Air-water solution processes in recently compacted soil. *Developments in Geotech. Eng. 2000, Bangkok, Thailand.* pp. 303–310.
Bruggeman, J.R., Zanger, C.N. & Brahtz, J.H.A. (1939) *Notes on Analytical Soil Mechanics.* Tech. Mem. No. 592, Washington, D.C., U.S. Bureau of Reclamation.
Gidigasu, M.D. (1974) Degree of weathering in the identification of lateritic materials for engineering purposes – A review. *Eng. Geol.,* 8 (3), 213–266.
Gidigasu, M.D. & Dogbey, J.L.K. (1980) The importance of strength criterion in selecting some residual gravels for pavement construction. *7th Regional Conf. Africa Soil Mech. & Found. Eng., Accra, Ghana.* Vol.1, pp. 300–317.
Hilf, J.W. (1948) Estimating construction pore pressures in rolled earth dams. *2nd Int. Conf. Soil Mech. & Found. Eng.* Vol. 3, pp. 234–240.
Mitchell, J.K., Hooper, D.R. & Campanella, R.G. (1965) Permeability of compacted clay. *J. Soil Mech. & Found. Eng. Div. ASCE.,* 91 (SM4), 41–65.
Rodda, K.V., Perry, C.W. & Roberto Lara, E. (1982) Coping with dam construction problems in a tropical environment. In: Engineering and Construction in Tropical and Residual Soils. *ASCE Geotech. Div. Spec. Conf., Honolulu, Hawaii.* pp. 695–713.
Terzaghi, K. (1958) Design and performance of Sasumua Dam, Paper No. 6522. *Proc. Instn. Civ. Engrs.,* Vol. 9, pp. 369–388.

United States Environmental Protection Agency (USEPA) (1989) *Requirements for Hazardous Waste Landfill Design, Construction and Closure.* Seminar Pub., EPA 625 4-89/022, Washington, USA.

Wesley, L.D. (2010a) *Geotechnical Engineering in Residual Soils.* Hoboken, USA, Wiley.

Wesley, L.D. (2010b) *Fundamentals of Soil Mechanics for Sedimentary and Residual Soils.* Hoboken, USA, Wiley.

Chapter 5

Steady and unsteady flow of water and air through soils – permeability of saturated and unsaturated soils

V.K. Garga & G.E. Blight

5.1 DARCY'S AND FICK'S LAWS OF STEADY STATE FLOW

In the middle years of the nineteenth century, Darcy showed, by means of experiment, that the steady state rate of flow of liquid (water) is proportional to the gradient of pressure head in the direction of flow, i.e., the loss of pressure head per unit of distance in the direction of flow. Darcy's law can be written as:

$$v = -ki, \quad \text{or} \quad k = -v/i \tag{5.1}$$

where v is the velocity of flow with dimensions [m/s]
    k is called the coefficient of permeability

The negative sign shows that the pressure or pressure head decreases in the direction of flow. The pressure head is the pressure [N/m^2] divided by the unit weight of water [N/m^3], i.e., [N/m$^2 \cdot$ m^3/N $=$ m], hence

i, the gradient of the pressure head has a unit of [m/m], i.e., it is dimensionless.

Thus the units of k are [m/s], i.e., velocity.
    The unit weight of water is denoted by γ_w.
    If p is the pressure and z is a distance measured in the direction of flow.

$$i = \frac{p/\gamma_w}{z}$$

If equation 5.1 is expressed as:

$$\frac{dm}{dt} = -D_c \frac{dp}{dz} \tag{5.2}$$

D_c is called the diffusion coefficient.
where dm/dt is the mass of fluid per unit of area normal to the direction of flow per unit time, with units of [kg $\cdot$ m^{-2} s^{-1}],
        dp/dz is the gradient of pressure with distance in the direction of flow, with units of [N m$^{-2} \cdot$ m^{-1}] $=$ [kg m s$^{-2} \cdot$ m$^{-2} \cdot$ m^{-1}] $=$ [kg m^{-2} s^{-2}], therefore
        D_c has units of [s].

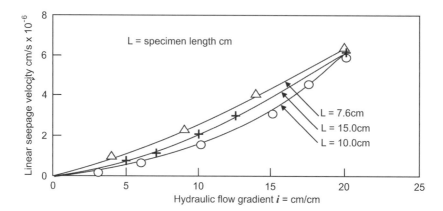

Figure 5.1 Observed relationships between flow velocity and flow gradient i.

Equation 5.2 is known as Fick's law, which applies to all fluids, whether compressible or incompressible, i.e., whether liquids or gases.

The equation of state for air that can be used to account for its compressibility is:

$$pV = \frac{R\theta}{m_a} \cdot m, \quad \text{or} \quad m = \frac{m_a pV}{R\theta} \tag{5.3}$$

where p is the air pressure above atmospheric pressure,

V is the volume of air,
R is the universal gas constant $= 8.313\,kPam^3/Kmol$,
θ is the absolute temperature, [K],
m_a is the molecular mass of air $= 0.02897\,kg/mol$,
m is the mass of air per unit of area moving in the direction of flow, and
$K = °C + 273$ (e.g., $20°C = 293K$).

It is easy to be misled by equations 5.1 and 5.2 into thinking that the constant of proportionality k or D is constant for all values of pressure p. This is not so either for water or air flow. Figure 5.1 is a set of data showing that the flow velocity v through a saturated soil is not strictly directly proportional to the flow gradient i. The soil was a clayey sand residual from weathered granite, saturated and subjected to the seepage of water in a triaxial cell. Figure 5.1 shows that the linear seepage velocity did not increase in direct proportion to the increase in seepage gradient, although the increase was approximately proportional.

Figure 5.2 shows relationships between pressure gradient and mass velocity for both air and water flow through the same fine-pored ceramic disk (Blight, 1971). In the case of air flow, the ceramic was dry, while it was completely saturated for the test on water flow. The resultant flow curves are very similar to those shown in Figure 5.1. For both fluids, flow does not increase strictly linearly with increasing pressure gradient, but over a small range of gradients, both Darcy's and Fick's laws are reasonably accurate.

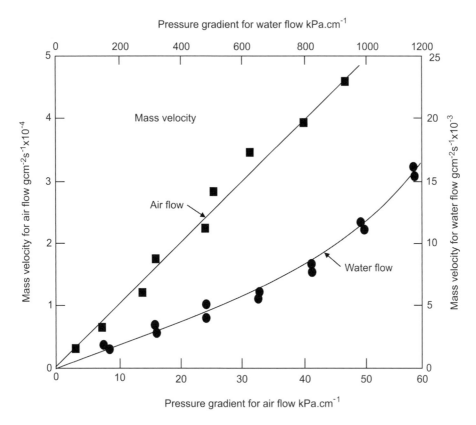

Figure 5.2 Comparison of air and water flow under steady state conditions through the same porous ceramic disc, in terms of mass velocity versus pressure gradient.

5.2 DISPLACEMENT OF WATER FROM SOIL BY AIR

Air can enter a saturated soil only by displacing water from the pores. For displacement to occur, the air pressure at entry to the soil must be large enough to overcome the capillary forces retaining water in the pores. Therefore, there will be an air pressure threshold below which the soil is impervious to air. Once the threshold or air entry pressure is exceeded, the interconnected coarser pores will drain and a limited air permeability will be established. As the air pressure is increased, an increasing proportion of soil pores will drain and the air permeability will increase progressively until a balance is reached between the air pressure and the capillary stresses retaining water in the soil.

In layered soils the threshold pressures for flow parallel to and normal to the laminations will differ. Air flow will be initiated through the coarser zones of the soil, usually parallel to the laminations. Unless the air pressure rises above the threshold value for flow across the laminations, the soil will be permeable to air flow in one direction only.

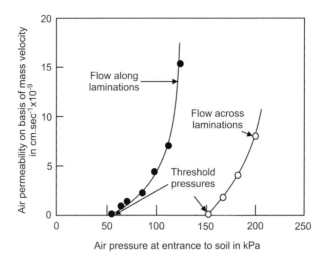

Figure 5.3 Air permeability of initially saturated specimen of tailings.

Figure 5.3 shows the air flow characteristics typical of a layered soil. The threshold or air entry pressure for flow across the laminations is three times that for flow along the laminations. Also, as the air pressure at entrance to the soil rises and drainage of pores proceeds, the permeability to flow along the laminations increases very rapidly. In a practical situation this rapid increase in permeability may prevent the air pressure in the soil from reaching the threshold value for flow across the laminations. Hence the soil may remain permeable to air flow in one direction only, as appears to be the case in Figure 5.3. The measurements shown in Figure 5.3 represent a series of steady-state conditions. The soil was allowed to drain fully under each increment of air pressure before the permeability was measured. The unsteady flow of air through a soil from which water is being actively displaced would be governed by an equation of the form of equation 5.5a (derived below). This cannot be solved without introducing the relationship between air pressure p and degree of saturation S. (Because the void ratio e, degree of saturation, S and water content w are related by $eS = wG$, Figure 5.3 is closely linked to the SWCC curves shown in Figures 1.16b and c. The curves shown in Figure 5.3 are also subject to hysteresis if the pressure of the permeating air is reduced.)

5.3 UNSTEADY FLOW OF AIR THROUGH PARTLY SATURATED AND DRY SOILS

Firstly, consider an elemental volume of fixed size $dx\,dy\,dz$ in an unsaturated soil through which air at constant temperature is flowing in the z-direction only. The boundaries of the element are fixed in space, and the soil in the element has a porosity n and a degree of pore space saturation S.

If $M=$ the mass of air entering the element in the z-direction in unit time, then using Fick's law:

$$M = (1 - S)nD_c\frac{\partial p}{\partial z}dx\,dy \tag{5.4a}$$

and using Darcy's law:

$$M = k\frac{m_a}{R\theta}(1 - S)np\frac{\partial p}{\partial z}dx\,dy \tag{5.4b}$$

If $m=$ the mass of air contained or stored in the element at any time, then for mass continuity of the air:

$$\frac{\partial m}{\partial t} = M - \left(M + \frac{\partial M}{\partial z}dz\right) = -\frac{\partial M}{\partial z}dz$$

From the equation of state for air, equation 5.3:

$$m = \frac{m_a}{R\theta}(1 - S)np\,dx\,dy\,dz$$

and hence from equation 5.4a:

$$\frac{\partial}{\partial t}(1 - S)np = \frac{D_cR\theta}{m_a}\frac{\partial}{\partial z}(1 - S)n\frac{\partial p}{\partial z} \tag{5.5a}$$

From equation 5.4b:

$$\frac{\partial}{\partial t}(1 - S)np = k\frac{\partial}{\partial z}(1 - S)np\frac{\partial p}{\partial z} \tag{5.5b}$$

If one is concerned with air flow through relatively rigid soils in which the volume is sensibly constant with time and if the air is not actively displacing water from the pores of the soil, n and S can be regarded as constants. In this case, equations 5.5a and b reduce to

$$\frac{\partial p}{\partial t} = \frac{D_cR\theta}{m_a}\frac{\partial^2 p}{\partial z^2} \tag{5.6a}$$

$$\frac{\partial p}{\partial t} = k\frac{\partial}{\partial z}\left(p\frac{\partial p}{\partial z}\right) \tag{5.6b}$$

Equation 5.6a is identical with the linear Terzaghi equation for the one-dimensional consolidation of a saturated soil with $D_cR\theta/m_a$ replacing c_v. Equation 5.6b, however, is nonlinear.

If equation 5.6a describes, with sufficient accuracy, the unsteady flow of air through dry rigid soils, and through soils whose behaviour approximates this condition, the large number of established solutions of the equation (and its two and three dimensional forms) that have been worked out for consolidation and seepage problems

in saturated soils, become available. Moreover, the steady state form of equation 5.6a, the Laplace equation

$$\frac{\partial^2 p}{\partial z^2} = 0 \qquad (5.7)$$

will also apply to the steady-state flow of air through dry rigid soils and soils approximating to this condition, i.e., soils in which air is not either displacing, or being displaced by water.

5.4 UNSTEADY FLOW OF AIR THROUGH UNSATURATED SOIL

An unsaturated soil that contains a proportion of continuous air voids will be permeable to air. If the air pressure at entry to the soil is not greatly in excess of the capillary pressures retaining water in the pores, air flow will take place with little displacement of water. Any displacement that does occur will, however, alter the air permeability of the soil.

Curve A in Figure 5.4 shows the results of an air pressure build-up test on a specimen of clay compacted 2% wet of the Proctor optimum water content. The degree of saturation of the specimen varied between 94% and 98% during the test. The amount of pore space available for the passage of air was, therefore, very limited. Nevertheless, the experimental equalization curve deviates very little from the curve predicted by the Terzaghi theory.

Curve B in Figure 5.4 shows typical experimental consolidation curves for two compacted clays saturated by back pressure before consolidation and consolidated against a back pressure to maintain complete saturation. The experimental degrees of

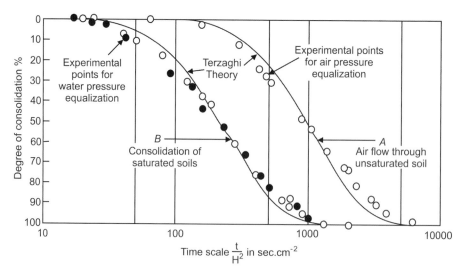

Figure 5.4 Line A: data for unsteady flow of air through unsaturated soil specimen. Line B: data for consolidation of two saturated soil specimens.

consolidation deviate from the theory by a maximum of 7% which is very similar to the deviation of Curve A. The application of Fick's law to unsteady air flow through unsaturated soils is thus subject to a degree of error similar to that of applying Darcy's law to the unsteady flow of water through a saturated soil.

5.5   MEASURING PERMEABILITY TO WATER FLOW IN THE LABORATORY

The coefficient of water permeability k is very much affected by the structure and macro-structural features of a natural soil (shrinkage cracks, fissures, layering, alternating coarse and fine layers, occasional coarse layers, etc.). As a result of this sensitivity, small scale measurements of permeability (usually laboratory measurements) tend to be unreliable because a small scale specimen is unlikely to contain a macro-feature, and if it does, the true overall effect of a series of such macro-features will not be correctly simulated. Hence, in general, larger scale field measurements are preferred. However, at the early design stage for a new project, there may be no alternative to measuring permeabilities in the laboratory.

There are two basic types of laboratory apparatus for measuring the coefficient of permeability – rigid walled and flexible walled permeameters. Originally, permeameters consisted of a simple cylinder of metal or rigid plastic material which contained the particulate material to be tested. This is the rigid walled permeameter. However it was realized that leakage between the wall and the sample could badly affect measurements, resulting in apparent permeabilities that may be considerably greater than true values. The solution adopted was to measure the permeability on a specimen contained in a triaxial cell where the flexible triaxial membrane, forced against the specimen by the cell pressure, effectively prevents leakage down the side wall. This is the flexible walled permeameter. For specimens containing sharp particles, however, such as crushed gravels, or for very large particles such as waste rock, it is difficult to avoid punctures in the flexible membrane and to overcome this, a hybrid rigid-flexible wall consisting of a rigid cylinder, lined with resilient sheet rubber can be used. In testing highly pervious large-particled materials a steel cylinder lined with a resilient material, such as a relatively impervious closed cell foam sheet (e.g., flexible polyurethane foam) can be used to prevent side wall leakage.

5.6   OBSERVED DIFFERENCES BETWEEN SMALL SCALE AND LARGE SCALE PERMEABILITY MEASUREMENTS

Day & Daniel (1985) and Daniel (1987) conducted comparative field and laboratory measurements of permeability on two clays. Test ponds were constructed in the field, and samples were later retrieved from the test liners for laboratory measurements. Measurements of seepage rate were made for the pond as a whole, and by means of single and double ring infiltrometers. Tests using both rigid and flexible walled permeameters were made on block and tube samples of the clay compacted *in situ*, and also on samples compacted in the laboratory. Effective confining stresses in the laboratory were about 100 kPa and seepage gradients ranged from 20 to 200. Day

and Daniel found that values of permeability deduced from seepage losses from the ponds were 900 to 2000 times larger than permeabilities measured in the laboratory, but only 1.2 to 1.9 times larger than field infiltrometer measurements. This work has recently been summarized by Daniel & Koerner (2007).

Chen & Yamamoto (1987) also carried out a comparison of field and laboratory permeability measurements, using infiltrometers and porous probes *in situ*, and flexible-walled permeameters in the laboratory. For the laboratory tests, effective stresses were about 200 kPa and the seepage gradient was 180. They found field permeabilities were 10 times larger than laboratory values. Elsbury *et al.* (1990) made a comparison of field and laboratory permeability measurements on a highly plastic clay. They found that double ring infiltrometer tests gave slightly lower permeabilities than did seepage rates from a test pond. Also, compaction in the field with a vibratory roller resulted in a clay with a permeability ten times larger than one compacted using the same roller without vibration. Permeabilities measured in the laboratory used seepage gradients of 20 to 100 and effective stresses of 15 to 70 kPa. Permeabilities measured in the field proved to be between 10 000 and 100 000 times greater than values measured in the laboratory.

Pregl (1987) has stated that a permeability measured in the laboratory serves as an index of material quality but is not directly related to the permeability of a prototype lining in the field. The permeability in the field will always be less than that measured in the laboratory (according to Pregl) because the seepage gradient used in laboratory tests is usually of the order of 30 whereas that in the field approximates to unity. Also, the Darcy coefficient of permeability is not constant with seepage gradient (see Figure 5.1). However, the majority of observational evidence shows that field–measured permeabilities far exceed those measured on laboratory specimens.

It is apparent from these studies that there are several possible reasons why a permeability measured in the field may differ from one measured in the laboratory:

- A large area exposed to seepage is more likely to contain defects in the form of cracks and more permeable zones than is a small area.
- If the Darcy coefficient of permeability is not constant with flow gradient, the use of different seepage gradients in the field and laboratory will result in different field and laboratory values.
- A similar remark applies to effective stresses. A specimen subjected to a high effective stress can be expected to show a lower permeability than a similar one with a low effective stress.

Nevertheless, it is possible to obtain reasonable agreement between field and laboratory permeability tests, as shown by Table 5.1 which compares the results of the ponding tests (shown later in Figure 5.10a) with a number of laboratory test results.

All laboratory permeability tests referred to in Table 5.1 were of the constant head flexible wall triaxial type performed on 100 mm diameter specimens. The average effective stress was kept at 3 kPa for all tests and the seepage gradient at unity. This stress was the lowest value that could be controlled reliably in the laboratory and was similar to the effective overburden stress in the pond tests of 0.5 m (only 4 to 5 kPa).

Table 5.1 compares the permeability values measured in the laboratory on specimens compacted to the same dry density as the upper 150 mm of soil in the ponds.

Table 5.1 Comparison of field and laboratory permeability tests.

Range of Mean Permeability from Pond Test (cm/s × 10^{-6})	Range of Mean Permeability from Laboratory Tests (cm/s × 10^{-6})
59 to 81	37 to 93 (63 to 115% of field test values)

Hence this set of measurements shows that it is possible, with correct interpretation and correct testing, to estimate *in situ* permeabilities reasonably closely from the results of laboratory tests.

However, it must be noted that the permeability of this granite soil is higher than the permeabilities that were considered by, e.g., Day & Daniel (1985). It is noted from the literature that discrepancies between field and laboratory permeabilities generally appear to increase as the soil becomes less permeable, and presumably therefore, discontinuities and defects may play a greater role in modifying the permeability of large volumes of the soil.

5.7 LABORATORY TESTS FOR PERMEABILITY TO WATER FLOW

The two main types of test for permeability to water are the constant head or constant gradient test and the falling head or reducing gradient test. As mentioned earlier, permeability values determined in the laboratory do not necessarily represent the *in situ* behaviour of residual soils. This is particularly true for undisturbed soils where the relatively small size of laboratory samples is inadequate to incorporate the various geological discontinuities, e.g., permeable and impermeable fissures and veins and other relict structures, present in a weathered profile, permeable layers and shrinkage cracks in a transported profile or a compacted layer. It should be noted that Darcy's law is only partially valid and that the coefficient of permeability of a soil can vary considerably with the flow gradient (Figure 5.1), It is therefore important to use a similar flow gradient in the laboratory to that likely to occur in the field.

Conventional permeability tests in the laboratory may however be applicable to compacted soils and more uniformly structured natural soils *in situ*, especially where permeability is determined on both horizontally and vertically trimmed samples. It is then possible to estimate the overall, or effective permeability for uniformly textured soils. Laboratory tests, unlike field tests, also have the advantage of providing an indication of the variation in the coefficient of permeability with changes in effective stress. These data are often important for the design of earthworks and are generally not available from field tests. Constant head permeability tests, carried out in a triaxial apparatus coupled with pore pressure dissipation tests to measure the coefficient of consolidation c_v, are particularly useful in this regard. Such tests permit the determination of permeability at various effective stresses as the sample undergoes consolidation (or swelling) and its relation to compression and void ratio (Tan, 1968; Garga, 1988).

In constant head tests the permeability is measured directly by maintaining a small pore pressure differential (10–20 kPa or 1 to 2 m of water head) across the sample and by applying Darcy's law when a steady state flow rate is achieved. The coefficient

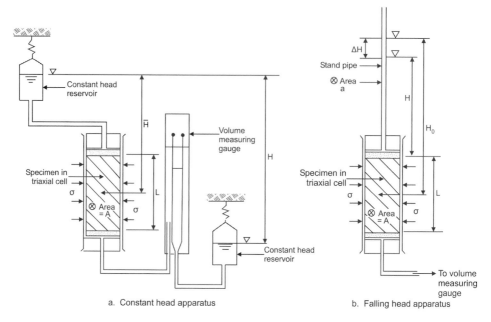

a. Constant head apparatus

b. Falling head apparatus

Figure 5.5 Layout of laboratory constant and falling head permeability tests.

of permeability can be determined by the following expression which is simply the defining equation (5.1) for Darcy's law expressed in directly measurable quantities:

$$k = \frac{q_\infty L}{A \Delta H} = \frac{v}{i} \tag{5.1a}$$

where q_∞ = steady state rate of flow (volume per unit time),
   A = area of cross-section of the sample, therefore
   $q_\infty/A = v$,
   L = length of sample,
   ΔH = constant differential head across the sample,
   $\Delta H/L = i$.

The layout of the apparatus for the constant head test is shown in Figure 5.5a, in which $H - \overline{H} = \Delta H$, and that for the falling head test is shown in Figure 5.5b.

For the falling head test, the water flowing through the sample is supplied from the open standpipe of cross-sectional area a. If the head falls by ΔH in time ΔT the rate of flow is $v = a\Delta H/\Delta t$ which equals AkH/L. Written differentially,

$$-a\frac{dH}{H} = \frac{Ak}{L} \cdot dt$$

Integrating between times $t = 0$ and $t = t$, and heads H_0 and H gives:

$$k = \frac{aL}{At} \cdot \ln\left(\frac{H_0}{H}\right) = 2.3\frac{aL}{At} \cdot \log_{10}\left(\frac{H_0}{H}\right) \tag{5.2b}$$

where $H_0 = H$ at $t = 0$ $H = H$ at $t = t$.

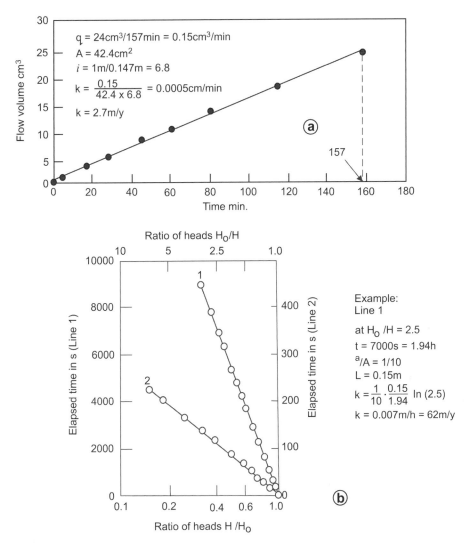

Figure 5.6 Typical results for permeability tests on soil: (a) constant head test, (b) falling head test. Note that $\ln(H_0/H) = -\ln(H/H_0)$, either ratio may be used.

In less pervious soils or waste materials, it may take a very long time for q to attain the steady state value q_∞. It may also take a long time for an accurately measureable quantity of flow to accumulate. For this reason, falling head tests are usually preferred for less pervious soils Not only is it not necessary to wait for equilibrium flow, but the falling head can be magnified by increasing the ratio A/a of sample area to standpipe area.

Figure 5.6a shows typical results for a constant head permeability test on a tailings specimen, in which the cumulative flow has been plotted against elapsed time. The flow gradient was 10 kPa over a length of 0.147 m, i.e., i = 6.8 and k = 2.7 m/y.

Table 5.2 Typical ranges of permeability to water flow for various soil types.

Coefficient of Permeability k (cm/s)														
	10^2	10	1.0	10^{-1}	10^{-2}		10^{-3}	10^{-4}	10^{-5}	10^{-6}		10^{-7}	10^{-8}	10^{-9}
Types of soil		Clean gravel		Clean sands, clean sand and gravel mixtures				Very fine sands, silts, mixtures of sand, silt & clay					Homogeneous clays	

Figure 5.6b shows results for falling head permeameter tests on two tailings specimens. Here, the ratio of the heads (H/H$_0$) is plotted to a logarithmic scale against the elapsed time t. k can then be calculated from equation 5.2b.

Table 5.2 shows the range of coefficients of permeability to be expected for various types of soil. The values are given in cm/s. To convert these values to m/y, multiply by 315 360. For example, 1×10^{-6} cm/s $= 0.32$ m/y.

5.8 MEASURING PERMEABILITY TO AIR FLOW

Permeability to air can also be measured either by constant or falling head methods. However, because air is a compressible fluid, it is difficult to measure the volume of air passing through a laboratory specimen as the volume of the throughflow has to be measured at constant pressure and temperature. It is much simpler to use the falling head principle (Blight, 1977), in which the volume of the permeating air supply is kept constant and the decline in pressure as the air passes through the soil is measured.

Differentiating the equation of state for air (equation 5.3) partially with respect to t:

$$\frac{\partial m}{\partial t} = \frac{V m_a}{R\theta} \cdot \frac{\partial p}{\partial t}$$

Hence, from equation 5.2:

$$D_c \frac{\partial p}{\partial z} = \frac{V m_a}{R\theta} \cdot \frac{\partial p}{\partial t} = D_c \cdot \frac{p}{d}$$

where d is the length of seepage path, i.e., the length of the test specimen, and A is its superficial area. It is assumed that the air flow exhausts into air at atmospheric pressure, i.e., reduces from p to zero.
Rearranging:

$$\frac{\partial p}{p} = \frac{D_c A R\theta}{d V m_a} \cdot \partial t$$

Integrating:

$$\frac{V m_a d}{D_c A R\theta} \ln(p) = t + \text{constant}$$

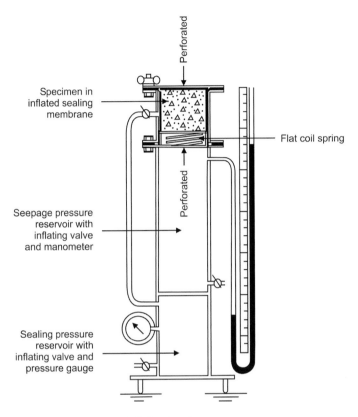

Figure 5.7 Falling head air permeameter.

When $t=0$, $p=p_o$ $\therefore$ constant $= \dfrac{Vm_ad}{D_cAR\theta}\ln(p_o)$ or $t = \dfrac{Vm_ad}{D_cAR\theta}\ln\left(\dfrac{p_o}{p}\right)$

Finally, the diffusion coefficient is given by:

$$D_c = \frac{Vm_ad}{AR\theta t}\ln\left(\frac{p_o}{p}\right)\ [\mathrm{s}] \tag{5.8}$$

The values of m_a and R are given below equation 5.3 on page 120.

With these values inserted (and kPa converted to $\mathrm{kg\,m^{-1}s^{-2}}$), equation 5.8 becomes:

$$D_c = 3.48 \times 10^{-3} \cdot \frac{Vd}{A\theta t}$$

Note that

θ (in K) $=$ °C $+ 273$, e.g., $20°C = 293K$

If m_a is in $[\mathrm{kg\,m^{-2}}]$, V in $[\mathrm{m^3}]$, d in [m] and t in [s], the units of D_c will be [s].

Figure 5.7 shows the construction of the permeameter which is conveniently made out of a length of 150 mm diameter steel pipe. The specimen is at the top, sealed at the

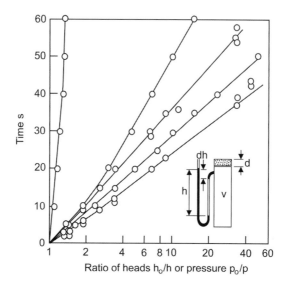

Figure 5.8 Experimental relationships between time and head or pressure in measurement of coefficient of air diffusion (Equation 5.8).

sides by an inflated latex rubber sleeve, held in position against the perforated top plate by a flat coil spring. The side seal inflating air reservoir is at the base of the permeameter and the percolating air reservoir is above it, and is inflated through the valve, shown lower right. The falling pressure in the constant volume of percolating air is measured by means of a manometer, pressure transducer or pressure gauge.

Figure 5.8 shows some typical relationships between time and the ratio of heads h_0/h or pressures p_0/p (plotted logarithmically, equation 5.8). Note that the relationship is usually linear, but in some cases is not completely linear.

5.9 METHODS FOR MEASURING WATER PERMEABILITY *IN SITU*

The most common techniques to measure water permeability in the field involve some form of either constant head or falling head testing in unlined or lined (i.e., cased) boreholes. It is very common to obtain an estimate of k values by performing simple falling head tests in the drill stem at various depths as the drilling proceeds. In special cases, permeability testing can also be carried out using inflatable packers to isolate specific zones for testing, or by installing sealed hydraulic piezometers at different depths.

Many near surface soils have sufficient cohesion to permit a testing hole to be opened up with a manual or machine operated auger, without the need of a casing and to remain stable during permeability testing. A continuous-flight auger drill can be used successfully in unsaturated soils where speedy drilling may be necessary. Other methods for drilling in cohesive soils e.g., undisturbed residual soils, include wash boring, percussion-hammer drilling and rotary drilling. Whatever the method used, it is important that the inside surface of the hole used for permeability testing be free of

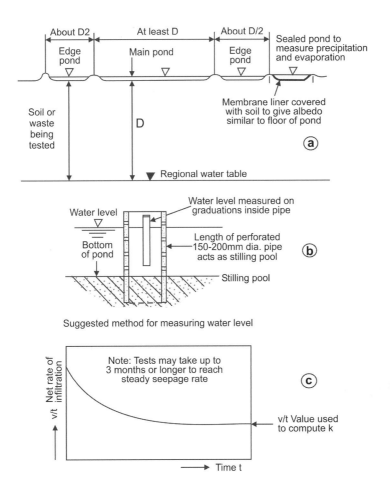

Figure 5.9 (a) Layout of ponding test to measure permeability for vertical flow *in situ*. (b) Method of shielding water level measurements from effects of wind. (c) Variation of infiltration rate with time after filling pond.

loose or remoulded (smeared) material. Smear must be removed by brushing the sides of the hole or surging it with water.

The most frequently used direct measurements of in-situ permeability can be divided into two major groups: those which feed water into the ground and those which extract water. The feed-in tests may be used above or below the water table, while the extraction tests can only be conducted below the water table.

5.9.1   Permeability from surface ponding or infiltration tests

Ponding tests are suitable for measuring the coefficient of permeability for vertical flow and are thus appropriate for obtaining permeability values for the design and checking the permeability of compacted clay liners. The layout and geometrical requirements for a ponding test are shown in Figure 5.9a.

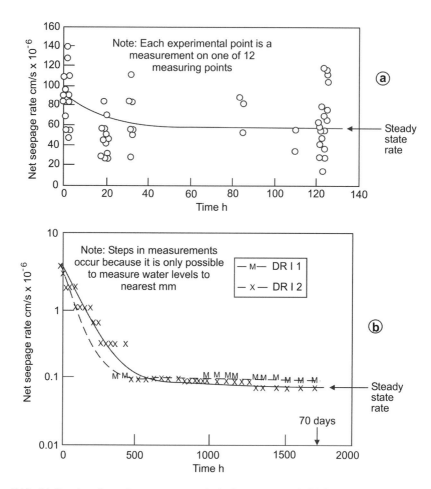

Figure 5.10 (a) Results of ponding test on residual silty granite soil, (b) Results of two side-by-side double ring infiltrometer tests on compacted clayey residual granite soil.

The main pond, on which the seepage measurements are made is surrounded by edge ponds (or moats) so that the effects of essentially vertical flow are measured. It is essential that the position of the regional water table be known and that the dimensions of the ponds be related to the depth of the water table. The ponding test referred to in Figure 5.10a consisted of four ponds surrounded by a moat. One of the four was lined and served as an evaporation measuring pond. Each pond had four water level observation points. The scatter evident in Figure 5.10a resulted from the difficulty of measuring small changes of water level accurately and also difficulty in compensating the measured seepages for evaporation losses which also depend on measuring small changes of water level. Temperature changes also affected the accuracy of the measurements by causing the plastic pipes (used as small stilling ponds against wind effects) to change in length. Thus an increase in temperature caused an apparent increase in seepage rate, and vice versa.

A subsidiary pond with an impervious liner is used to allow for the effects of precipitation and evaporation during the course of the test. The geomembrane liner must be covered with soil or waste similar to the floor of the prototype reservoir or pond to give a similar reflective albedo. This is absolutely essential, as the rate of evaporation is often of the same order as the seepage rate. As an alternative to separately measuring the evaporation rate, a layer of lubricating oil can be used to cover the surface of the water to eliminate evaporation. Seepage rates may be of the order of mm/day and it is extremely difficult to measure seepage accurately. The measuring system must be protected from the effects of temperature and wind if accurate results are to be achieved. Figure 5.9b shows a suggested system to do this. After filling the ponds the net infiltration v = (change in water level of main pond – change in water level of sealed pond) should be observed and plotted against time t (see Figures 5.9c and 5.10a and b). Plate 5.1 shows a pond layout with perimeter moats and lined evaporation pond. Obviously, any replenishment of water in the ponds either by rain precipitation or other means must also be taken into account. The calculated coefficient of permeability is based on the net rate of infiltration once this has reached a steady value.

The single or double-ring infiltrometer test is a smaller and cheaper variant of a full-scale ponding test (see Figure 5.11). For the double ring test a pair of concentric sheet metal rings are set and sealed into the ground surface and filled with water. The outer ring serves the same purpose as the edge pond and reduces edge effects. To give reasonable values of permeability, the diameter of the outer ring should not be less than 1200 mm and that of the inner ring, 600 mm.

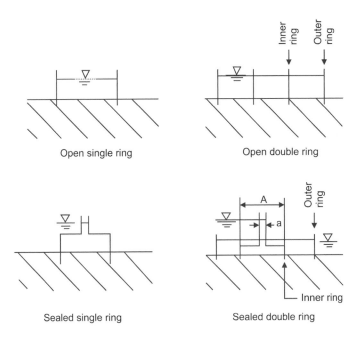

Figure 5.11 Open and sealed single and double ring infiltrometers.

Plate 5.1 Set of four moated seepage test ponds with lined pond (for assessing evaporation) closest to camera.

The two forms of ring infiltrometers popularly used are the "open" and "sealed" forms. Four variations are illustrated in Figure 5.11. Open rings are less desirable because with a low conductivity soil, it is difficult to separate the drop in water level of the pond caused by seepage from changes caused by evaporation and rainfall.

With sealed rings, however, very low rates of flow can be measured. Single-ring infiltrometers allow flow to spread laterally beneath the ring, complicating the interpretation of test results. Single rings are also susceptible to the effects of temperature variation: as the system heats up, the whole of it expands and as it cools down, the whole system contracts. This can lead to erroneous measurements when the rate of flow is small.

The sealed double-ring infiltrometer has proved the most successful. The outer ring forces the infiltration from the inner ring to be closer to one dimensional. Covering the surface virtually eliminates the effects of evaporation and rainfall, and the use of small diameter standpipes make more accurate flow measurements possible. With a standpipe, the flow rate can be measured more accurately. If the plan area of the inner ring is A, and the cross-sectional area of the standpipe is a, the drop in level is magnified by the ratio A/a. One of the most difficult problems is to prevent the water contained in the ring from breaking out beneath the edge of the embedded ring. This is a particular problem in long-term tests that may take weeks to complete. The authors have found that the best solution is to prepare a groove in the soil surface to receive the ring. This should be at least 50 mm deep. The ring is then put in place and sealed by pouring plaster of Paris into the groove on both sides of the ring, up to the level of the soil surface.

The average coefficient of permeability of the stratum between the soil surface and the regional water table at depth D is given by:

$$k = \frac{v}{t} \cdot \frac{D}{D+d} \tag{5.9}$$

where d is the average depth of water in the pond during the test.

Note that d only becomes significant if D is not much greater than d. For example, if $d = D$, $k = 1/2(v/t)$, if $d = D/10$, $k = 10/11(v/t)$.

Figure 5.10b shows the results of a pair of sealed double ring infiltrometer tests carried out on a compacted clay liner. Note that the results are repeatable from test to test with much less scatter in the measurements than in Figure 5.10a. Also note the length of time taken to achieve a steady flow rate.

5.9.2 Permeability from borehole inflow or outflow (USBR, 1951 & 1974)

Variable Head Tests

If the water table is close to the surface, the permeability may be estimated by observing the rate of rise of water in a borehole which penetrates the water table. This method suffers from the disadvantage that the permeability measured is primarily that for horizontal flow. Thus the presence of a highly permeable horizon such as the pebble marker or a sandy layer may have an overriding effect on the results.

If the borehole does not intercept the water table, measurements can be carried out using the same principle. Now, however, it is necessary to fill the borehole with water and observe the fall of water level with time as it seeps into the surrounding soil. The rate of fall in the water level should be observed for several hours, refilling the hole when necessary, as initially the flow gradients due to gravity will be augmented by suction in the soil around the hole. If the soil surrounding the hole is not thoroughly wetted, a true permeability cannot be measured.

The methods of analysis (Hvorslev, 1951; Schmidt, 1967) consist of first determining the basic time lag T, for which either of the two methods shown in Figure 5.12 may be employed. The coefficient of permeability k may be obtained from:

$$k \text{ (or } k_h \text{ if soil is anisotropic)} = \frac{A}{FT}$$

where:
A = cross-sectional area of the standpipe,
F = appropriate shape factor shown in Figure 5.13 (F has dimension of length),
k = isotropic permeability,
k_h = horizontal permeability (in anisotropic soil),
T = basic time lag.

Note that Figure 5.12 is drawn for an artificially raised water level in the borehole, whereas Figure 5.13 is drawn for a water level in the borehole that has been lowered by pumping.

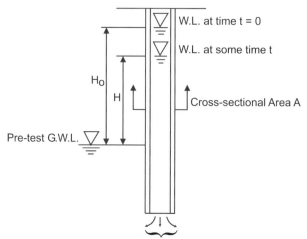

Zone allowing flow has shape factor F
Definition of symbols for falling head tests

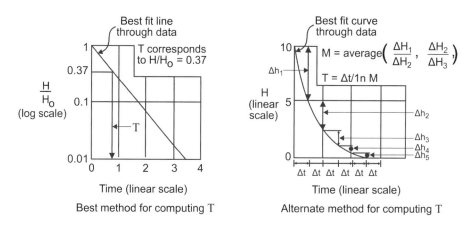

Figure 5.12 Calculation of basic time lag, T, for variable head tests.

Where the soil is anisotropic, the ratio of $k_h/k_v = m^2$ must be estimated*, or obtained from laboratory tests. However, it should be noted that the error in evaluating the permeability of the soil due to error in selection of m^2 is less than the inherent error in a falling head test. (*Note that in this section, m denotes a dimensionless ratio of horizontal to vertical permeability, not the unit of metres or a mass.)

Constant Head Tests

These analyses (Hvorslev, 1951; Schmidt, 1967) can be used for any feed-in test where the inflow during a test under a constant head becomes constant over time, i.e., when steady state flow conditions are achieved or approached. The analysis consists of the following steps.

i = isotropic conditions: $k_h = k_v = k$
a = anisotropic conditions: $k_h \neq k_v$

$$k \text{ or } k_h = \frac{A}{(F \cdot T)}$$

Note: Flow direction shown for falling head tests for clarity;
  "A/F"values also applicable for rising head tests

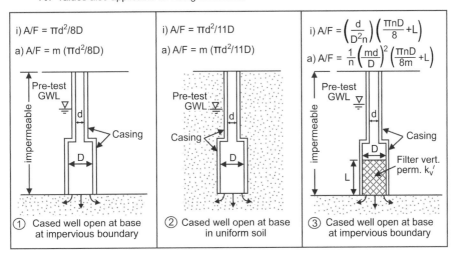

i) A/F = $\pi d^2/8D$

a) A/F = m ($\pi d^2/8D$)

① Cased well open at base
   at impervious boundary

i) A/F = $\pi d^2/11D$

a) A/F = m ($\pi d^2/11D$)

② Cased well open at base
   in uniform soil

i) A/F = $\left(\dfrac{d}{D^2n}\right)\left(\dfrac{\pi nD}{8}+L\right)$

a) A/F = $\dfrac{1}{n}\left(\dfrac{md}{D}\right)^2\left(\dfrac{\pi nD}{8m}+L\right)$

③ Cased well open at base
   at impervious boundary

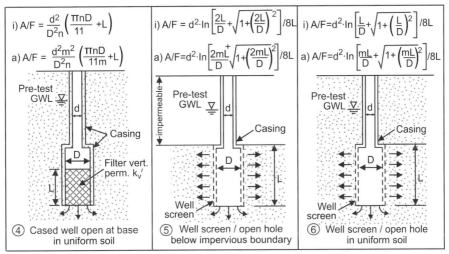

i) A/F = $\dfrac{d^2}{D^2n}\left(\dfrac{\pi nD}{11}+L\right)$

a) A/F = $\dfrac{d^2m^2}{D^2n}\left(\dfrac{\pi nD}{11m}+L\right)$

④ Cased well open at base
   in uniform soil

i) A/F = $d^2 \cdot \ln\left[\dfrac{2L}{D}+\sqrt{1+\left(\dfrac{2L}{D}\right)^2}\right]/8L$

a) A/F = $d^2 \cdot \ln\left[\dfrac{2mL}{D}+\sqrt{1+\left(\dfrac{2mL}{D}\right)^2}\right]/8L$

⑤ Well screen / open hole
   below impervious boundary

i) A/F = $d^2 \cdot \ln\left[\dfrac{L}{D}+\sqrt{1+\left(\dfrac{L}{D}\right)^2}\right]/8L$

a) A/F = $d^2 \cdot \ln\left[\dfrac{mL}{D}+\sqrt{1+\left(\dfrac{mL}{D}\right)^2}\right]/8L$

⑥ Well screen / open hole
   in uniform soil

Definitions: $k_m = \sqrt{k_v k_h}$; $m = \sqrt{k_h / k_v}$; $n = k_v' / k_v$

  where k_v = vertical permeability of soil/rock mass
       k_h = horizontal permeability of soil/rock mass
       k_v' = vertical permeability of filter in casing
       T is termed the basic time lag
       See Figure 6.7 for best method to determine representative value of T

Figure 5.13 Shape factors for variable head tests.

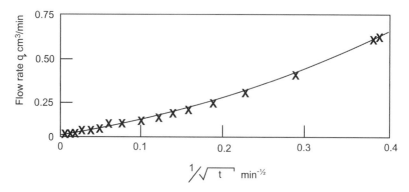

Figure 5.14 Typical plot of q versus $1/\sqrt{t}$ from *in situ* constant head permeability tests.

Determination of the steady state conditions.

An approximate value of the steady flow rate can be found from observed changes in flow rate with time, as follows: If H denotes the constant height of the water in the test hole above the base of the test zone, q denotes the infiltration flow rate, and t denotes the infiltration time, then q at steady state can be obtained from a plot of q versus $\log(1/\sqrt{t})$, or q versus $1/\sqrt{t}$ as t becomes large. A typical plot of this kind (Garga, 1988) is shown in Figure 5.14.

Determination of the effective head at test zone, H_c.

For all cases except the packer test, this is the constant head of water above the test zone. In the case of packer tests, the height of the column of water above the test zone is adjusted for head losses in the water hose and couplings as well as for any additional pressure head supplied by the pump.

Determine the shape factor F for a given test configuration from Figure 5.15, then:

$$k = \frac{q_c}{FH_c} \tag{5.10}$$

where q_c is the constant flow under steady state conditions.

The analysis of in-situ constant head permeability tests conducted in sealed hydraulic piezometers is also common practice. This analysis takes into account the compressibility of the soil and the resultant volume change as an excess (or deficit) pressure head is applied. A graph of the rate of flow q versus $1/\sqrt{t}$ is plotted as the test progresses (see Figure 5.14). Because $1/\sqrt{t}$ reduces as t increases, the steady state value of q is approached at $1/\sqrt{t} = 0$. It should be noted that it is not necessary to obtain a continuous record of the flow for the entire duration of the test. It is sufficient to monitor the flow rate over small time intervals periodically as the test proceeds. If t_1 and t_2 are the times (from commencement of the test) over which the flow rate is measured, then q can be plotted against $2/(\sqrt{t_1} + \sqrt{t_2})$. At large times, the expression for the coefficient of permeability reduces to:

$$k = \frac{q_\infty}{F\Delta h} \tag{5.10a}$$

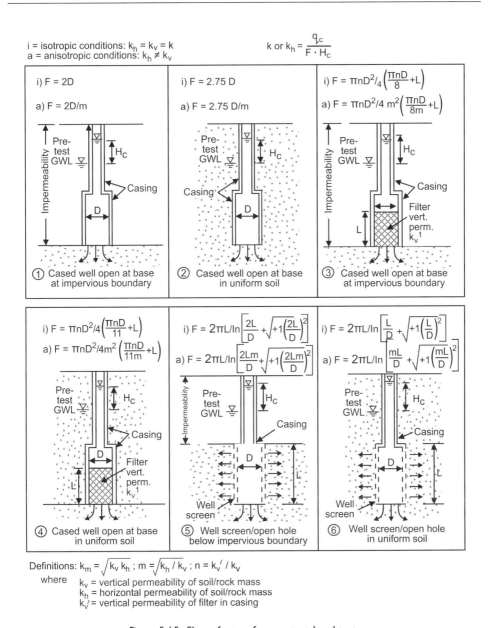

i = isotropic conditions: $k_h = k_v = k$
a = anisotropic conditions: $k_h \neq k_v$

$$k \text{ or } k_h = \frac{q_c}{F \cdot H_c}$$

---

① i) $F = 2D$

a) $F = 2D/m$

Pre-test GWL ∇ / H_c / Casing / D

① Cased well open at base at impervious boundary

---

② i) $F = 2.75\,D$

a) $F = 2.75\,D/m$

Pre-test GWL ∇ / H_c / Casing / D

② Cased well open at base in uniform soil

---

③ i) $F = \pi n D^2/_4 \left(\frac{\pi n D}{8} + L \right)$

a) $F = \pi n D^2/4 \, m^2 \left(\frac{\pi n D}{8m} + L \right)$

Pre-test GWL ∇ / H_c / Casing / Filter vert. perm. $k_v{}^1$ / L

③ Cased well open at base at impervious boundary

---

④ i) $F = \pi n D^2/4 \left(\frac{\pi n D}{11} + L \right)$

a) $F = \pi n D^2/4m^2 \left(\frac{\pi n D}{11m} + L \right)$

Pre-test GWL ∇ / H_c / Casing / D / Filter vert. perm. $k_v{}^1$ / L

④ Cased well open at base in uniform soil

---

⑤ i) $F = 2\pi L/\ln \left[\frac{2L}{D} + \sqrt{+1\left(\frac{2L}{D}\right)^2} \right]$

a) $F = 2\pi L/\ln \left[\frac{2Lm}{D} + \sqrt{+1\left(\frac{2Lm}{D}\right)^2} \right]$

Pre-test GWL ∇ / H_c / Casing / D / L / Well screen

⑤ Well screen/open hole below impervious boundary

---

⑥ i) $F = 2\pi L/\ln \left[\frac{L}{D} + \sqrt{+1\left(\frac{L}{D}\right)^2} \right]$

a) $F = 2\pi L/\ln \left[\frac{mL}{D} + \sqrt{+1\left(\frac{mL}{D}\right)^2} \right]$

Pre-test GWL ∇ / H_c / Casing / D / L / Well screen

⑥ Well screen/open hole in uniform soil

Definitions: $k_m = \sqrt{k_v k_h}$; $m = \sqrt{k_h / k_v}$; $n = k_v{}' / k_v$

where k_v = vertical permeability of soil/rock mass
         k_h = horizontal permeability of soil/rock mass
         $k_v{}'$ = vertical permeability of filter in casing

Figure 5.15 Shape factors for constant head tests.

where q_∞ = flow rate as t becomes large (or $1/\sqrt{t}$ approaches zero),
Δh = constant head applied during the test,
F = shape factor.

The shape factors for cylindrical tips (Olson & Daniel, 1981), with length L and diameter D, are shown in Figure 5.16.

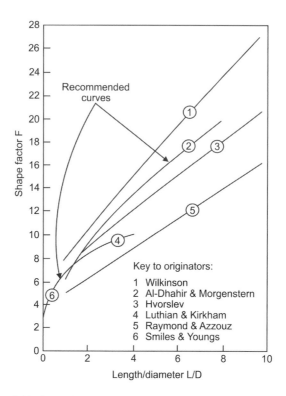

Figure 5.16 Shape factors for cylindrical tips (Olson & Daniel, 1981).

5.10   ESTIMATION OF PERMEABILITY FROM FIELD TESTS

Often only a rough preliminary estimate of the soil permeability is required, in which case useful results can be obtained from a simple soak-away test in a test-pit. The test-pit is filled with water, and the fall in water level is recorded over a period of several days. The hole must be covered over with a metal or plastic sheet to prevent evaporation losses and surrounded by a mound to prevent surface water from running in. The measurements can then be analyzed approximately to assess the order of magnitude of the permeability. Figure 5.17 shows a set of data and a calculation for a soak-away test performed on a clayey silt residual from a mud-rock. The object of the test was to see if a low-permeability clay liner would be required at the site (required permeability 0.1 m/y). The result of 30 m/y showed that a liner would be required.

The determination of field permeability by Matsuo *et al.*'s (1953) method has been widely applied to compacted soils in earth embankment construction. The test appears to work well in soils with permeability in the range of 10^{-4} to 10^{-6} cm/s (30 to 0.3 m/y). This simple method consists of excavating of a large rectangular test pool of width B and length L. The sides of the excavation may be sloped if required. The flow rate q necessary to maintain a constant level H in the pool is monitored. The seepage flow in this case is a three dimensional flow. In order to obtain a two dimensional estimate

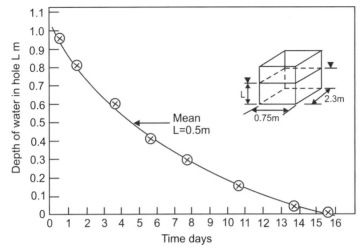

Approximate analysis by case in Figure 5.15 (see caption)
D = (0.75 + 2.3)/2

$$A/F = \left(\frac{0.75 + 2.3}{2}\right)^2 \ln \left\{ \frac{2 \times 0.5}{(0.75 + 2.3)/2} + \left[1 + \left\{ \frac{2 \times 0.5}{(0.75 + 2.3)/2} \right\}^2 \right]^{\frac{1}{2}} \right\} / 8 \times 0.5$$

A / F = 0.45 m   T = 5.5 days

A / FT = 0.082 m/d = 30 m/y (= 95x10^{-6} cm/s)

Figure 5.17 Observed soak-away curve for soak-away test in a test pit. (Approximate analysis according to case 5 in Figure 5.15 "open hole below impervious boundary").

of the flow, the pool is next enlarged to twice the initial length ($L_1 = 2L$). The new flow rate, q_1 required to maintain the same constant level H is noted. By subtracting the two flow rates, the effect of flow near both ends of the lateral cross-section can be eliminated, and the average discharge per unit length may be calculated as follows:

$$q_{ave} = \frac{q_1 - q}{(L_1 - L)} \tag{5.11}$$

The range of *in situ* permeability coefficient, depending on whether the flow is perpendicularly downwards or horizontal, may be obtained from the following simple expressions:

$$k_v = \frac{q_{ave}}{B - 2H} \quad \text{and} \quad k_h = \frac{q_{ave}}{B + 2H} \tag{5.12}$$

Figure 5.18 shows the results of a series of tests using Matsuo *et al.*'s method to explore possible anisotropy in permeability. As shown by the diagram, the calculated permeability proved to be constant regardless of the ratio of wetted side area A_s to base area A of the pit. It could thus be concluded that in this case the permeability was isotropic for all intents and purposes.

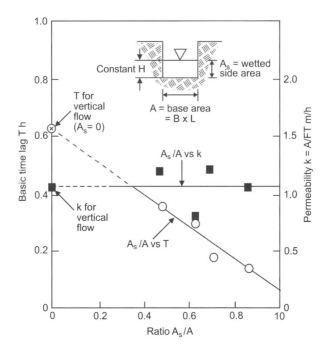

Figure 5.18 Variation of basic time lag T with ratio of side area A_s to base area A.

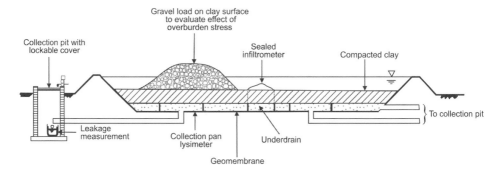

Figure 5.19 Large-scale permeability test by means of a test pad.

5.11   LARGE-SCALE PERMEABILITY TESTS USING A TEST PAD

The most realistic way of testing the permeability of a compacted clay layer is to incorporate a test pad, or a number of such pads into the prototype installation. The pad can be constructed in accordance with the design and specifications for the prototype. If tests on the pad (or pads) are successful, it (or they) can be incorporated into the prototype and monitored on a long term basis as a means of checking on the long-term performance of the clay liner. Figure 5.19 shows a schematic layout for a typical test

Table 5.3 Permeability of weathering profiles in igneous and metamorphic rocks (After Deere & Patton, 1971).

Soil zone	Qualitative permeability
organic top soils	medium to high
mature residual soil and/or colluvium	low (generally medium or high in lateritic soils if pores or cavities present)
young residual or saprolitic soil	medium
saprolite	high
weathered rock	medium to high
sound rock	low to medium

pad (USEPA, 1989). The dimensions of such a test pad should be generous, and at least 25 m × 25 m or 31.6 m × 31.6 m (to give an area of 1000 m^2).

5.12 PERMEABILITY CHARACTERISTICS OF RESIDUAL SOILS

Despite the enormous influence of seepage on slope stability, design of foundations for dams, excavations and underground openings, the geotechnical literature provides very limited information on the permeability of residual soils. The variation in the macrofabric of a weathering profile of a residual soil can result in large variations in permeability, both laterally and with depth. Generalizations of the "typical" values of permeability for various types of residual soils can therefore be misleading and must be avoided. Lumb (1975) and Blight (1988) have indicated that the variation in the test results from given sites of residual soil in Hong Kong and South Africa can be of the same order of magnitude as the variation from site to site. Typical weathering profiles of residual soils presented by Lumb (1962), Deere & Patton (1971), Vargas (1974), Blight (1988) and others clearly indicate the variations in grain size, void ratio, mineralogy, degree of fissuring, and the characteristics of the fissures that will affect permeability values from site to site. Table 5.3 shows some qualitative permeabilities of weathering profiles in igneous and metamorphic rocks. Tables 5.4 and 5.5 show some values of permeability measured in undisturbed residual soils, both *in situ* and in the laboratory.

The methods used to determine the permeability of residual soils in practice both in the field and in the laboratory, are similar to those used for transported soils. The most common methods in the field are constant head and variable head permeability tests in boreholes, auger holes and test pits (Brand & Phillipson, 1985). Brand & Phillipson's review of international practice in testing of residual soils clearly indicates a strong preference for in-situ permeability testing. The limitations of small sized laboratory test samples to include the micro- and macro-structural variations encountered in the field are clearly recognized.

The permeability of a saprolitic soil is controlled to a large extent by the relict structure of the material. Most of the flow takes place along relict joints, quartz veins, termite and other biochannels. De Mello *et al.* (1988) describe problems experienced with termite channels in the foundation of a 30 m high earth dam. Blight (1991) has

Table 5.4 Permeability values measured in residual soils of Brazilian dam foundations.

Parent rock	Residual soil	Permeability (m/y)	Type of test
Basalt	Mature residual and saprolitic	90 to 3×10^3	Variable head in piezometers and infiltration in boreholes
Basalt	Mature residual and saprolitic	0.03 to 30	Infiltration and pumping in boreholes
Gneiss	Mature residual and saprolitic	30 to 150	Infiltration and pumping in boreholes
Gneiss	Mature residual and saprolitic	1500	Infiltration and pumping in boreholes
Gneiss	Mature residual (porous clay) saprolitic	70	Infiltration in pit
Gneiss	Saprolitic	30	Infiltration in pit and pumping
Gneiss	Saprolitic	30	Variable head laboratory permeameter
Migmatite	Saprolitic	3 to 1000	Infiltration in boreholes

(After Costa Filho & Vargas Jr., 1985).

Table 5.5 Permeabilities of residual soils derived from granitic and gneissic rocks.

	Parent rock	Permeability (m/y)	Comments
Saprolitic soil	Granite	$600–125 \times 10^3$	Laboratory test
(young residual soil)	Granite	0.15–1000	
	Granodiorite	0.15–175	Consolidation tests
	Granodiorite	3	Variable head permeability tests
	Quartz-diorite	3–100	Field and laboratory tests
	Gneiss	6–150	Laboratory tests normal to schistosity
	Gneiss	3–60	Laboratory tests normal to schistosity
Mature residual	Granite	6–120	Laboratory tests
	Granite	0.15–60	Laboratory tests
	Granite	30–60	Micaceous layers in gneiss

(After Costa Filho & Vargas Jr., 1985).

Table 5.6 Field test method capability matrix (After O'Rourke et al., 1977)

KEY:
— Refers to single test
— Refers to stage test

RATING:
4 is most favourable
1 is least favourable

	HOLE PREPARATION COST	EQUIPMENT COST	PERFORMANCE COST	OPERATION TIME	OPERATION EASE	EASE OF ANALYSIS	ACCURACY WITH DEPTH SHALLOW <50m	ACCURACY WITH DEPTH DEEP >50m	ACCURACY WITH PERMEABILITY RANGE HIGH k>30m/y	ACCURACY WITH PERMEABILITY RANGE LOW k<30m/y	GEOLOGICAL SENSITIVITY HOMOGENEOUS	GEOLOGICAL SENSITIVITY STRATIFIED	GEOLOGICAL SENSITIVITY COMPLEX	PERMEABILITY DISCRIMINATION VERTICAL	PERMEABILITY DISCRIMINATION LATERAL (AREAL)
FALLING OR RISING HEAD TEST	3	4	4	4/2	4/2	4	1	1	3	1	2	1/2	1	1/4	1
CONSTANT HEAD TEST	4	3	4	3/2	4/2	4	2	2	3	2	3	1/2	1	1/4	1
PACKER TEST WITH CALIBRATION	4	2	2	2	1	4	3	1	4	4	4	4	4	4	1
PACKER TEST WITH PRESSURE TRANSDUCER	4	2	2	2	1	4	4	4	4	4	4	4	4	4	1
WELL PUMP TEST, EQUILIBRIUM ANALYSIS	1	1	1	1	2	3	4	4	4	2	4	3	3	1	4
WELL PUMP TEST, NON-EQUILIBRIUM ANALYSIS	1	1	1	1	2	1	3	3	4	3	4	3	3	1	4

reported on the effects of termite channels that were allowing effluent to leak from a series of evaporation ponds. In both cases the presence, or significance of these channels had been missed during the site investigation, and represented a very difficult repair problem. Termite channels have also caused a very carefully set out and levelled reference measurement base to go out of level, with the reference monuments settling by up to 155 mm over a period of 11 years (Watt & Brink, 1985, Figure 6.11).

In summary, because permeability is governed by macro-scale features, it usually cannot be reliably assessed by laboratory tests on undisturbed samples, as the scale of these is too small. The most reliable way is to assess the permeability by means of fairly large-scale field tests. If the water table is low, these can take the form of ponding tests or infiltration tests into test pits. Pumping tests from test pits or holes can be used if the water table is close to surface. Considering the multiplicity of forms of field permeability test available, it is useful to have a guide to the choice of the most suitable test for given circumstances and situation. Table 5.6 (O'Rourke, et al., 1977) is very appropriate for this purpose, especially as it enables an overall figure of merit to be assigned to available methods. It was designed to apply to borehole tests, but the concept and the rating system can easily be adapted as a means of comparison of pit or pond tests.

REFERENCES

Blight, G.E. (1971) Flow of air through soils. *J. Soil Mech. & Found. Div.*, ASCE, 97 (SM4), 607–624.

Blight, G.E. (1977) A falling head air permeameter for testing asphalt. *Civ. Eng. South Africa*, June, 123–126.

Blight, G.E. (1988) Construction in tropical soils. *2nd Int. Conf. Geomech. Tropical Soils. Singapore.* Vol. 2, pp. 449–467.

Blight, G.E. (1991) Tropical processes causing rapid geological change. In: *Quaternary Engineering Geology*, British Geological Society, Eng. Geol. Spec. Pub., No. 7, pp. 459–471.

Brand, E.W. & Phillipson, H.B. (1985) *Sampling and Testing of Residual Soils. A Review of International Practice.* Hong Kong, Scorpion Press.

Chen, H.W. & Yamamoto, L.D. (1987) Permeability tests for hazardous waste management with clay liners. In: Geotechnical and Geohydrological Aspects of Waste Management, Lewis, Boca Raton, USA, pp. 229–243.

Costa Filho, L.M. & Vargas Jr. E. (1985) Hydraulic properties. Peculiarities of Geotechnical Behaviour of Tropical Lateritic and Saprolitic Soils, Progress Report (1982–1985). Sao Paulo, Brazil, Brazilian Society of Soil Mechanics. pp. 67–84.

Daniel, D.E. (1987) Earthen liners for land disposal facilities. *ASCE Spec. Conf. on Geotechnical Practice for Waste Disposal, ASCE, New York.* pp. 21–39.

Daniel, D.E. & Koerner, R.M. (2007) *Waste Containment Facilities.* 2nd Edition. Virginia, USA ASCE.

Day, S.R. & Daniel, D.E. (1985) Hydraulic conductivity of two prototype clay liners, *ASCE J. of Geotech. Eng.*, 111 (8), 957–970.

Deere, D.V. & Patton, F.D. (1971) Slope stability in residual soils. *Pan American Conf. Soil Mech. & Found. Eng. Puerto Rico.* Vol. 1, pp. 87–170.

de Mello, L.G.F.S., Franco, J.M.M. & Alvise, C.R. (1988) Grouting of canaliculae in residual soils and behaviour of the foundations of Balbina dam. *2nd Int. Conf. on Geomech. in Tropical Soils, Singapore.* Vol. 1, pp. 385–390.

Elsbury, B.R., Daniel, D.E., Sraders, G.A. & Anderson, D.C. (1990) Lessons learned from compacted earth liners. *ASCE J. Geotech. Eng.*, 116 (11), 1641–1659.

Garga, V.K. (1988) Effect of sample size on consolidation of fissured clay. *Canadian Geotech. J.*, 25 (1), 76–84.

Hvorslev, M.J. (1951) *Time-lag and Soil Permeability in Groundwater Observations.* Vicksburg, USA, US Waterways Experiment Station, Bulletin No. 36.

Lumb, P. (1962) The properties of decomposed granite. *Géotechnique*, 12 (2), 226–243.

Lumb, P. (1975) Slope failures in Hong Kong. *Quart. J. Eng. Geol.*, 8, 31–65.

Matsuo, S., Hanmachi, S. & Akai, K. (1953) A field determination of permeability. *3rd Int. Conf. Soil Mech. & Found. Eng., Zurich.* Vol. 1, pp. 268–271.

Olson, R.E. & Daniel, D.E. (1981) Measurement of the hydraulic conductivity of fine-grained soils. *Permeability and Groundwater Contaminant Transport.* ASTM, STP746, pp. 18–64.

O'Rourke, J.E., Essex, R.J. & Ranson, B.K. (1977) *Field Permeability Test Methods with Applications to Solution Mining.* Washington, DC, US Dept. of Commerce, Nat. Tech. Inf. Serv. Report No. PB-272452.

Pregl, O. (1987) Natural lining materials. *Int. Symp. on Process, Technology and Environmental Impact of Sanitary Landfills. Cagliari, Italy.* Vol. 2, paper 26, pp. 1–7.

Schmidt, W.E. (1967) Field determination of permeability by the infiltration test. *Permeability and Capillarity of Soils*, ASTM, STP 417, pp. 142–158.

Tan, S.B. (1968) *Consolidation of Soft Clays with Special Reference to Sand Drains.* PhD thesis, University of London.

United States Environmental Protection Agency, Office of Research and Development (1989) Requirements for hazardous waste landfill design, construction and closure. USEPA/625/4-89/022, Cincinnati, USA.

US Bureau of Reclamation (1951) *Permeability Tests Using Drill Holes and Wells.* Denver, CO, Geology, US Dept. of the Interior. Report G-97.

US Bureau of Reclamation (1974) *Earth Manual.* 2nd edn. Water and Power Resource Service, Denver, CO, US Dept. of the Interior.

Vargas, M. (1974) Engineering properties of residual soils from South-Central Region of Brazil. *2nd Int. Congr., Int. Assoc. of Eng. Geol.,* Sao Paulo, Brazil. Vol. 1, pp. 5.1–5.26.

Watt, I.B. & Brink, A.B.A. (1985) Movement of benchmarks at the Pienaarsriver survey base. In: Brink, A.B.A. (ed) *Engineering Geology of Southern Africa.* Pretoria, South Africa, Building Publications. Vol. 4, pp. 199–204.

Chapter 6

Compressibility, settlement and heave of residual soils

R.D. Barksdale & G.E. Blight

6.1 COMPRESSIBILITY OF RESIDUAL SOILS

Many methods have been used to assess the compressibility of residual soils. *In situ* methods have included the standard penetration test, the pressuremeter test and forms of plate loading tests. Laboratory methods have been based on the oedometer and triaxial compression tests.

All residual soils behave as if overconsolidated to some degree. Their compressibility is relatively low at low stress levels. Once a threshold, yield stress or equivalent preconsolidation stress has been exceeded, the compressibility increases. In most cases, the stress range will be such that the soil will remain within the pseudo-overconsolidated range of behaviour and compressibility will be low.

Figure 6.1 shows two typical oedometer compression curves for a weathered andesite lava from two depths of the same profile as that represented by Figure 1.13. The yield stress is not very clearly defined, but correlates reasonably well with both

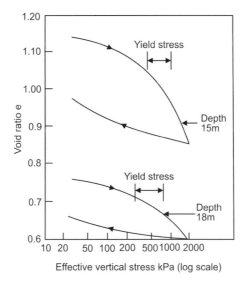

Figure 6.1 Typical oedometer curves for specimens from a profile of residual andesite lava.

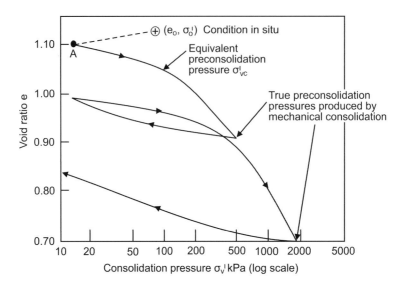

Figure 6.2 Typical consolidation curve for residual andesite lava, showing the equivalent preconsolidation pressure and the establishment of true preconsolidation pressures by subsequent mechanical overconsolidation.

depth and initial void ratio. Note the large variation of initial void ratio that may occur over a relatively small vertical distance as a result of variable composition and decomposition of the parent rock. Figure 6.2 shows typical characteristics of the compression curve for a residual andesite lava in more detail. In particular, once the equivalent preconsolidation stress has been exceeded, a true preconsolidation stress can be established by mechanical consolidation, and a residual soil will then behave similarly to a transported soil.

The equivalent preconsolidation pressure is a measure of the strength of the inter-particle, or inter-mineral crystal bonds remaining in the soil after weathering. It is therefore reasonable to expect that σ'_{vc} would increase with depth in the profile. Figure 6.3 shows the variation of σ'_{vc} with depth for three profiles of andesite lava. These data show that because the degree of weathering decreases with depth, σ'_{vc} tends to increase with depth and the increase is roughly linear.

In a transported soil profile, lateral stresses are related to the overconsolidation ratio, and increase with increasing overconsolidation. As a residual soil weathers and decomposes, the minerals swell, but simultaneously lose material by leaching, internal erosion of ultrafine particles, etc. Hence, unless the end-products of weathering are expansive, it is reasonable to expect that lateral stresses in a residual soil profile will be less than the overburden stress. In other words, K_0, the at rest pressure coefficient, will be less than unity, even though the soil may exhibit a pseudo-preconsolidation or yield stress in a consolidation test.

An estimate of the lateral stress in a soil profile can be obtained by measuring the suction in an undisturbed specimen taken from the profile, and comparing this with

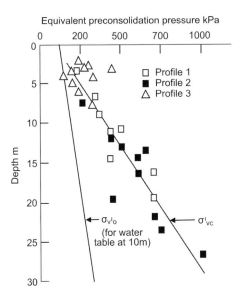

Figure 6.3 Variation of equivalent preconsolidation stress in three adjacent profiles of residual andesite lava.

the overburden stress. If σ'_s is the mean effective stress in the soil after sampling (which equals the suction in the unconfined specimen), it can be shown (Blight, 1974) that

$$R_s = \frac{\sigma'_s}{\sigma'_{V0}} = K_0 - A_s(K_0 - 1) \tag{6.1}$$

where σ'_{v0} is the effective overburden stress and A_s is the A-parameter for the change in pore pressure in the soil specimen that results from releasing the shear stress $(\sigma'_{v0} - \sigma'_h)$ during sampling (σ'_h is the horizontal effective stress). As A_s is usually small and positive,

$$R_s = K_0 \text{ (approximately)} \tag{6.1a}$$

Figure 6.4 shows that there is roughly a linear relationship between σ'_s and σ'_{v0}, for the profile referred to in Figure 6.3. Also, the value of R_s in the profile is about 0.3. The mean value for the well known Jaky expression for K_0 is

$$K_0 = 0.9(1 - \sin \varphi') = 0.38 \tag{6.2}$$

while the mean value of K_0 measured in triaxial tests in the laboratory is 0.42. Thus there is a strong indication that the value of K_0 for the profile is less than 0.5, which confirms the expectation deduced by considering the weathering process and products. While data of this sort appear to be available for only one profile (of andesite) it is not unreasonable to assume that K_0 in other profiles of non-expansive residual soil will similarly be less than unity, even though the soils behave as if overconsolidated.

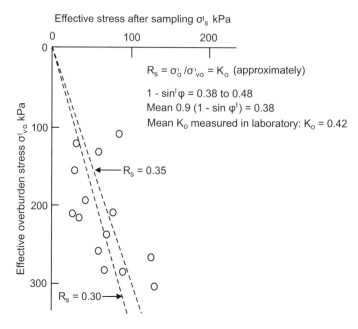

Figure 6.4 Indirect estimates of K_o in a residual soil profile, the profile illustrated by Figure 1.13.

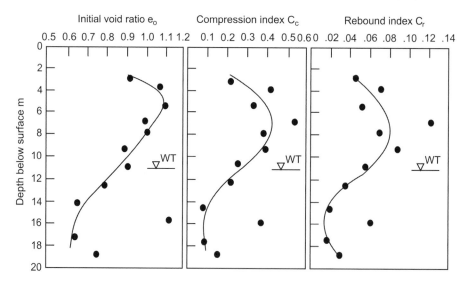

Figure 6.5 Variation of initial void ratio, compression index, rebound index and preconsolidation pressure with depth in profile of residual weathered andesite lava.

Figure 6.5 shows the variation of void ratio e_0, compression index C_c and rebound index C_r in a profile of residual lava. There is a general tendency for all three parameters to decrease with depth (with a local anomaly at about 16 m) which indicates that the degree of weathering decreases with depth. The strong similarity in the shapes of the

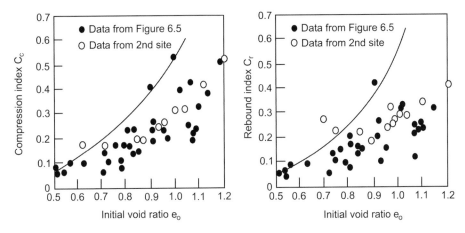

Figure 6.6 Correlations between initial void ratio and compression and rebound indices (Blight & Brummer, 1980). The solid lines are the approximate upper bounds for the data from Figure 6.5.

three curves suggests that both C_c and C_r should correlate with void ratio as indeed, Figure 6.6 shows they do. The data for a second andesite profile about 5 km away shows that the relationships in Figure 6.6 are of fairly general applicability and that the void ratio of the weathered material gives a good indication of its (rather variable) compressibility.

Similar data are shown in Figure 6.7 for a norite gabbro that has weathered to a profile (similar to the centre profile illustrated in Figure 3.2) consisting of a black highly expansive clay (or cotton soil) (usually from 1.5 m to 3 m in depth) overlying friable silty sands which grade with depth into very soft rock (usually between 3 and 9 m). These measurements by Hall *et al.* (1991) are very similar to those shown in Figure 6.6.

6.2 THE PROCESS OF COMPRESSION AND SWELL IN UNSATURATED SOILS

In terms of the unsaturated effective stress variables $(\sigma - u_a)$ and $(u_a - u_w)$ (e.g., Blight, 1965) the processes of heave and collapse are illustrated by Figures 6.8 and 6.9. Figure 6.8 shows the relationship between $(\sigma - u_a)$ and $(u_a - u_w)$ for the swelling process in an unsaturated soil. If $(\sigma - u_a)$ is kept constant, the equivalent of a constant total stress in the field, a reduction of the suction $(u_a - u_w)$ resulting from increasing moisture, will cause the soil to swell along lines such as AB in Figure 6.8. If the soil has a stable fabric or grain structure, swell will continue until $(u_a - u_w) = 0$ and the soil becomes saturated (at B). Thereafter, swell will continue along BC′, if σ is further reduced (u_a does not now exist, as the soil is saturated). If, however, the soil has an unstable grain structure, collapse settlement may occur once $(u_a - u_w)$ falls below a critical value for the value of $(\sigma - u_a)$ being carried and the soil will swell along the path from D to E and then collapse to F. Once the grain structure has stabilized, swell

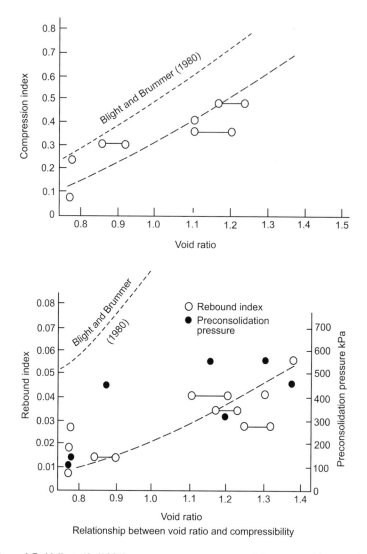

Figure 6.7 Hall *et al.*'s (1991) measurements on a saprolitic norite gabbro profile.

may resume along FG, if $(u_a - u_w)$ continues to be decreased, and the soil will behave normally.

Figure 6.9 illustrates the results of a constant volume swell process in which $(u_a - u_w)$ is reduced by increasing the moisture content of the soil, and $(\sigma - u_a)$ is adjusted to prevent settlement or swell from occurring. The paths traced out in the plane of zero volumetric strain represent contours of constant effective stress and experiments have shown that they can be followed regardless of whether the soil is expansive or collapsing. At constant effective stress, collapse will not occur, regardless of the value of $(u_a - u_w)$, as shown by the example for a collapsing sand in Figure 6.9.

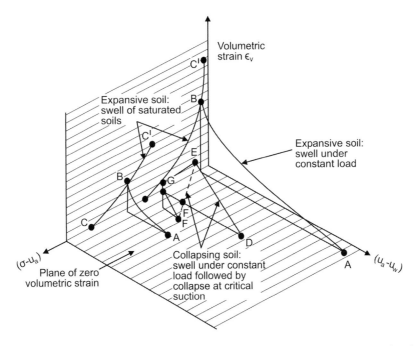

Figure 6.8 Three-dimensional stress-strain diagram for the swell of a partly saturated soil under constant isotropic load.

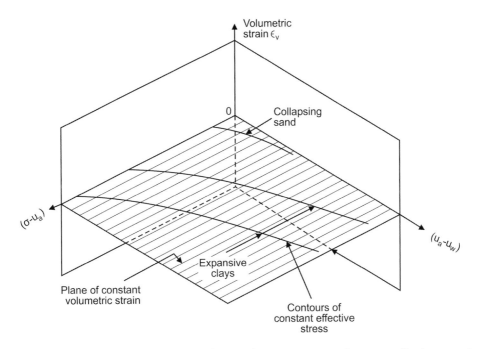

Figure 6.9 Three-dimensional stress-strain diagram showing contours of constant effective stress in partly saturated soils.

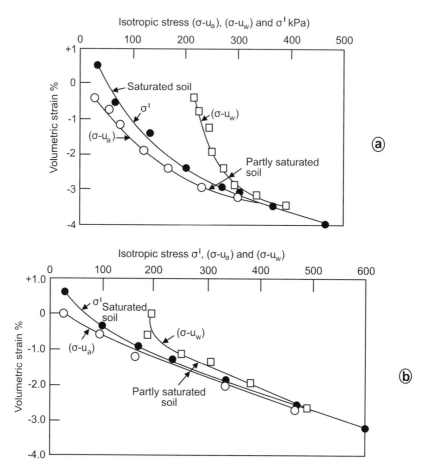

Figure 6.10 Typical curves relating volumetric strain to the effective stress components $(\sigma - u_a)$ and $(\sigma - u_w)$ for isotropic compression of samples from an expansive clay profile. The diagrams also show the compression curves for the saturated soil. Initial water contents were: a: 21.4%, b: 22.5%.

Figure 6.10 shows detailed isotropic compression curves for two pairs of expansive soil samples from a profile of weathered mudrock. Each diagram shows the compression curves for an unsaturated sample in terms of the applied stress $(\sigma - u_a)$ and the resultant values of $(\sigma - u_w)$ as the soil is compressed at constant natural water content. It also shows the compression curve for the second sample, once it has been saturated to an effective stress of 30 kPa. The equation relating a change in volumetric strain to changes of σ', $(\sigma - u_a)$ and $(\sigma - u_w)$ can be written, in terms of the Bishop parameter χ as

$$\Delta\varepsilon_v = C\{\Delta(\sigma - u_a) + \chi\,\Delta(u_a - u_w)\} = C\Delta\sigma' \tag{6.3}$$

where C is the compressibility. See section 1.14.1.

Solving for χ, in the equation $\Delta\sigma' = \Delta(\sigma - u_a) + \chi\,\Delta(u_a - u_w)$, gives

$$\chi = \frac{\Delta\sigma' - \Delta(\sigma - u_a)}{\Delta(u_a - u_w)} \tag{6.3a}$$

For the initial values of σ', $(\sigma - u_a)$ and $(u_a - u_w)$, (equal to $(\sigma - u_w) - (\sigma - u_a)$), for Figure 6.10a,

$$\chi = \frac{\sigma' - (\sigma - u_a)}{(\sigma - u_w) - (\sigma - u_a)}$$

$$\chi = \frac{65 - 25}{215 - 30} = 0.22$$

and similarly for Figure 6.10b

$$\chi = \frac{51 - 25}{200 - 15} = 0.14$$

6.3   BIOTIC ACTIVITY

This brief introduction would be incomplete without mention of the activities and effects on residual soil compressibility of termites and other burrowing insects. Termites are very common in tropical and sub-tropical areas where residual soils occur, and their activities may significantly modify the compressibility of soils. There are two ways in which termite activities may affect the settlement of structures: In an area infested with termites either presently or in the past, the soil profile may be riddled with termite channels, thus materially increasing both the macro void ratio of the soil as well as its compressibility and permeability. Unfortunately, this effect does not appear to have been quantified, but it is an effect that should be recognized when examining, sampling and testing soil profiles. It is important to note that termite channels may be present in the soil even though there is no sign of termite mounds on the surface. The termites may have left the area decades, centuries or millennia ago, but the effects of their tunnels and tunnelling remain in the soil.

Because termites carry soils fines to the surface, large objects such as boulders and shallow foundations tend to be under-mined and carried downwards. In the case of a foundation, the settlement may be sufficient to cause structural distress. The following are a few examples of distress caused by termite activity:

- Partridge (1989) has reported that termite activity may result in the formation of a collapsible soil structure which may be subject to severe settlement if loaded during a dry period of the year and subsequently wetted by infiltrating water.
- In Johannesburg, one of a pair of augered cast *in situ* piles 10 m deep, supporting a pile cap, started to settle as load came on to it, with the result that the pile cap rotated. A 750 mm diameter auger hole was drilled next to the pile to determine the

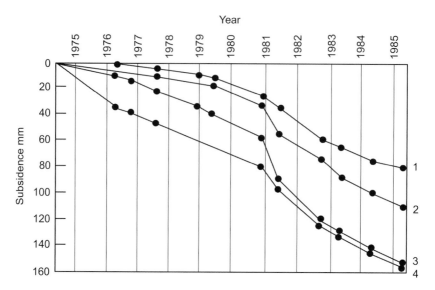

Figure 6.11 Subsidence of various surface monuments as a result of termite activity. (After Watt & Brink, 1985).

cause of the settlement. It was discovered that the pile had been founded directly above the food storage chamber of a termitary. This highly compressible spherical structure, about 500 mm in diameter was responsible for the settlement.

• In 1974 a precise survey base was constructed at Pienaarsriver, South Africa, by Finnish geodesists (Watt & Brink, 1985). It was intended for the calibration of electro-optical and electronic distance measuring instruments to an accuracy of 1 part in 10 million. The monuments supporting the calibration marks were mass concrete blocks measuring 1 m square in plan and founded at depths of 2.5 to 3 m on a yellow to reddish brown very stiff to very soft rock consistency silty sand, residual from the *in situ* decomposition of dolerite. Within two months it was found that certain of the monuments had settled by as much as 6 mm and had tilted slightly. This small movement could have been caused by shrinkage of the concrete, hence measurements were continued until 1976 when the precise distances were set out. At this time settlements of up to 16 mm had occurred. The movement continued and by 1980 it was found that the distance between the zero and 432 m marks had shortened by 12 mm (1 in 36 000). The base was quite clearly unable to meet the accuracy requirements of 1 in ten million and had to be abandoned. Figure 6.11 shows settlement records for some of the monuments, indicating that settlements of more than 150 mm had taken place over a period of 11 years.

There was abundant surface evidence of termite activity in the area and test pits showed the existence of subterranean cavities, channels and food stores. There was evidence on surface that termites were actively transporting soil from below and depositing it on surface. Hence it is important to look for and, if found, record the presence of termites or termite channels in the soil.

6.4   MEASURING THE COMPRESSIBILITY OF RESIDUAL SOILS

A thorough foundation investigation must be conducted in residual soils since they often vary erratically both with lateral position and depth. Test borings, advanced by either machine or hand, should be placed no further apart than about 20 m for commercial buildings and about 45 m for large industrial facilities. The groundwater table should be measured at the time the test boring is made and at least 24 hours later. (See section 3.4.3.) These observations should be confirmed by means of carefully installed standpipe piezometers, observed over the ensuing months.

The following are the most commonly used *in situ* test methods to assess the compressibility of residual soils. As indicated previously, laboratory oedometer tests are also used very extensively. In selecting a test, the convenience, economy and good control over moisture and stress conditions of the oedometer test have to be considered in relation to the advantages (though at greater cost) of *in situ* tests.

6.4.1   The conventional plate load test

The plate load test is carried out by applying load to a rigid plate and measuring the resulting vertical deformation (Figure 6.12). Since the plate load test is performed *in situ*, soil disturbance effects, which are important in residual soils, are reduced. Elastic moduli are calculated from the equation

$$E = \frac{qB(1 - v^2)}{\rho_0} \cdot I_f \tag{6.4}$$

In which q is the stress applied to the plate, B is its width, ρ_0 is the corresponding settlement under the plate surface, v is Poisson's ratio (usually taken as 0.2 to 0.3), and values of I_f are given in Table 6.1. The elastic moduli obtained from the plate load test are usually greater than those determined from laboratory consolidation tests even on specimens hand trimmed from block samples. Some studies have also found elastic moduli derived from plate load tests to be greater than those obtained from screw plate tests (see section 6.4.3). For a plate of breadth B, the plate load test gives a modulus of elasticity representative of the soil located within a depth of B to at most 2B beneath the plate. In general, the plate load test should be carried out at the level of the bottom of the footing or below this. Plate load tests should only be conducted in the desiccated crust at the ground surface if the zone of actual footing influence lies primarily in that stratum. Frequently, plate load tests are performed at several depths in different strata. Before conducting a plate load test, the unsaturated soil beneath the plate should be soaked if it may become wetted during periods of prolonged wet weather. Also, careful soaking of the soil should be carried out if the test is intended to model deformation behaviour beneath the water table. Soaking should be continued for a sufficiently long period of time to wet the soil for a depth of at least 1.5B below the bottom of the plate. This may take several days or even weeks to occur (see section 8.4).

Test pit

Usually the plate load test is performed in a pit excavated down to the desired level. However, elastic half-space theory, which is frequently used to calculate the modulus

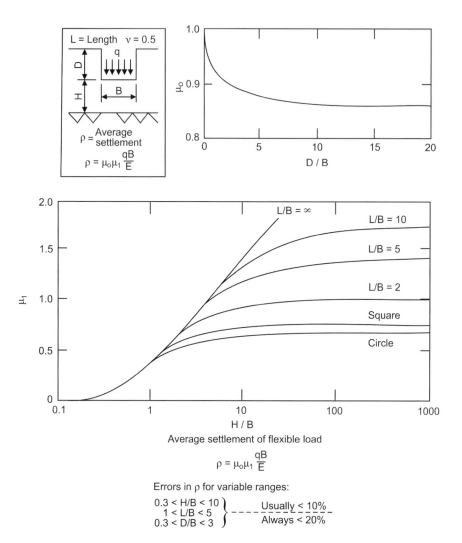

Figure 6.12 Elastic settlement calculation method for homogeneous soil: Rigid layer and foundation embedment (Christian & Carrier, 1978).

of elasticity, assumes the load is applied at the surface of a soil mass of wide lateral extent. The equation given in Table 6.1 which can be used to reduce the plate load test data, makes these assumptions.

If elastic half space theory is used to reduce the data, an unconfined plate load test should usually be performed. To perform this type of test, the soil overburden adjacent to the plate should be removed for a distance of at least 1.5B and preferably 2B away from the edge of the plate, where B is the diameter or least dimension of the plate. For a confined test, the soil surcharge should be left as close as possible to the plate. The effect of embedment should then be considered in calculating the modulus of elasticity, as discussed below.

Table 6.1 Influence factors I_f for computing immediate surface settlements ρ_0 of shallow footings.

		$\rho_0 = qB(1 - v^2)I_f/E$ (I_f for flexible footing)			
Shape of the footings in plan		Centre(λ)	Edge(λ)	Corner(λ)	Rigid footing
Circular		1.00 (1.27)	0.637 (0.81)	—	0.785
Rectangular	L/B	—	—	—	—
	1.0	1.12 (1.37)	0.76 (0.93)	0.56 (0.69)	0.815
	1.5	1.36 (1.35)	—	0.68 (0.67)	1.01
	2.0	1.52 (1.36)	1.12 (1.00)	0.86 (0.76)	1.12
	2.5	1.68 (1.39)	—	0.84 (0.69)	1.21
	3.0	1.78 (1.37)	—	0.89 (0.68)	1.30
	4.0	1.96 (1.37)	1.52 (1.06)	0.98 (0.68)	1.43
	5.0	2.10 (1.35)	1.68 (1.08)	1.05 (0.68)	1.55
	10.0	2.56 (1.22)	2.10 (1.05)	1.28 (0.61)	2.10

Notes: 1. Assumptions – pressure q is applied at the surface of an isotropic, homogeneous, elastic semi-infinite soil mass; 2. B = diameter of circular footing and minimum dimension of rectangular footing; q = net increase in applied pressure; 3. Elastic constants: E = modulus of elasticity; and v = Poisson's ratio; 4. ρ (flex, rect.) = $\lambda \cdot \rho$ (rigid, rect.), where λ = numbers given in parentheses (e.g., for L/B = 1.0, 1.37 × 0.815 = 1.12); 5. For square or rectangular footings, edge of foundation is midpoint of long side.

Plate size and type

Either circular or square rigid plates can be used as loading plates as well as cast-in-place concrete foundations. A desirable plate size (B) is 0.8 m with the minimum reliable size being about 0.3 m. The reliability of the plate test results increases as the plate size approaches the actual size of the footing. Tests on three different size plates can be used to help extrapolate settlements to those of the full size foundations. (Note that soil variability may obscure the effect of differing plate sizes, see Figure 6.16.)

Deformation measurement

Plate settlement should be measured using at least 2 dial indicators (or other types of displacement measuring devices such as LVDTs or precise leveling) with measurements diametrically opposite each other and at equal distances from the centre of the plate. Three displacement measuring devices are preferable to two. If three devices are used, space them 120° apart. The measuring devices should be capable of reading at least to the nearest 0.02 mm. These gauges should be tightly clamped to an independent reference beam supported on each side at least 1.5 m from the plate. (See Figure 6.13 which shows a typical arrangement for a plate load test.)

Load application

Load is usually applied to the plate by a hydraulic jack. The jack can react against either a platform supporting a dead load or a portable lightweight truss held down by helical anchors or tension piles (as illustrated in Figure 6.13). It is desirable for a load cell, proving ring, or other measurement device to be used to measure the applied load accurately. If the jack is slightly out of vertical, friction on its piston may give a false indication of the load. The applied load must be carefully aligned perpendicular

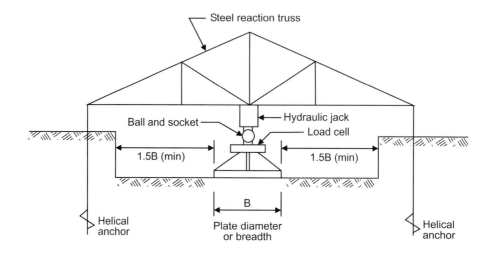

Steel reaction truss

Ball and socket

Hydraulic jack
Load cell

1.5B (min)

1.5B (min)

Helical
anchor

Helical
anchor

B

Plate diameter
or breadth

Notes: 1. Conduct plate load test at several levels below the
bottom of the footing.
2. Test depth of influence is only 1 to at most 2 times
the plate diameter or width.
3. Plate deformations should be measured with at least
2 dial indicators or LVDTs from an independant
reference beam extending at least 2.4m from
the plate on each side. Alternatively precise
leveling to an accuracy of ± 0.1mm may be used.

Figure 6.13 Plate load test set-up.

to the plate and applied through a ball and socket to minimize the detrimental effects of eccentric loading.

To obtain good contact between the plate and the soil surface, a quick setting grout is placed between the plate and the soil. Plaster of paris (dehydrated gypsum), either alone or mixed with a clean, fine sand or a quick setting commercial grout can be used to minimize bedding errors and help in levelling the plate. The grout bed should be as thin as practical and be allowed to cure before testing. Great care should be taken in levelling the plate and applying a perpendicular load at its centre.

Load test

The plate should in general be loaded to no more than the anticipated foundation design pressure to avoid excessive shear strains. Load should be applied in 5 equal increments allowing time for the settlement after each load increment to stop before applying the next increment. To determine the immediate settlement, measurements should be taken as soon as possible after applying the load for each increment. Measurements can be timed to a logarithmic scale, e.g., 1, 2, 4, etc. minutes after each load application. As the test progresses, plate settlement should be plotted as a function of the logarithm of time. The final test load should be maintained until the settlement has been constant for at least 2 hours. After reaching the maximum load, unloading should be made in at least

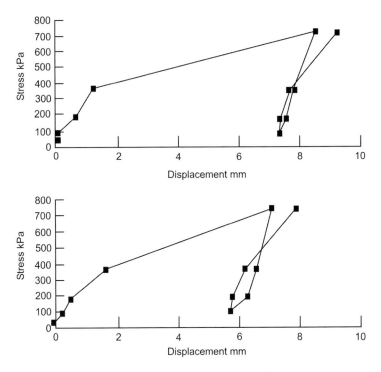

Figure 6.14 Stress-displacement curves for cross-hole plate loading tests on a residual andesite lava soil.

two decrements. Application of at least a second load-unload cycle is recommended, as shown in Figure 6.14. The first loading is usually affected by disturbance to the soil caused by excavating and setting the plate and anchors in position. The design is then based on parameters measured for the second load-unload cycle.

Primary consolidation settlement

For some saturated and/or clayey residual soils sufficient time may not be available, from a practical standpoint, to allow the delayed settlement to occur completely. As a practical expedient in testing these soils, the plate load test can be performed measuring only the instantaneous deformation. The modulus of elasticity calculated using the instantaneous deformation can then be corrected to consider, approximately, delayed settlement using one-dimensional consolidation test results. For each load increment in the one-dimensional consolidation test (or selected increments in the field plate load test), determine the immediate settlement as a percent of the total settlement (i.e., immediate plus primary). Do not include secondary compression in determining the total settlement. Plot the immediate settlement, expressed as a percent of total settlement, as a function of applied pressure. Now determine the average value of immediate settlement, expressed as a function of total settlement, over the pressure

range applied by the foundation. Reduce the calculated modulus of elasticity to reflect the effect of delayed settlement.

Unsaturated residual soils usually have instantaneous deformations greater than 60 to 70% of the total settlement. For example, assume that the immediate modulus of elasticity is 9500 kPa, calculated from the plate load test results. Also assume that for this soil the average instantaneous deformation is 75% of the total deformation (excluding secondary compression). The modulus of elasticity corrected for time-dependent settlement is then equal to $0.75 \times 9500\,\text{kPa} = 7125\,\text{kPa}$. This correction technique can also be used for correcting triaxial test values of E when the test has been performed too rapidly to allow delayed deformation to occur. (See section 6.4.5.)

If the relative amount of immediate settlement compared to total settlement varies significantly over the range of applied load, the immediate settlement measured for each load increment can be corrected by dividing the measured immediate settlement by the ratio of immediate to total settlement for that increment as determined from the consolidation test.

Modulus of elasticity

The modulus of elasticity is calculated from the theory of elasticity. For a plate load test performed in the unconfined condition (i.e., no surcharge near the sides of the plate), the equation given at the head of Table 6.1 can be used to calculate the modulus of elasticity. Use the appropriate influence factor I_f from Table 6.1 to account for the shape of the plate. For a confined test, use the elasticity approach summarized in Figure 6.12 which considers the depth of embedment used in the plate load test.

Soil disturbance

In performing a plate load test in a pit, the soil near the bottom of the plate undergoes the most disturbance because of loosening due to both the excavating and stress release. This disturbance can be minimized by excluding the deformation over a depth of about 0.5B beneath the plate. Good results have been reported in using this technique. The modulus of elasticity, however, must be backcalculated using appropriate theory and special instrumentation must be used to measure soil deformation at 0.5B depth below the plate.

6.4.2 The cross-hole plate load test

The cross-hole plate load test is a variant of the conventional plate loading test. Two plates are used and are jacked horizontally against the sides of a test hole or trench. For reasons of greater safely, the test should only be performed after the installation of side support in the test hole or trench. The load versus horizontal compression curve is measured, and the lateral compression is taken as half of the measured total extension of the jack.

The test is very convenient, as it can be used to measure elastic moduli at several depths in the same hole. The measured modulus is for horizontal compression of the soil, as is the case with the better-known pressuremeter test. The modulus for horizontal

Table 6.2 Set of typical cross-hole plate loading test results.

Hole No	Depth m	Soil Horizon	Field Consistency	E_h MPa	Stress Range kPa
TPI	2.5	Res. Andesite	Firm	38.4	92–367
				8.1	367–736
TP2	2.96			30.5	46–367
				10.6	367–736
TP4	2.8			36.1	46–367
				27	367–736
TP7	2.4			40.5	46–367
				10.2	367–737

compression does, however, appear to be not dissimilar to that for vertical compression at the same depth.

Figure 6.14 shows two stress-displacement curves recorded for cross-hole plate bearing tests on a weathered andesite lava.

Based on the assumption that each of the plates moves the same distance, a drained modulus of elasticity for the soil can be calculated using an expression by Bycroft (1956).

$$E'_h = \frac{(7 - 8v)(1 + v)\text{Pav} \cdot \pi R}{16(1 - v)\rho_h} \tag{6.5}$$

where Pav is the pressure on the plate [kPa],
  ρ_h is the movement of the plate under load [m],
  R is the radius of the plate [m],
  v is Poissons's ratio taken as 0.2 or 0.3,
  E'_h is the elastic modulus in the horizontal direction [kPa or MPa).

A set of typical results for a set of cross-hole plate bearing tests is given in Table 6.2.

6.4.3 The screw plate load test

The screw plate load test is a form of *in situ* plate load test that is performed at different depths beneath the surface. The screw plate consists of a single turn of a helical auger with a projected area of a full circle. Screw plate diameters typically vary from 100 to 300 mm. The screw plate test is performed by screwing the plate into the soil, either by hand or machine, using a relatively simple jack and reaction system to apply load to the plate. When frictional resistance to turning the plate becomes too large, a hole with the same diameter as the plate is augered out, and the plate is screwed below the bottom of the hole. The modulus of elasticity is then back-calculated from the load-settlement curve. Valuable information concerning the rate of consolidation and the drained or undrained shear strength can also be estimated from screw plate test results. The screw plate test is described in detail, for example, by Selvadurai *et al.* (1980) and Smith (1987a, 1987b). The principle of the screw plate is shown in Figure 6.15.

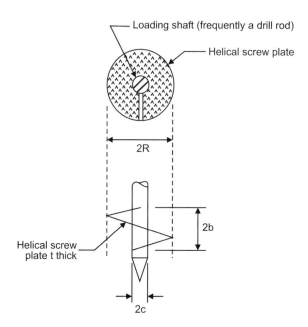

Figure 6.15 Screw plate test – Helical screw plate geometry.

Screw plate geometry

The screw plate geometry, illustrated in Figure 6.15, should have approximately the following ratios to minimize soil disturbance: $c/R = 0.125$; $b/c = 0.25$; $t/R = 0.02$

where: R = radius of screw plate,
         $b = \frac{1}{2}$ of screw pitch,
         c = radius of loading shaft,
         t = screw plate thickness.

A tungsten tipped cutting edge on the screw plate is desirable to minimize wear caused by abrasion of the soil.

Test reactions

The test plate and loading shaft can react against a reaction frame or truss held down by dead load or helical tension anchors screwed into the ground, or by a rotary drilling rig, or a heavily loaded vehicle.

Screw plate installation

Installation of the screw plate is usually accomplished by augering to within 400 mm of the desired test elevation. The screw plate is then advanced to the elevation of the test at a penetration rate per turn equal to the pitch of the helical screw.

Use of a rotary drill rig and hollow stem auger sufficiently large to allow inserting the screw plate shaft into the hollow stem decreases the time required for performing a test. Using tension anchors to hold down the drill rig can substantially increase the

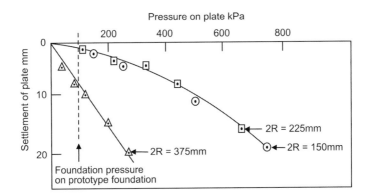

Figure 6.16 Results of screw-plate bearing tests on sand residual from weathered granite. All tests were at the same depth of 3 m.

reaction capacity. After augering the hole, the hollow stem drill rod is clamped. The inner loading shaft, which is attached to the screw plate by a hexagonal connection, is then advanced using the rotary feed on the drill rig.

Load test

The load test is conducted by applying load to the shaft which connects to the screw plate. A hydraulic jack is used to apply the load which is measured either by a calibrated proving ring or load cell. The combined deflection of the screw plate and elastic short-ening of the loading shaft are measured using two dial gauges or LVDTs placed 180° apart. The dial gauges or LVDTs can easily be clamped magnetically to the loading shaft. Relative movement is measured between the loading shaft and a fixed reference beam which should be at least 3 m in length. Compression measurements are taken from the time the drill rod is released from the chuck of the drill rig.

The screw plate loading procedure is to first apply a load equal to 20% of the estimated failure load and then remove it. This procedure reduces effects of soil dis-turbance and seats the plate. Two load/unload cycles to the design pressure are then applied, and, finally, the screw plate is loaded to failure and, if practical, unloaded.

For low plasticity soils, the load rate should be about 0.25 to 1.25 mm per minute. In residual soils where primary consolidation is important, each load increment should if possible be left on until at least 70%, and preferably 100%, of primary settlement has occurred.

Typical stress-settlement curves for a set of three screw plate tests are shown in Figure 6.16. (Two of the curves coincide.) These curves illustrate the difficulty often experienced in interpreting *in situ* tests on residual soils. As the depth of influence of a loaded foundation increases with increasing size, it would be expected that the settlements recorded on plates of increasing size would, at the same applied pressure, increase in proportion to their lateral dimensions. The set of results in Figure 6.16 does not show this trend, as a result of variability of the soil from the site of one

plate test to another. Variability of this sort must be expected when testing residual soils, as it affects the reliability of the settlement predicted for the prototype structure.

Elastic modulus

The elastic modulus is calculated from the screw plate test results using the following general formula which is based upon the theory of linear elasticity:

$$E_{sp} = \frac{\lambda \Delta qR}{\rho} \tag{6.6}$$

where E_{sp} = modulus of elasticity obtained from the screw plate test [kPa],
   Δq = net increase in average stress applied to the screw plate [kPa],
   ρ = measured screw plate deflection [m],
   R = radius of screw plate [m],
   λ = a constant depending upon depth of screw plate below the surface, method of plate installation, and plate rigidity.

For a deep screw plate located greater than 12 to 16 screw plate radii below the surface, a value of $\lambda = 0.65$ can be used if the hole is augered out to within 400 mm of the screw plate. $\lambda = 0.75$ can be used if the plate is screwed into the soil without any removal of soil. For a shallow screw plate test performed at a depth of less than 4 screw plate radii below the surface, $\lambda = 1.0$ can be used for either a cleaned or disturbed hole. For intermediate depths, values of λ can be interpolated.

6.4.4 The pressuremeter test

The pressuremeter test offers an excellent *in situ* method for evaluating the modulus of elasticity from the surface down to great depths. Use of the pressuremeter reduces, but does not eliminate, soil disturbance, as compared to tests performed on undisturbed, thin-walled tube samples. The shear strength of the soil and the *in situ* horizontal pressure can also be estimated by using the pressuremeter. The widely-used Menard pressuremeter test is illustrated in Figure 6.17. (See also Baguelin *et al.*, 1978; Finn *et al.*, 1984; Mair & Wood 1987). Other forms of pressuremeter include the self-boring pressuremeter pioneered by Cambridge University (e.g., Hughes *et al.*, 1977).

Disadvantages of the pressuremeter include the use of relatively expensive equipment which is quite sensitive to equipment calibration and lack of operator expertise. Pressuremeter test results are also influenced by the method of installing the device, test procedures, and the method of interpretation of the test results. Although the modulus of elasticity is measured in the horizontal direction, this does not appear to be an important disadvantage for residual soils.

The pressuremeter test can be visualized as a generalized form of lateral plate load test. The pressuremeter, which is cylindrical in shape, is inserted into a vertical borehole. The pressuremeter is expanded radially by increasing the fluid pressure inside the device. Cables and/or pipes are used to lower the pressuremeter down the hole and to connect it to instrumentation located at the surface. Since the contact surface of the pressuremeter with the soil consists of a rubber membrane, a uniform radial pressure

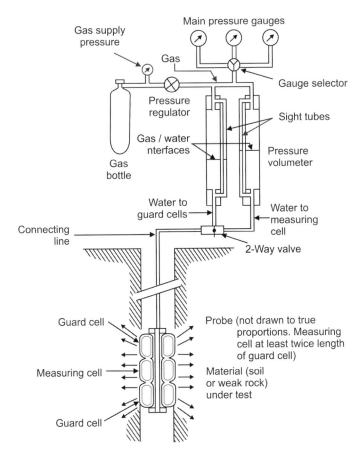

Figure 6.17 Principle of Menard pressuremeter test.

is applied to the sides of the hole over a vertical length of about 300 to 800 mm, depending on instrument design.

Either gas or water pressure is used to expand the pressuremeter laterally and hence expand the borehole in the radial direction. The expansion of the hole is measured indirectly by either determining the change in volume of the fluid in a calibrated chamber or else by deformation gauges arranged inside the measuring cell and measuring the increase of radius of the cell wall. For a given radial expansion pressure, the change in diameter of the hole can be related, through theoretical and empirical approaches, to the horizontal modulus of elasticity of the soil in the vicinity of the pressuremeter.

Hole preparation

The Menard type pressuremeter test should be performed in a smooth-sided hole. In residual soils, pushing a thin walled tube sampler ahead of a machine-augered bore hole has been found to work well. To obtain good results in residual soils, the cavity

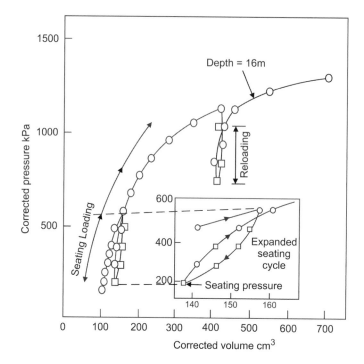

Figure 6.18 Unload – reload cycle in a pressuremeter test to reduce effects of drilling disturbance and stress release.

should first be expanded, i.e., the pressuremeter is seated to the stress level applied by the foundation at working load. The stress exerted by the pressuremeter is then reduced to a small seating value, and again increased back to the working stress level. The reload stress-volume change (or cavity radius) curve (see Figure 6.18) should be used for calculating elastic moduli. The load-unload-reload sequence takes out some effects of soil disturbance resulting from advancing the hole.

Equipment calibration

Pressuremeter test results and hence the modulus of elasticity determined from them are quite sensitive to equipment calibration and test correction effects. System calibration must take into consideration:

1 **System compliance.** System compliance includes both decrease in membrane thickness and also changes in the volume of pressure transmission tubes, etc. which both become increasingly important as the test pressure increases.
2 **Pressure effects.** Pressure loss must be accounted for due to membrane and sheath resistance to expansion. Also, differences in pressure between the pressuremeter and the pressure-measuring system must be considered if the pressure is measured at the surface. Figure 6.19 shows a typical measured relationship between expansion pressure and volume (note the large values of pressure required.)

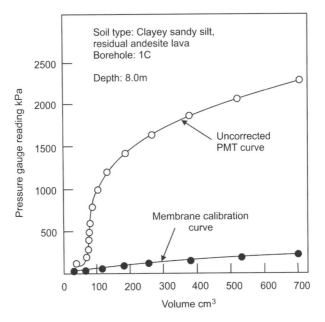

Figure 6.19 Typical pressuremeter test result with membrane calibration curve.

The membrane has a finite resistance to expansion and this has to be subtracted from the measured expansion pressure. Figure 6.20 shows the corrected expansion curve and its interpretation:

- In phase I, the soil is being re-compressed to its original *in situ* radial lateral pressure.
- In phase II the soil compresses linearly with increasing pressure.
- In phase III, the soil begins to fail and deforms plastically until it reaches the limit of volume change that can be accommodated by expansion of the membrane.

The pseudo-elastic phase, the steepest slope of the load curve, is then used to calculate the pressuremeter elastic modulus E_M for the soil from the equation:

$$E_M = 2(1 + v)V\Delta p/\Delta V \tag{6.7}$$

where v is Poisson's ratio, usually taken as $1/3$,
   $\Delta p/\Delta V$ is the slope of the (reloading) pressuremeter curve,
   V is the cavity or expansion volume for the pressure range over which E_M is being measured.

The recycled modulus is typically 2 to 2.5 times larger than the modulus for first loading.

Figure 6.21 shows a set of corrected pressuremeter curves measured at various depths in a profile of weathered andesite lava. The diagram shows how the shear

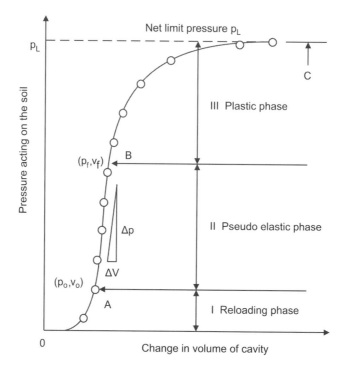

Figure 6.20 Phases in the course of a pressuremeter test.

strength, related to the net limit pressure p_L (C in Figure 6.20), and the stiffness of the soil increase as the depth increases and the degree of weathering of the soil correspondingly decreases. As emphasized earlier, the properties of residual profiles may vary widely from one hole to another. Figure 6.22 illustrates a typical profile of pressuremeter moduli measured by Menard pressuremeter tests in different holes in a profile of saprolitic residual diabase. The diagram shows the effect of the increasing stiffness of the soil with depth that was also illustrated by Figure 6.21, as well as the variation from borehole to borehole and the variability down the depth of each borehole.

6.4.5 Slow cycled triaxial tests

Multi-stage, slow cycled consolidated, undrained triaxial tests performed on undisturbed specimens offer a practical approach for evaluating the modulus of elasticity of residual soils when *in situ* tests are not feasible. An important advantage of slow cycled triaxial tests is that conventional triaxial testing equipment can be used. Also, high quality thin wall tube samples can be tested, although hand trimmed block samples should give the best results (i.e., closest to the *in situ* modulus of elasticity). The use of a modulus of elasticity that is close to the *in situ* value results in smaller estimated settlements which usually agree better with measured actual values. Studies have

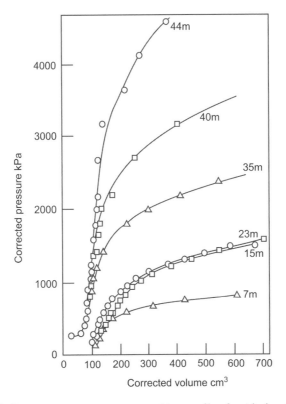

Figure 6.21 Pressuremeter curves measured in a profile of residual andesite lava.

found that cycled triaxial tests give a higher modulus of elasticity than conventional, one-dimensional consolidation tests.

Details of test

The cycled triaxial test usually employs two complete load-unload cycles for each of three effective confining pressures. An alternative method is to do one load unload-reload cycle on each of several specimens, each at a different effective confining stress, or a number at the same confining stress, depending on the availability of specimens. The effective confining stress for the first load cycle is often the nominal "bedding" value which is used to remove bedding errors from the test. The elastic modulus is then determined for the initial portion of the reload cycle. Figure 6.23 shows a typical result of such a test performed on a specimen of weathered andesite lava, using an initial bedding stress for the first loading cycle.

Caution should be exercised to apply a deviator stress $(\sigma_1 - \sigma_3)$ well below failure so as not to damage the usually delicate structure of the residual soil. Effective confining pressures should be selected which bound the stresses expected to exist beneath the completed structure. For many conventional building loadings, effective confining

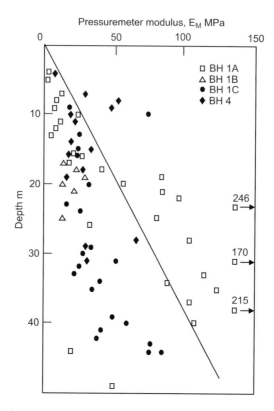

Figure 6.22 Variation of pressuremeter modulus of residual andesite lava with depth in 4 adjacent test holes.

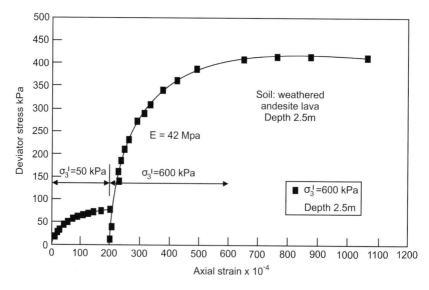

Figure 6.23 "Bedding cycle" triaxial test in which specimen is bedded in under a low confining stress, unloaded and then reloaded under a higher confining stress.

Table 6.3 Typical stress ranges for slow cycled triaxial testing of silty sand and sandy silt residual soils.

Effective Applied Stresses kPa	Soil Consistency					
	Stiff (N*= 8 to 15)		Firm (N = 5 to 8)		Soft (N = 3 to 4)	
σ_3	$\sigma_1 - \sigma_3$	σ_1	$\sigma_1 - \sigma_3$	σ_1	$\sigma_1 - \sigma_3$	σ_1
35	40	75	30	65	20	55
70	50	120	40	110	30	100
140	100	240	70	210	60	200

*N = SPT "N" value.

pressures of 35, 70 and 140 or 170 kPa are often suitable for use in the multi-stage triaxial test.

The deviator stress for the bedding stress should be selected to cause a maximum shear stress $\frac{1}{2}(\sigma_1 - \sigma_3)$ in the specimen no greater than about 15% of the maximum failure shear stress. An acceptable maximum deviator stress should be estimated using past experience for the material to be tested. It is better, however, to carry out a trial test or tests to assist in deciding the range of stresses to be used.

After selecting appropriate shear strength parameters, the failure envelope should be plotted using the same scale for the normal stress (abscissa) and shear stress (ordinate) axes. For each effective confining pressure to be used in the multi-stage test, a failure circle should be drawn on the plot limiting the maximum shear stress to no more than about 30% of the estimated failure stress. Table 6.3 illustrates typical stress ranges suitable for residual silty sands and sandy silts.

At least two complete load-unload cycles of testing should be carried out for each of the three effective confining pressures used in the test. The specimen should be tested in a saturated condition if, during the structure's design life, the soil may reach a high degree of saturation. A consolidated, undrained triaxial test is used to obtain a result in effective stress terms.

After initially applying each confining pressure, pore pressures should be measured as a function of time to determine when primary specimen consolidation is complete. Upon reaching the end of primary consolidation, the deviator stress load-unload cycles may be applied, using an axial strain rate that will result in at least a 90% equalization of pore pressure in the test specimen.

The problem of deciding on a suitable testing time can be approached theoretically by means of consolidation theory (e.g., Gibson & Henkel, 1954; Bishop & Henkel, 1962), but the theoretical solution assumes that the drained surfaces are 100% efficient which they are not (e.g., Blight, 1963) and for this reason, an experimental solution was developed. This appears in Figure 7.24 and will be described here, as the technique is needed in Chapter 6, although it fits better into Chapter 7 on shear strength measurement. The input data consist of the specimen dimensions and the drainage conditions.

Figure 7.24 shows (upper diagram) experimentally established equalization curves B and D for drained triaxial shear tests equipped either with all-round filter paper

drains or with double end drainage. These have been drawn in terms of the degree of drainage versus the time factor $T = c_v t_f / H^2$, where t_f is the time to reach failure, or the end of the test, and H is half the sample height. The lower diagram shows a similar experimental equalization curve G for undrained tests with all-round drains. It will be noted that the equalization curve for drained tests with all-round drains is almost identical with that for undrained tests with all-round drains. Theoretically, if the time factor T_{90} for 90% equalization in tests with all-round drains is 0.04, T_{90} for tests with double end drainage, in which the minimum drainage path length is twice as long should be 4 times larger, i.e., T_{90} should be 0.16. In fact, the experimental value is 20 times larger, i.e., 0.8. This is because the theory assumes that the peripheral all-round filter paper drains are 100% efficient, whereas their efficiency is only about 20%. (c_v = coefficient of consolidation in cm^2/s or m^2/y.)

Figure 7.25 shows the information from Figure 7.24 redrawn in the form of a simple chart for which the degree of equalization is 95%.

As far as pore pressure equalization is concerned, drained tests with double end drainage and undrained tests without circumferential filter paper drains behave alike. The time to failure for 95% degree of drainage can be calculated from:

$$t_f = \frac{1.6H^2}{c_v} \tag{6.8a}$$

Drained and undrained tests with double end drainage and circumferential drains (all-round drains) behave alike and the time to failure can be calculated from:

$$t_f = \frac{0.07H^2}{c_v} \tag{6.8b}$$

To convert a time to failure to a suitable rate of strain it is usually advisable to do a trial test to establish the strain at failure or the required range of strains for the particular material being tested.

Modulus of elasticity

For each confining pressure, the modulus of elasticity should be determined from the slope of each unload cycle. Usually the modulus of elasticity for each subsequent load-unload cycle is greater than for the previous cycle at a lower effective stress. The rate of increase usually becomes less with increasing numbers of cycles. The modulus of elasticity is calculated as the change in deviator stress during unloading divided by the smaller of the change in height of the specimen or the change in axial distance between displacement measurement points on the specimen, expressed as axial strain, i.e., $E = \Delta(\sigma_1 - \sigma_3)/\Delta E$ (axial).

Usually the modulus of elasticity obtained from the last unload cycle is used in design. To aid in selecting design values of the modulus of elasticity, all of the measured elastic moduli for a particular site should be plotted on log-log graph paper. The modulus of elasticity is plotted on the vertical axis and confining pressure on the horizontal axis. A curve is then fitted (by eye or least squares) to the data points obtained for each specimen (i.e., the modulus of elasticity measured at the three confining pressures). Each curve (which represents the results for one specimen) is then

labelled including a complete soil description of the specimen, depth of sampling, and the observed value of standard penetration resistance, cone tip resistance, or any other related information.

6.4.6   Comparisons of different methods of assessing elastic modulus for residual soils

Very few comparisons have been published of the various available methods of assessing compressibility or elastic moduli for residual soils. One of the few such comparisons was made by Jones & Rust, (1989) for a saprolitic weathered diabase. Their comparison, illustrated in Figure 6.24a, shows that the Menard pressuremeter, plate bearing test and oedometer test give comparable results for E provided that rebound curves are used for the plate loading and oedometer tests. The self-boring pressuremeter, that should give the lowest disturbance to the soil, also gives the highest values of E. Also note from Figure 6.24b that the N value from the Standard Penetration Test (SPT) in the same profile has a very similar trend with depth. From this it may be deduced that SPT results in residual soils may also be used to obtain estimates of E. For this profile, the correlation is:

$$E = 1.6N \quad \text{MPa} \qquad (6.9)$$

A popularly used correlation is

$$E = N \quad \text{MPa} \qquad (6.9a)$$

which is rather more conservative, in terms of calculated settlement, than equation 6.9.

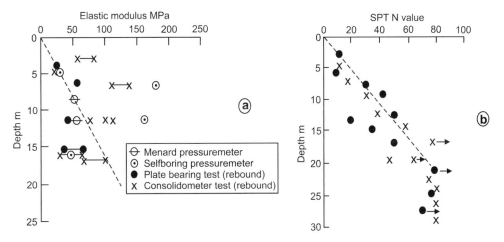

Figure 6.24   (a) Comparison of E values derived by four different methods for a weathered diabase profile. (b) Variation of standard penetration test N with depth in the same profile. (After Jones & Rust, 1989).

6.5   SETTLEMENT PREDICTION CALCULATIONS FOR RAFT AND SPREAD FOUNDATIONS

Brand & Phillipson's (1985) survey shows that rafts and spread foundations have been used to transfer load to residual soil in a number of countries, including Brazil, Hong Kong, India, Nigeria, Singapore, South Africa and Sri Lanka. These types of foundation are also widely used in the United States (e.g., Barksdale et al., 1982) and Australia (Moore & Chandler, 1980).

As stated earlier, residual soils behave as if overconsolidated, the degree of overconsolidation depending on the degree of weathering. Williams (1975) found that conventional oedometer tests could successfully be used to predict settlement on residual soils provided the Schmertman (1955) corrections were applied, as well as the Skempton & Bjerrum (1957) corrections for overconsolidation.

Barksdale et al. (1982) used a number of methods to predict the settlement of a large water tower founded on a weathered biotite gneiss. They found that methods based on in situ tests could considerably overestimate the settlement whereas those based on laboratory tests could give excellent predictions.

Working in the same geographical area, Wilmer et al. (1982) reported on five settlement studies of spread footings on residual soils. Using conventional one-dimensional oedometer tests, they concluded that calculated settlements for residual soils would normally be about 30% in excess of measured movements.

Barksdale et al. have observed that differential settlement on residual soils may be as much as 75% of total settlement, but averages less than 25% of total settlement.

The prediction of settlement is a critical aspect of foundation design. As described above, settlements of structures on residual soils are usually determined using one of the following two techniques:

- one-dimensional consolidation test results and stress distribution theory, or
- elastic theory which uses a modulus of elasticity measured in the field or laboratory (section 6.4), or determined from empirical correlations with field measurements.

Accurate settlement predictions are difficult to perform for structures founded on residual soils. Soil disturbance effects together with testing errors in both the laboratory and field and lack of knowledge of residual soil characteristics are important factors which help to account for actual settlements frequently being less than predicted values by 30 to 50 percent.

Considerable progress has been made in recent years towards understanding the behaviour of residual soils. The use in settlement prediction methods of elastic moduli measured in situ has given improved results. Also, the measurement of axial deformation within the middle portion of a triaxial test specimen subjected to a small strain level and thus eliminating end-of-specimen errors, has been found to give more realistic results (i.e., higher elastic moduli) than using conventional methods for measuring overall specimen deflection. Nevertheless, the available settlement prediction approaches have not proved entirely successful for all soil and loading conditions. Local experience, developed by comparing predicted and measured settlements, is still needed to verify settlement calculation results.

6.5.1   Selection of settlement prediction methods

The analytical approach to predicting the settlement of residual soils must have a sound theoretical basis, be able to handle varying soil and foundation conditions, and be suitable for use with a variety of laboratory and field testing methods.

To satisfy these widely varying needs, an elasticity-based approach has been found useful. For shallow foundations this includes the strain influence diagram method originally proposed for sands by Schmertmann (1970). The strain influence diagram method has proved to be well-suited for predicting the settlement of shallow foundations on most residual soils. Elasticity-based methods are not suitable for predicting settlements of collapsible or swelling soils where the soil structure may change and volume changes are climate controlled, rather than load controlled. Elasticity methods should not be used with highly plastic residual soils which exhibit more than 40 to 50% delayed settlement, unless local experience has shown it to be a valid approach.

6.5.2   Strain influence diagram method

The strain influence diagrams used in the method are based on vertical strains measured in model studies and calculated using the finite element method. Vertical strain has been observed, in a number of laboratory and large scale field studies, to decay with depth more rapidly than indicated by Boussinesq-type stress distribution theory for a homogeneous soil. Both measurements made on model foundations and finite element analyses show that the maximum strain does not occur immediately beneath a rigid foundation as implied by the usually used stress distribution-based settlement prediction methods. Instead, the maximum strain is developed between about 0.5 and 1 times the diameter or width of the foundation as shown in Figure 6.25a. Therefore, the strain influence diagram approach agrees better with observed foundation behaviour than do the commonly used stress distribution-based settlement methods.

The strain influence diagram method is versatile and easy to use. The vertical settlement of a foundation is equal to the integral of the vertical strain over the depth influenced by foundation stresses beneath the point at which the settlement is to be calculated. Numerical integration of strain with depth is easily performed by dividing the depth influenced by the loading into sublayers of thickness Δz. The total

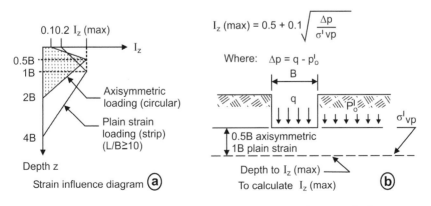

Figure 6.25 Strain influence diagram settlement calculation method.

foundation settlement is then the summation of the average vertical strain in each sublayer multiplied by its thickness Δz.

The strain influence diagrams used in calculating foundation settlement for rigid circular foundations on homogeneous soil are shown in Figure 6.25. To perform a settlement analysis using the strain influence diagram method, first divide the strata beneath the foundation into n convenient sublayers of thickness Δz_i. Sublayer boundaries are usually placed at the break in the strain influence diagram and where changes in the modulus of elasticity occur, such as at boundaries between different strata. Typically 2 to 4 sublayers are used in a settlement analysis. The foundation settlement (ρ) is then calculated using the following generalized equation:

$$\rho = C_1 \cdot C_2 \Delta p \sum_{i=1}^{i=n} \left(\frac{I_z}{E}\right)_i \Delta z_i \tag{6.10}$$

where
$\rho = $ total foundation settlement,
$n = $ total number of sublayers,
$C_1, C_2 = $ constants accounting for foundation depth and time effects,
$\Delta p = $ net increase in effective pressure applied by the foundation,
$\Delta z_i = $ variable increment of depth, $\Sigma \Delta z_i$ equals the depth of the strain influence diagram,
$I_z = $ average strain influence factor over the increment of depth Δz_i,
$E = $ average modulus of elasticity of the soil over the corresponding increment of depth Δz_i.

The correction factor C_1, used in equation 6.10 reduces the calculated settlement to account for the beneficial effect of foundation embedment:

$$C_1 = 1 - 0.5\left(\frac{p'_o}{\Delta p}\right) \tag{6.11}$$

where $p'_o = $ initial effective overburden pressure (refer to Figure 6.25b),
$\Delta p = $ net increase in effective foundation pressure, i.e., $\Delta p = (q - p'_o)$.

The correction factor C_1 should not exceed 0.5 for deeply embedded foundations. For foundations embedded less than 0.7 m, the use of $C_1 = 1.9$ is suggested.

The C_2 factor in equation 6.10 accounts for secondary compression of the soil. Secondary compression occurs over a period of time, primarily after excess pore pressures caused by the applied loading have dissipated. The constant C_2 is calculated from the expression:

$$C_2 = 1 + 0.2 \log_{10}(10t) \tag{6.12}$$

where t is the time in years after the load is applied.

The rate at which secondary compression occurs is related to the compressibility of the soil. As a result, density, soil structure, mineral composition of the soil, including mica content, and other variables all influence the rate of secondary settlement. Although equation 6.12 does not directly consider these factors, it appears to give fair estimates of secondary compression effects, based on somewhat limited verification for residual soils.

The elasticity-based solutions given in Table 6.1 can also be used to calculate settlement for the conditions described above and are particularly useful for uniform soil conditions.

Circular and rectangular strip footings on homogeneous, deep layer

Figure 6.26 gives strain influence factors for rigid circular footings, taking account of various depths and lateral distances The footings are assumed to rest on a deep soil layer having an approximately constant modulus of elasticity (i.e., stiffness) with depth. To avoid tilt, the centroid of the applied loading must coincide with the centroid of the foundation. For most accurate results, the depth to rock must be greater than about 3B beneath the bottom of the footings. The ground surface should be reasonably level. Figure 6.26 shows a cross section taken through the centre of the footing and the soil underlying it. All dimensions are in terms of the foundation diameter B. The

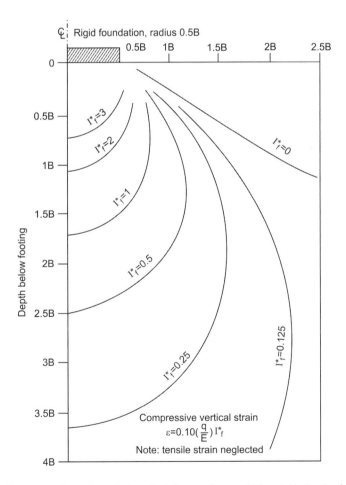

Figure 6.26 Contours of equal vertical strain influence factors I_f^* for rigid circular foundation: Deep homogeneous soil mass.

cross section shows contours of equal vertical strain influence factor I_f^* as a function of depth and radius from the axis of the footing.

Adjacent footings

The additive effect of adjacent footings can be estimated using superposition and the solutions given in Table 6.1. The settlement caused by adjacent footings can also be estimated using Figure 6.26. Determine the settlement by using the strain influence factors and a slightly modified form of equation 6.10.

$$\rho = 0.10 \cdot C_1 C_2 \Delta p \sum_{i=1}^{i=n} \left(\frac{I_f^*}{E} \right)_i \cdot \Delta z_i \tag{6.10a}$$

All terms in the above equation have been previously defined.

As shown on the contour influence diagram given in Figure 6.26, vertical strain caused by the applied loading dies out quite quickly with increasing lateral distance from the footing. For many problems, the effect of adjacent footings, when greater than 2B away, can therefore be neglected.

Footings at great depth

Table 6.1 can also be used for square, circular and rectangular footings resting near the surface of a deep layer of homogeneous soil. The generalized elasticity equation used to calculate settlement is given in the heading of the table.

Rectangular foundations: Generalized strain influence diagrams

Consider the determination of settlement of rigid, rectangular foundations of varying shape resting on a deep homogeneous stratum. Settlement can be estimated using the strain influence diagram method for rectangular foundations. As the length to width ratio, L/B, of the foundation becomes greater, the following quantities associated with the strain influence diagram increase:

- the dimensionless maximum depth of the influence diagram, z(max)/B.
- the dimensionless depth below the bottom of the footing to the maximum strain influence factor I_z^0 (max), z'/B and
- the value of the strain influence factor, I_z^0 at the bottom of the footing.

For a given value of the length to width ratio, L/B, of the foundation, determine from Figure 6.27 the value of the strain influence factor I_z^0 at the bottom of the foundation. this figure also shows the notation used for the influence diagram variables which are a function of L/B. The maximum depth, z(max)/B, of the diagram and depth to the peak value (z'/B) are obtained from Figure 6.28. Multiply the dimensionless values obtained from this figure by the foundation width (i.e., the minimum dimension of a rectangular foundation) to obtain the actual value of the variable. The settlement calculations are carried out using equation 6.10 in the same manner as previously discussed.

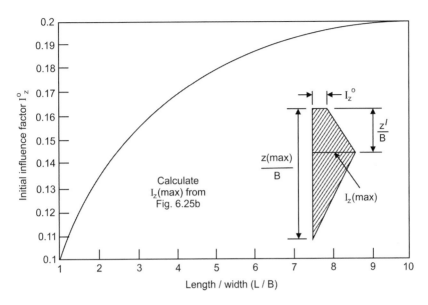

Figure 6.27 Strain influence factor I_z^o at base of rectangular foundation as a function of L/B.

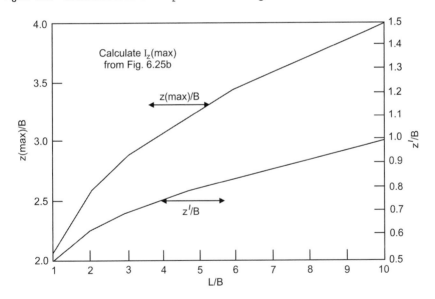

Figure 6.28 Dimensionless maximum depth, z(max)/B, and depth to peak influence factor, z'/B, as a function of L/B, for rectangular foundation.

Flexible circular, square and rectangular foundations, homogeneous deep strata

A flexible foundation is one that has a small stiffness when subjected to bending. A layer of fill or a thin, lightly reinforced concrete mat foundation can be considered as

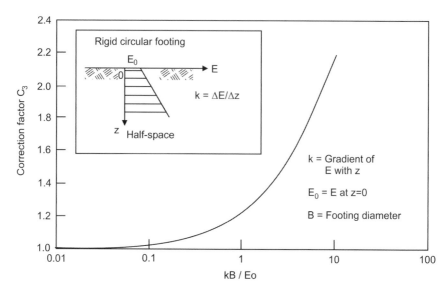

Figure 6.29 Correction factor C_3 for rigid circular foundation resting on a deep, homogeneous layer; modulus of elasticity increases linearly with depth.

flexible loadings. A centrally loaded, flexible foundation has a dish-shaped settlement profile with the greatest settlement occurring in the centre and the least settlement at the edges.

The solutions given in Table 6.1 can be used to calculate the settlement of flexible foundations resting on a deep layer of homogeneous soil. The strain influence diagram method can also be used for flexible foundations. First calculate the settlement for a rigid foundation having the correct shape using the strain influence diagram method. Then correct the results by multiplying the calculated rigid foundation settlement by the correction factors given in parentheses in Table 6.1. For either method of settlement calculation, superposition can be used to obtain settlements at locations on the foundation for which influence factors are not given in Table 6.1.

Circular rigid foundation – increasing stiffness with depth

The stiffness E of residual soils frequently increases with depth. The increase in stiffness is at least partly due to the degree of weathering becoming less and confining stress greater with increasing depth. For this type of soil stiffness profile, the vertical strain developed beneath a footing dies out more rapidly with increasing depth than for a homogeneous soil.

To approximate this condition, first consider a rigid, circular footing of diameter B underlain by a deep soil stratum whose stiffness E increases linearly with depth. The elasticity solutions for these conditions are given in Table 6.1 or the strain influence diagram approach for a homogeneous soil (Figure 6.25) can still be used, provided the calculated settlement is corrected using Figure 6.29. First calculate the settlement for a circular footing resting on a deep homogeneous strata. In these calculations use the soil stiffness at a depth beneath the foundation equal to its diameter B. Then multiply this settlement by the appropriate correction factor C_3 given in Figure 6.29.

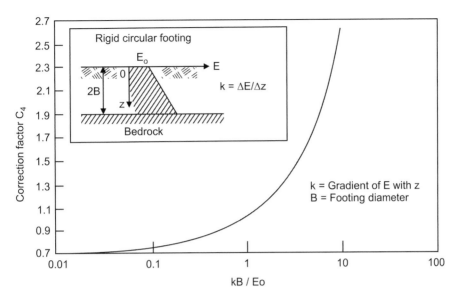

Figure 6.30 Correction factor C_4 for rigid circular foundation resting on layer 2B in depth; modulus of elasticity increases linearly with depth.

Figure 6.30, which is similar to Figure 6.29 is used to calculate the settlement when a rigid layer is at a depth of 2B beneath the foundation. The stiffness E increases linearly with depth below the circular, rigid foundation. The modulus of elasticity at a depth of B beneath the foundation is used. This calculated settlement is then multiplied by the appropriate correction factor C_4 given in Figure 6.30.

6.5.3 Menard method for calculating settlement of shallow foundations

The prediction of settlement is made via a semi-empirical equation:

$$\rho = \frac{(q - \sigma'_{vo})}{9 E_M} [2B_0 (\lambda_d B/B_0)^\alpha + \alpha \lambda_c B] \tag{6.13}$$

In this equation the first term in [ ] accounts for settlement caused by shear distortion, while the second term accounts for compressional settlement.

where q   = gross average bearing stress,
     σ'_{vo} = effective overburden stress at founding level,
     B_0 = a reference width, usually 600 mm,
     B   = width or diameter of footing where $B \geq B_0$,
     α = a rheological factor that depends on soil type and the ratio E_M/P_L (Table 6.5),
     p_L^* = net limiting pressure $= p_L - \sigma'_{vo}$,
     λ_d, λ_c = shape factors depending on the length/width ratio L/B of the foundation (Table 6.6).

Table 6.4 Menard's correction factors for foundation embedment.

D/B	Correction factor
0	1.2
0.5	1.1
1	1.0
>1	1.0

Table 6.5 Values of Rheological Factor α.

	Silt		Sand		Sand and Gravel	
Soil Type	E_M/p_L^*	α	E_M/p_L^*	α	E_M/p_L^*	α
Over-consolidated	>14	2/3	>12	1/2	>10	1/3
Normally consolidated	8–14	1/2	7–12	1/3	6–10	1/4
Weathered and/or remoulded		1/2		1/3		1/4

Table 6.6 The shape factors λ_d and λ_c.

L/B_o	Circle	Square	2	3	5	20
λ_d	1	1.12	1.53	1.78	2.14	2.65
λ_c	1	1.10	1.20	1.30	1.40	1.50

Equation 6.13 applies in cases where the depth of founding D is at least equal to the footing width B, i.e., D ≥ B. Where D < B the correction factors tabulated in Table 6.4 are applied.

The values of α for silt, sand and sand-gravel mixtures are tabulated in Table 6.5 and the shape factors are tabulated in Table 6.6.

6.6 SETTLEMENT PREDICTIONS FOR DEEP FOUNDATIONS

According to the Brand & Phillipson (1985) survey, deep foundations of various types are widely used in residual soils. Driven displacement piles and driven steel tube piles have been used in Brazil, but bored piles and caissons of various types appear to be more widely used in tropical soils. Hand-dug caissons are widely used in Hong Kong, with bored contiguous piles frequently used to support the sides of building excavations. Bored piles are also used in India, Nigeria and Singapore. Driven H piles and precast concrete piles are used in Singapore. In South Africa the commonest type of pile in residual soils is the bored cast *in situ* pile, although driven and driven displacement piles are also used. The situation in Sri Lanka is similar.

Pavlakis (1983, 2005) has had considerable success in predicting settlements of piles from the results of pressuremeter tests in weathered andesite lava. To predict

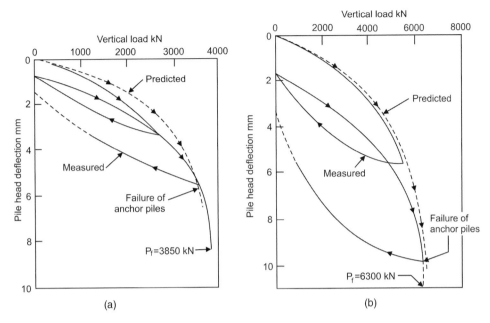

Figure 6.31 (a) Measured and predicted load-deflection behaviour for a single driven, cast-in-situ pile.
(b) Measured and predicted load-settlement behaviour of 2 driven, cast-in-situ piles under one pile cap.

the pile failure loads, he used standard Menard techniques (Menard, 1965 and section 6.5.3), while to predict the load-settlement curve, he followed the procedure of Sellgren, (1981). Sellgren's method for calculating settlements is described in section 6.6.1. Figures 6.31a and b are examples of the excellent agreement he obtained between measured and predicted load-settlement curves, in Figure 6.31a for a single driven cast-in-situ pile, and in Figure 6.31b for two piles loaded through a single pile cap.

6.6.1 Sellgren's method for predicting the settlement of piles

Sellgren (1981) suggested that the load displacement curves for single driven piles should be taken to be in the form of a hyperbola as follows (Figure 6.32).

$$S = \frac{aP}{1 - bP} \tag{6.14}$$

where $S =$ pile head settlement,
$P =$ vertical load on pile,
$\alpha =$ slope of initial part of hyperbola (see Figure 6.32),
$a = \tan \alpha$,
$b = 1/P_f$, where P_f is the failure load of the pile.

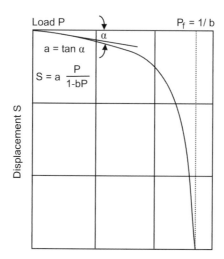

Figure 6.32 Load-displacement relationship for pile loading test proposed by Sellgren (1981).

Equation 6.14 thus becomes

$$S = \frac{aP}{1 - P/P_f} \tag{6.14a}$$

From the results of many pile tests, it has been found that factor "a" can be determined by equation 6.15

$$a = \frac{4\{1 + \beta/[\theta E_p D_p \cdot \tanh(\theta L_p)]\}}{\pi D_p \{\beta + \theta E_p D_p \tanh(\theta L_p)\}} \tag{6.15}$$

where D_p = pile diameter,
$\quad\quad L_p$ = pile length,
$\quad\quad \beta$ = $6E_M/(1 + v)$,
$\quad\quad \theta$ = $\sqrt{4B/E_p D_p}$,
$\quad\quad B$ = $4.17\ E_M$ for $v = 0.3$,
$\quad\quad E_M$ = pressuremeter modulus,
$\quad\quad E_p$ = Young's modulus for pile material.

For piles with square cross section, the term $1/B$ replaces $4/\pi D_p$ in equation 6.15, where B is the width of the pile.

6.7 MOVEMENT OF SHALLOW FOUNDATIONS ON RESIDUAL SOILS

The main movement problems with shallow, lightly loaded foundations on residual soils arise from seasonal or time-cumulative swelling, or shrinkage, or from collapse

of highly leached, high void ratio, unstable grain structures on wetting. By their very nature, these problems tend to occur in areas with well-defined wet and dry seasons, where seasonal or perennial soil water deficits occur, usually in combination with expansive or shrinkable residual clays, or highly weathered granites or loess-type collapsible soils.

The geotechnology of seasonal or long term heave or settlement is highly specialized, and of great importance, as the effects of expansive and collapsing soils necessitate many millions of dollars worth of repairs annually to homes throughout much of the world. Very often, the most extensive damage is caused to low-cost homes whose owners can least afford the cost of repairs. As an example, it is estimated that in South Africa alone, with its relatively small population of 50 million people, the annual cost of repairs to dwelling houses built on unsatisfactory foundation soils is the equivalent of 100 million US dollars per annum. This sounds small in terms of the cost per head of population, but in a country struggling to provide housing for its rapidly expanding population, repair costs represent about 15 000 fewer houses for the poor per annum, or 100 000 fewer people without a decent roof over their heads. Considering that the annual increase of population exceeds 3 million, this exacerbates an already impossible problem.

Solutions to these problems tend to be highly site- and country-specific. The reader is referred to the voluminous literature on the subject, and in particular to the series of international conferences on expansive clay soils held at various venues between 1965 and 1991. These have since been replaced by a series of conferences with the wider scope of unsaturated soils. Brief summaries of the approach to combating heave and collapse, used in various countries and references to literature have been given by Richards (Australia), Gidigasu (Ghana), Desai (India), UI Haq (Pakistan) and Blight (South Africa), collected by Brand & Phillipson (1985). Only a few examples of possible counter-measures will be given in this book.

6.7.1   Heave of expansive soils

Heave is commonly experienced with clayey residual soils, especially if the clay is smectitic. Damage to structures by the heave or expansion of clays occurs in many parts of the world, and is characteristic of arid and semi-arid zones (basically, the unshaded areas in Figure 1.3a). The structures most commonly damaged by heave (or sometimes shrinkage) are houses and other types of low-rise dwelling unit. Insurance policies usually specifically do not cover damage caused by ground movement, and hence the house-owner (either a private citizen or the state) has to bear the cost of repairs directly out of his/her pocket. Damage by heave can be severe, even so severe as to necessitate the demolition of the building. In other cases, repair costs have been as high as 60% to 70% of the present value of the structure. Usually there is no complete guarantee that the damage will not recur in the future, often as a result of a change in the garden layout, growth of trees or creepers, or a change in ownership resulting in a more or less irrigated garden.

As illustrated in Figure 6.8, heave occurs when the total stress on the soil $(\sigma - u_a)$ remains approximately constant, but the suction $(u_a - u_w)$ decreases because the soil gains in moisture content. The reasons that suction decreases are usually that the surface land-use has changed. For example, the suction in a soil profile will decrease

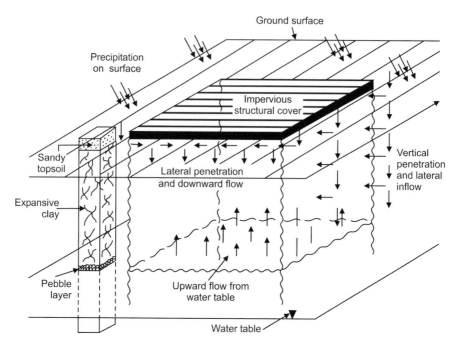

Figure 6.33 Moisture accumulation beneath an impervious structural cover. Possible mechanisms of moisture supply.

if the land is developed and its surface is covered by structures, pavements or irrigated gardens or lawns. All of these will have the effect of allowing water to accumulate in the soil, thus reducing the suction, and leading to swell. There will always be a seasonal component to heave. The soil profile will usually dry out to a certain extent over part of the year. Hence shrinkage will occur. This will, in turn, be reversed and heave will continue during the wetter part of the year. Figure 6.33 illustrates the process of moisture accumulation under an impervious structural cover, such as a house built on a slab-on-grade or a stiffened raft, or under a surfaced road.

In an idealized laboratory oedometer test to evaluate swell properties, a typical pair of tests results for nominally identical expansive soil specimens would be as shown in Figure 6.34. One specimen, compressed at constant natural water content would follow the path ABC. If the soil is wetted and allowed to swell under a very low total stress, it will swell along AA′ and on being compressed, will follow the path A′BC (Jennings & Knight, 1975). Figure 6.35 shows a set of real oedometer tests on a residual weathered shale. These tests show (with considerable irregularity) that swell will take place even if the soil is carrying considerable stresses (in this case up to 500 kPa), (see BB′ and CC′). DEF shows the path followed by a soil loaded and unloaded in an unsaturated state with no access to water. It is important to note that in both Figures 6.34 and 6.35, the compression paths AB, HC and DE are plotted in terms of total stress σ (or $\sigma - u_a$) with the suction being unknown. The paths A′BC or A′G, B′G etc. however, are plotted in terms of effective stress σ' because wetting the soil will have reduced the suction to zero, and at equilibrium, $u_w = 0$ and $\sigma = \sigma'$.

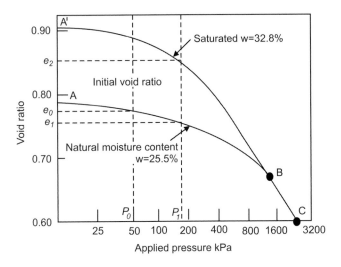

Figure 6.34 Compression curves for a heaving soil.

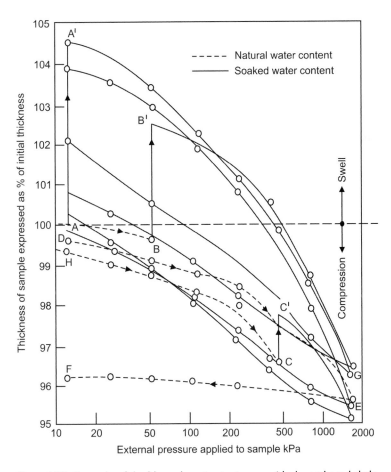

Figure 6.35 Example of double oedometer tests on residual weathered shale.

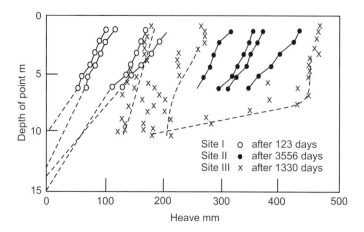

Figure 6.36 Heave of plates buried at various depths at a site near Kimberley, South Africa. (After Williams, 1991).

Other things being equal, the amount by which a soil profile will swell depends on the depth of expansive material in the profile. The lower extent of the expansive part of the profile is set by the water table, or by a change in the type of soil from a potentially expansive to an inert material. Similarly, the upper limit may be set by a non-expansive surface layer.

Deeply weathered mudrock and shale profiles occur extensively in many arid and semi-arid regions. Water tables in such areas may be extremely deep, depths of 30 to 50 m being quite common. Hence it is not unusual to find that the depth of potentially expansive soil is 30 to 50 m. As a result, amounts of heave can also be very large (hundreds of mm). Williams (1991) for example, has recorded surface heaves approaching 500 mm. At the same site, he has observed heaves of 200 mm at a depth of 10 m below surface (see Figure 6.36). At this site, the depth of potentially expansive material approaches 50 m. Usually, however, surface heaves are more moderate and seldom exceed 150 mm. Figure 6.37 shows a series of heave-depth curves for a group of houses supported on slabs-on-grade over a residual weathered shale foundation where the water table was at a depth of 25 m. The heave at a depth of 3 m varied from 110 mm to 180 mm which extrapolate to surface heaves of 180 to 220 mm. These are not unusually large heave movements.

The depth to which seasonal movement may occur can also be large in arid and semi-arid zones. There are usually clearly defined wet and dry seasons, with a short wet season of 4 to 5 months followed by a long dry season of 7 to 8 months. Profiles can dry out to depths of 15 to 20 m, which is also the depth to which soil can be desiccated by suction originating from tree roots. Figure 6.38 (Blight & Lyell, 1987) shows a case where the water table under annual summer crops was at a depth of 2 m below surface. Beneath an adjacent eucalyptus plantation, however, the water table had been drawn down, by evapotranspiration, to a depth of 21 m below surface. Hence, under the cropland, the depth of potentially expansive soil was 2 m, compared to 21 m if the plantation were to be felled and put to other, less water-demanding use. Rain at the start

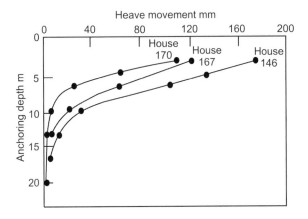

Figure 6.37 Movements of depth points beneath 3 houses founded on an expansive residual shale profile.

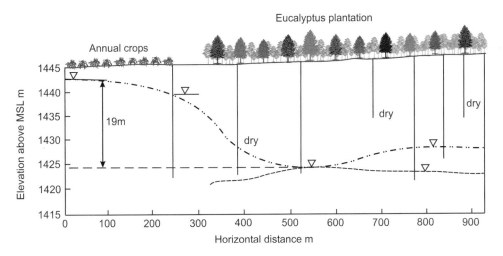

Figure 6.38 Depression of water table by evapotranspiration from a eucalyptus plantation.

of a wet season often occurs in the form of heavy thunder showers, and infiltration may deeply penetrate the profile down open surface shrinkage cracks in the soil. Pellissier (1991), for example, has found free water at the base of a pile in expansive clay at a depth of 7.5 m, shortly after rain, where no water table was found down to 16 m. It appears that the heave of this profile from 7.5 m downwards amounted to over 70 mm, showing that the soil must have been desiccated to well below 7.5 m. The pile-head had heaved extensively and the pile was suspected of having failed in tension. The pile was progressively exhumed and its load transferred to three jacks at the ground surface. Figure 6.39 shows the load on the jacks and the inferred load distribution down the length of the pile. The interesting thing about the load distribution is that it appears that there was virtually no friction on the shaft of the pile down its length to a depth of 4 m. The portion between 4 m and 5.5 m was subject to frictional uplift, and

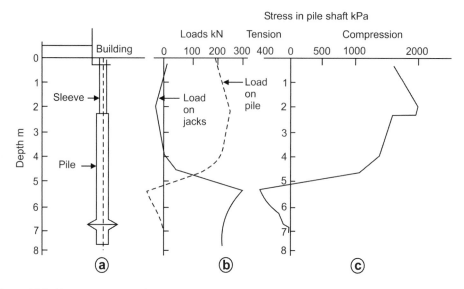

Figure 6.39 Measurements made on a pile in deep residual weathered mudrock: (a) Profile of pile. (b) Load on pile and jacks. (c) Stresses in pile shaft. (After Pellissier, 1991).

the length from 5.5 m to 6.7 m was acting to anchor the uplift. The upper 2.3 m of pile had been sleeved to reduce the friction, but the following 1.7 m must effectively have been out of contact with the surrounding soil as a result of seasonal drying. A similar observation has been made by Zeevaert (1980) relating to piles in Mexico City after a long dry season. In another case, where the water table was relatively shallow (8.5 m), Blight (1965) observed an almost immediate seasonal heave response to rain on a depth point anchored at 5.5 m indicating that seasonal desiccation occurred to below this depth.

Engineering solutions to counter the effects of heave are difficult to formulate. For heavy engineering structures, anchor piles or sleeved anchor piles (Blight, 1984) may be appropriate. For light structures, Williams (1980) and Blight et al. (1991) have reported considerable success in using surface sprinkling or ponding to pre-heave a site and reduce differential movements. Prewetting to pre-heave the profile is therefore another candidate solution. Stiffened rafts appear to be yet another popular solution for light structures (e.g., Pidgeon, 1980).

6.7.2 Prediction of heave in residual soils

In a comprehensive survey, Schreiner (1987) listed 39 published methods of predicting heave in expansive clay profiles and at this time, there are probably many more. The procedures, in vogue in various parts of the world, include completely empirical methods, usually based on indicator test results, methods based on void volume available to hold increased moisture, laboratory simulations of field processes, and a few applications of effective stress principles. The most rational methods combine effective stress principles, void volume considerations and water balance principles.

As an example of an empirical relationship, Brackley (1975) advanced the following completely empirical relationship between swell and soil properties, which was based on extensive laboratory tests on specimens of expansive clay deriving from four widely spaced sites. Of these, two of the clays were residual, one from an igneous rock and one from a shale, and two were alluvial clays. The relationship is:

$$\text{Swell \%} = \left(5.3 - \frac{1.47e}{\text{PI}} - \log_{10} p\right) \times (0.525\text{PI} + 4.1 - 0.85w_0) \tag{6.16}$$

where e $= in\ situ$ (original or unswelled) void ratio,
   p $=$ applied vertical stress [kPa],
   $w_0 = in\ situ$ (unswelled) gravimetric water content [%], and
   PI $=$ plasticity index of whole soil [%].

In a later publication, Brackley (1980) proposed the following simple semi-empirical equation:

$$\text{Swell \%} = \left(\frac{\text{PI}}{10} - 1\right) \log_{10}\left(\frac{s}{q}\right) \tag{6.17}$$

where s $=$ suction at centre of expansive layer [kPa], and
   q $=$ overburden plus foundation stress [kPa].

In Brackley's work, the suction was measured on undisturbed, unconfined specimens by means of psychrometers or pressure plates. He did not comment or draw attention to any discrepancy between the measurements, so the total suction must have been close to the matrix suction.

There must, however, be a caution applied to this type of equation: Empirical relationships used in geotechnical engineering are notorious for their inapplicability in conditions and climates that differ from those in which they were established and should not be used unless their applicability has been proved locally.

The generalized basis of a rational method of heave prediction is as follows:

- The initial and final effective stresses and hence the changes in effective stresses in the soil profile are estimated.
- Hence, from measured swell characteristics of the soil, heave movements are calculated, using methods similar to those used for calculating settlement.
- The rate of heave depends on the rate of accumulation of moisture in the soil profile under the changed surface conditions. The only rational way of estimating this appears to be by applying water balance principles. In an arid zone, the rate of heave is limited by the availability of water, and this, together with the available void volume governs the rate of penetration of the "heave front" into the soil. Initial effective stresses can be estimated either by *in situ* or laboratory suction measurements using psychrometric methods or by pressure plate (or preferably both methods), in the laboratory on undisturbed samples. The swell index of the soil can be measured in the laboratory, and at the same time, swelling pressure measurements can be made to determine initial effective stresses in the profile. (See e.g., section 8.5.)

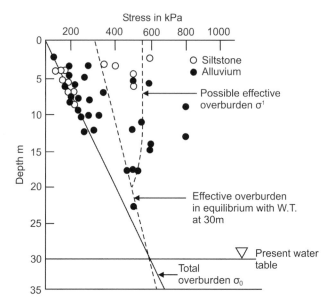

Figure 6.40 Effective stresses in a deep profile residual from siltstone (containing alluvium-filled channels) estimated from swelling pressure tests (Lethabo power station).

A significant difference between pressure plate and psychrometer measurements will indicate a significant component of solute suction. In this case it may be necessary to take account of the extra heave or swelling pressure caused by relief of the solute suction.

Figure 6.40 shows initial *in situ* effective stresses estimated by means of measured swelling pressures in a profile residual from siltstone, having a water table at 30 m depth (Blight, 1984). It will be noted that the data are erratic, and quite a lot of specimens showed no swelling pressure beyond the total overburden stress. Figure 6.41 shows measured values of the swell index C_s and its variation with depth at the same site. Clearly, from Figures 6.40 and 6.41, expansiveness of the soil decreases with increasing depth, as does the potential for expansion, in terms of moisture stress.

Estimating the final effective stresses in the profile is even more difficult than estimating initial effective stresses, as these are dependent on the long-term soil micro-climate. If the surface will be sealed, and if it is unlikely that much water will be contributed to the soil via leaking sewers, waste water soak-aways, etc., the approach first suggested by Russam & Coleman (1961), and developed by Aitchison & Richards (1965) can be used. The Russam-Coleman/Aitchison-Richards diagram (Figure 6.42) relates the Thornthwaite moisture index T to the equilibrium suction at depths of 450 mm and 3 m below the centre of a paved area. To obtain the complete profile of final effective stresses, it is necessary to know, or assume the suction at some other depth, possibly from the position of the water table. In Figure 6.42, if the soil water suction is expressed as $-u_w = \gamma_w h$, with h being the height above a free water surface in cm, then $pF = \log_{10} h$. The Thornthwaite climatic index is given by $T = (1000D - 60d)/E_p$

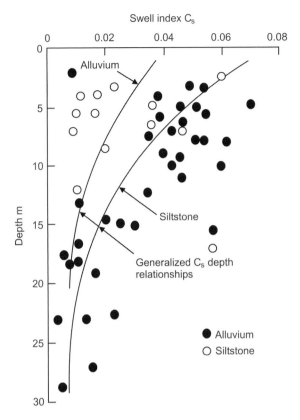

Figure 6.41 Relationship between swell index and depth for residual siltstone profile (containing alluvium-filled channels).

where D = annual infiltration into the soil,
     d = annual evapotranspiration from the soil, and
     E_p = annual potential evaporation (= annual A-pan evaporation).

In the case from which Figures 6.40 and 6.41 were derived, the site of a power station, observations after the power station was in operation showed that there were so many leaks and spillages of water at the surface, that the final suction profile corresponded to a state of continual downward percolation of water under a unit seepage gradient, i.e., the suction was zero at all depths down to the water table.

As stated above, the rate of heave depends on the rate of accumulation of moisture in the soil profile under the new surface conditions. This in turn, depends on the water balance for the site. The water balance can be written as:

$$\text{Rainfall} + \text{leakage} - \text{runoff} - \text{evapotranspiration}$$
$$= \text{infiltration into soil} \; - \; \text{seepage into the water table} \tag{6.18}$$

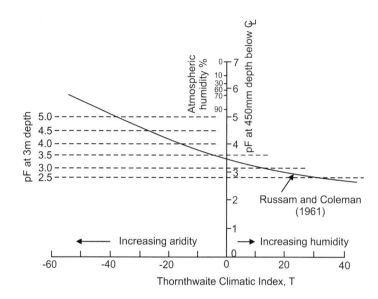

Figure 6.42 Russam-Coleman/Aitchison-Richards diagram relating equilibrium suction under pavement surfaces to Thornthwaite Index T.

In this equation, leakage would include irrigation water for lawns and flower beds and leakage from plumbing and sewers, etc. In the case of the power station, it included leaks from cooling tower ponds, blow-down from boilers, etc. It is obviously not easy to estimate an accurate water balance, but reasonable estimates for all of the terms can be obtained, if necessary, by means of a Delphi process (i.e., a discussion, leading to agreement on numerical design values, among a group of people knowledgeable about the subject).

Once the rate of infiltration into the soil has been estimated, the time for heave to occur can be calculated from the availability of water and the time necessary to fill the air-filled pore space in the profile and the additional pore space created by swelling. Where the lateral extent of a structure subject to heave is of the same order as the depth of expansive material, water accumulates in the profile by ingress through the surface and then by lateral flow into the soil under the structure. The typical shape of the time versus heave curve in this case is ogival or S-shaped. Figure 6.43 shows a typical ogival time-heave curve for a house built on a slab-on-grade where the depth of expansive soil was initially 5 m and the house measured 13.2 m × 7.1 m. The figure also shows the rise of ground water level as water accumulated in the profile. When the area subject to heave has lateral dimensions that are large in comparison with the depth of expansive soil and the lateral dimensions of the structures, the only way in which water can accumulate in the profile is by entering through the soil surface and migrating vertically downward followed by lateral movement (see Figure 6.33). For this situation, it is assumed that the infiltration enters the soil as a sharply delineated wetting front and hence that heave proceeds from the top down. Because the more expansive soil is usually located near the top of the profile (see Figure 6.36), the

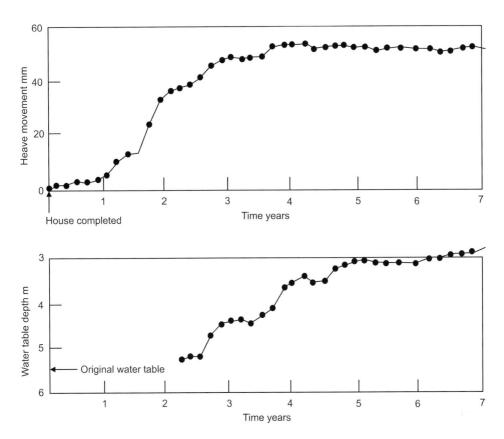

Figure 6.43 Ogival or S-shaped time-heave curve for an isolated building as moisture accumulates in the soil as a result of a change in land use (development as a housing estate). The effect of seasonal variations can be seen. The lower diagram shows the influence on the water table level.

resulting time-heave curve is concave-up, and does not have the convex-up S-shape. This is illustrated by Figure 6.44 which shows:

- the simplified characterization of the site as well as the assumed water inputs and outputs and including the variable surface cover; and
- the resultant concave-up time-heave relationship. Because of the various uncertainties, this is shown as a zone, rather than a single line.

Figure 6.45 shows a selection of time-heave curves observed for the profile referred to by Figures 6.40 and 6.41 that exhibit the concave-up shape. Note that the curves starting at the beginning of year 3 should be translated upwards to coincide with those starting 2 years earlier as the heave for both sets of curves started at the same time. Figure 6.45 also shows the large variability that may, and usually does, occur from point to point over a large site.

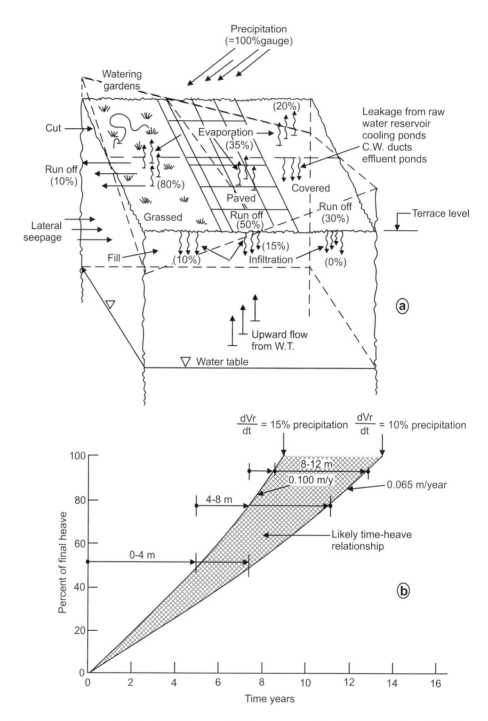

Figure 6.44 (a) Elements of a water balance model for a power station site. (b) Calculated time-heave curves for site water balance.

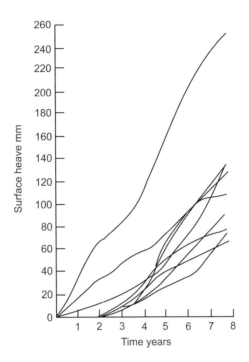

Figure 6.45 Time-surface heave relationship observed at various points on the site of Lethabo power station.

6.8 COLLAPSE OF RESIDUAL SOILS

The phenomenon of collapse settlement occurs in two types of residual soils. The first are loess or loess-like soils usually consisting of ancient wind-blown sands that have been lightly cemented at the points of contact of the soil particles. The second type consists of unusually highly weathered and leached soils residual from acid rocks, such as granites, that contain a large proportion of quartz and micas. As a result of leaching and loss of mineral material, the residual soil becomes a silty or clayey sand with a high void ratio and an unstable collapsible grain structure.

Ancient wind-blown sands

Extensive areas of Southern Africa are covered with a blanket of wind-blown sands of Quaternary age. Because of changes of climate in the fairly recent past, these dune sands that were deposited under desert conditions, now exist as fossil dunes in semi-deserts, e.g., as Kalahari semi-desert grassland or even under savanna conditions. Because of the present moister climate, the sands have partly weathered *in situ* and now contain a few per cent of silt and clay. Originally deposited with a loose wind-blown structure, their structure has become collapsible as a result of the presence of the clay (Knight, 1961). Figure 6.46 shows Knight's classic sketch of the fabric of a collapsing ancient windblown sand. Figure 6.47 (McKnight, 1999) shows the variation with depth of the

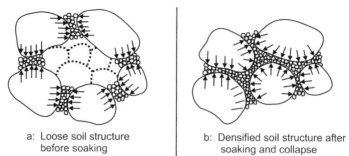

a: Loose soil structure
before soaking

b: Densified soil structure after
soaking and collapse

Flocculated clay particles consolidated between and around points of contact of sand grains

Unconsolidated flocculated clay particles

Figure 6.46 Knight's (1961) sketch of the mechanism of collapse of a collapsing sand.

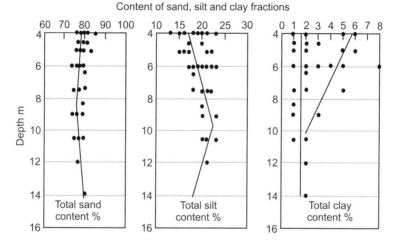

Figure 6.47 Variation of sand, silt and clay contents of a collapsing sand profile with depth. (After McKnight, 1999).

percentages of sand, silt and clay constituents of a collapsing sand profile in southern Mozambique. This shows that the sand and silt contents of the ancient wind-blown material vary from 92–100% at the surface to 97–100% at depth, with the clay content varying from up to 8% towards the surface to 2% at depth. As mentioned previously, leached weathered granites may also have collapsing properties. Figure 1.7 showed that increased annual rainfall over long periods results in greater leaching of weathered granite and therefore a higher void ratio and a greater tendency to be collapsible.

The characteristics of a collapsing soil have been contrasted with those of an expansive soil in Figures 6.8 and 6.9, and are further illustrated by Figure 6.48. Figure 6.48a illustrates the considerable strength (of about 275 kPa) of a collapsing soil at a confining stress of $(\sigma - u_a) = 15$ kPa and a suction of 480 kPa. When the soil was wetted,

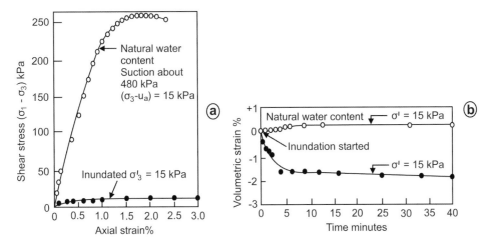

Figure 6.48 Triaxial shear tests on collapsing sand at natural water content and after inundation. a: deviator stress vs axial strain. b: volumetric strain vs time.

the effective stress fell to 15 kPa and the strength was only 12 kPa. Figure 6.48b shows that the soil actually swells by a small amount if wetted under a very small confining stress (1.5 kPa). If it is wetted while carrying 10 times that stress (15 kPa), it suffers a relatively sudden collapse settlement of 2%. Hence a collapsing soil can have a high strength and be relatively incompressible at low water contents (or high suctions) but loses strength and becomes compressible if the suction is reduced by an increasing water content.

Figures 6.49a and b show the differences between the compression characteristics of an unsaturated collapsing sand at its *in situ* water content, and compression of the sand after inundation, plotted in terms of $(\sigma - u_a)$ and $(\sigma - u_w)$ for the unsaturated sand and σ' for the saturated sand. The difference in compressibility of the sand when unsaturated and saturated is very striking. It will also be noted, in comparison with similar data for an expansive clay shown in Figure 6.10, that the compression curves for the saturated soil do not lie between the curves for $(\sigma - u_a)$ and $(\sigma - u_w)$. Although the unsaturated soil must obey the effective stress equation, the Bishop χ parameter cannot be evaluated by applying equation (6.3)a. However, as noted by Blight (1967) and shown in Figures 6.49a and b, the suction $(u_a - u_w)$ remains almost constant as the unsaturated collapsing sand is compressed. Therefore to a first approximation, $\Delta(u_a - u_w) = 0$ and equation (6.3) reduces to

$$\Delta\varepsilon_v = C\Delta(\sigma - u_a) \tag{6.3b}$$

Thus the compression of the unsaturated sand is controlled almost entirely by $(\sigma - u_a)$ and χ is either zero or does not exist in these circumstances.

Although collapse cannot be predicted on the basis of effective stress considerations, collapsing soils behave as effective stress-controlled materials both before and after the collapse takes place (see Figure 6.9). Depending on the water content of the soils, collapse may take place progressively, and not manifest as a sudden settlement

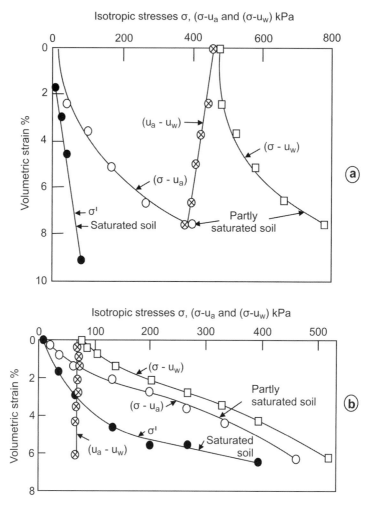

Figure 6.49 (a and b) Typical curves relating volumetric strain to the effective stress components $(\sigma - u_a)$ and $(\sigma - u_w)$ for isotropic compression of samples from a collapsing sand profile. The diagrams also show the compression curves for the saturated soil. Initial water contents were: a: 8.3 per cent, b: 9.6 per cent.

at all. An example of this is shown in Figure 6.50 (Wagener, 1985). If the soil of this example had been loaded at a time when its water content was high, it would have settled without collapse. If, on the other hand, it had been loaded at a low water content and was then subsequently wetted, it would have heaved slightly if loaded to less than 100 kPa, and collapsed if loaded beyond 200 kPa. There can thus be a continuous spectrum between expansive and collapsing behaviour.

The amount of collapse settlement that occurs depends on the initial void ratio of the soil and its water content, as well as the applied stresses. Under light foundation loads of 100 to 300 kPa, collapse settlements of up to 10% of the profile depth seem common, while settlements of up to 15% of profile depth have been reported.

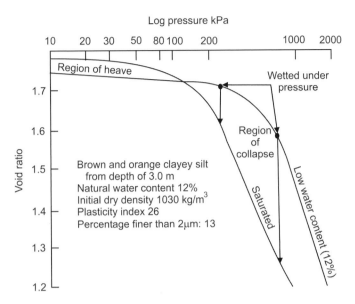

Figure 6.50 Reaction of a loose clayey silt (a residual weathered quartzite) to loading at natural water content, and after saturation.

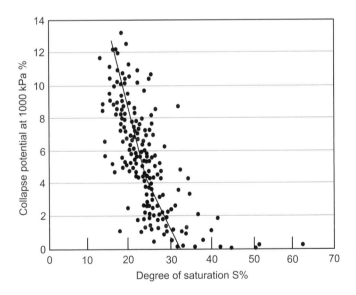

Figure 6.51 Collapse potential vs degree of saturation (McKnight, 1999) for the collapsing sand profile shown in Figure 6.47.

Figure 6.51 shows data collected by McKnight (1999) relating the degree of saturation to the collapse potential (at 1000 kPa) for the collapsing profile referred to in Figure 6.47. The collapse potential decreased from about 12% at a degree of saturation of 16% to almost zero at 32%.

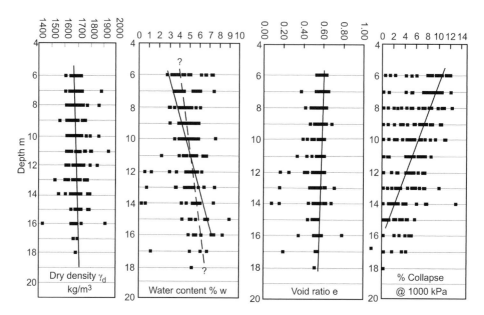

Figure 6.52 Variation in physical parameters with depth for profile of collapsing sand in Mozambique (McKnight, 1999).

Figure 6.52 (McKnight, 1999) summarizes the distribution with depth of *in situ* dry density, water content, void ratio and collapse potential at 1000 kPa for the profile shown in Figures 6.47 and 6.51. It is particularly noteworthy that the dry density was, on average, almost constant with depth, as was the water content and the void ratio. If the values $\gamma_d = 1700\,\text{kg/m}^3$ and $e = 0.55$ are chosen to represent the profile, the particle solid relative density (or specific gravity), $G = (1+e)\gamma_d/\gamma_w = 2.64$. The water content at saturation would then have been $w(\text{sat}) = e/G = 20.8\%$. Thus in terms of Figure 6.52, the profile was far from saturation, even at a depth of 18 m. (The profiles start at 6m because site levelling would bring the surface down by 6 m.)

Predicting collapse settlements

The amount of collapse that will occur is usually predicted on the basis of oedometer tests (e.g., Jennings and Knight, 1975) that attempt to simulate the process of loading followed by wetting that is likely to occur in the field. There appears to have been relatively little attempt to improve existing methods in recent years, even though the problem of collapse is wide-spread.

Combating the effects of collapse settlement

Structures most likely to be affected by collapse are those founded at or near the surface, e.g., roads, housing and slab-on-grade floors to framed structures. Stiffened rafts have been used with apparent success, but relatively little research seems to have been done on the reasons for success of this type of foundation. To design the raft, it is usual to assume that collapse will occur in form of a "soft spot" with a nominal diameter of up

to 2 m, that can occur anywhere under the raft (Tromp, 1985). Pile and pier solutions of various types have also been used (e.g., Schwartz and Yates, 1980).

The most widely used *in situ* treatment for collapsing sands is to attempt to densify them by compaction, i.e., to pre-collapse them. One of the difficulties in adopting this approach is that not much is known about the depth to which a soil profile will collapse under load, and therefore the depth to which a potentially collapsible profile should be densified. Refuge is usually taken in the old rule of thumb for settlement, i.e., that collapse will extend to a depth of 1.5 times the least lateral dimension of a foundation. For roads and extensive hard-standing areas (e.g., container yards), it is the practice to specify compaction of the top 0.5 m to 90% of Modified AASHTO density and the next 0.5 m down to 85% of Modified AASHTO density. This has some basis in the observation that traffic loading on roads built over collapsing sands has been found to produce a densification to depths of just over 1 m (Knight and Dehlen, 1963).

The practices used to achieve densification are limited only by the imagination of the designer or contractor. The following are a few examples of methods that have been tried:

- Removal and compaction. The soil is excavated, its moisture content adjusted, and it is then re-compacted in place to the specified densities. Obviously, this form of treatment can usually only be applied to limited areas, e.g., the plan area of a house or a small building. Because of the fine, predominantly single-sized nature of these sands, it is often difficult to compact beyond 93% of Mod AASHTO density.
- Densification by rolling or pounding the surface: Various forms of roller have been used, e.g., vibrating smooth wheeled rollers and impact (square or polygonal) rollers, with or without prior watering. Surface pounding by drop-weight or dynamic compaction has also been used. Success reported with all methods has been very variable. An interesting comparative study was reported by Jones and van Alphen (1980) who compared the effects of rolling a potentially collapsible profile with a heavy vibrating roller, with and without prior watering (the water penetrated to a depth of 300 to 500 mm), rolling with a loaded earth-moving motorscraper having wheel loads of 18 tons, and pounding with a 4 ton concrete block dropped from heights of up to 8 m. The results observed were as follows:
  - none of the methods produced effects that penetrated to a depth of more than 400 mm;
  - initially the roller and the loaded scraper loosened the soil, apparently by partly breaking down its fabric. Only large numbers of passes eventually caused a measurable, but small densification;
  - there was no difference in the behaviour of the watered and unwatered areas, and
  - ponding loosened the soil and re-compaction did not re-densify it.

Weston (1980) in a comprehensive review of the effects of rolling on collapsing sands for road construction has also reported mixed success in densifying collapsing sands *in situ*. Figure 6.53 reproduces some of Weston's data. Figure 6.53a shows the effect of compacting a section of road-bed with 20 to 30 passes of a heavy vibrating roller. Although some densification was achieved, there was relatively little change

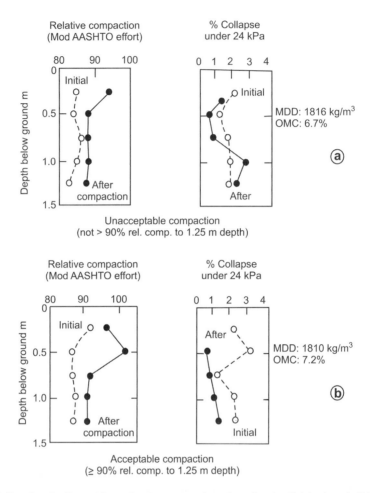

Figure 6.53 Results of rolling trials on the situ compaction of a collapsing Kalahari sand. (After Weston, 1980).

in the collapse settlement of the sand before and after rolling. Figure 6.53b shows results for a second section of the same road. Here, compaction was more successful in reducing the collapse potential of the sand. Weston concluded the following:

- The maximum relative compaction achieved by surface rolling is unlikely to exceed 90% of Modified AASHTO at depths of 0.5 to 1 m below the ground surface. The compactability is strongly influenced by moisture and clay content.
- There is no evidence to suggest that relative compactions in the top 0.5 m layer must be greater than 90% of Modified AASHTO for satisfactory road performance. In other words, the compaction results shown in Figure 6.53, although apparently disappointing were satisfactory.

Strydom (1999) reported a very interesting series of compaction tests on collapsing clayey sandy silt in which surface settlement was measured, as well as the rate of

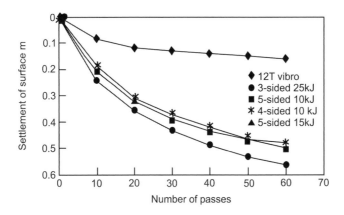

Figure 6.54 Settlement vs number of passes, for each compactor (Strydom, 1999).

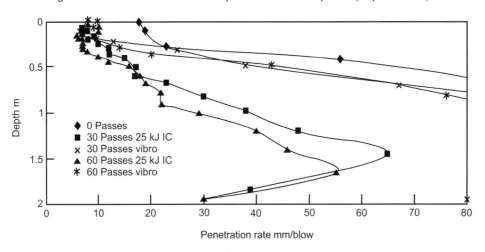

Figure 6.55 Variation in DCP count with depth: Vibro compactor vs 25 kJ compactor.

penetration of a cone penetrometer in mm of penetration per blow of a falling weight. An array of rollers or compactors was used, namely:

12 ton vibrating roller,
4-sided single drum impact compactor (10 kJ/blow),
5-sided dual drum impact compactor (10 kJ/blow),
5-sided dual drum impact compactor (15 kJ/blow),
3-sided dual drum impact compactor (25 kJ/blow).

(Energy per blow = Wh [Nm = J] in Figure 4.1b which shows a 4-sided roller.)

Figure 6.54 shows the settlement produced with increasing numbers of passes by the various compactors. There is no doubt from this that drum impact compactors were vastly superior to the vibrating compactor for this usage. There was, however, relatively little to choose between the impact rollers on the basis of energy per blow.

Figure 6.55 shows the relationship between falling weight penetration rate (DCP = Dynamic Cone Penetrometer) and depth for either 30 or 60 passes of the

vibrating roller and the 25 kJ impact compactor. Comparison with the zero pass line shows that all rollers immediately reduced the surface penetration rate from 18 mm/blow to 10 mm/blow. The vibrating compactor had little effect on the penetration rate. The 25 kJ/blow impact compactor reduced the penetration rate down to 1.5 m. Below this, the soil appears to have been stiffer to begin with. There was little benefit in increasing the passes to above 30.

REFERENCES

Aitchison, G.D. & Richards, B.G. (1965). A broad-scale study of moisture conditions in pavement subgrades throughout Australia. In, Moisture Equilibria and Moisture Changes in soils Beneath covered Areas. (G.D. Aitchison. ed.) Butterworths, Sydney, Australia, pp. 184–232.

Baguelin, F., Jezequel, J.F. & Shields, D.H. (1978) *The Pressuremeter and Foundation Engineering*. Aedermannsdorf, Germany, Trans Tech Publications.

Barksdale, R.D., Bachus, R.C. & Calnan, M.B. (1982) Settlements of a tower on residual soil. *Eng. & Const. Tropical & Residual Soils, ASCE Geotech. Divn. Spec. Conf.*, Honolulu, Published, American society of Civil Engineers, New York, USA. pp. 647–664.

Bishop, A.W. & Henkel, D.J. (1962) *The Measurement of Soil Properties in the Triaxial Test*. London, Edward Arnold.

Blight, G.E. (1963) The effect of non-uniform pore pressures on laboratory measurements of the shear strength of soils. *Laboratory Shear Testing of Soils*. ASTM STP No. 361, pp. 173–191.

Blight, G.E. (1965) The time-rate of heave of structures on expansive clays. In: Ed. G.D. Aitchison *Moisture Equilibria and Moisture Changes in Soils Beneath Covered Areas*. Butterworths, Sydney, Australia. pp. 78–87.

Blight, G.E. (1967) Effective stress evaluation for unsaturated soils. *J. Soil Mech. & Found. Eng. Div., ASCE*, 93 (SM2), 25–148.

Blight, G.E. (1974) Indirect determination of *in situ* stress ratios in particulate materials. *Eng. Found. Conf. Subsurface Exploration for Underground Excavation and Heavy Construction*, Henniker, USA. pp. 350–365.

Blight, G.E. (1984) Power station foundations in deep expansive soil. *Int. Conf. Case Histories in Geotech. Eng.*, St. Louis, USA. Vol. 1, pp. 353–362.

Blight, G.E. & Brummer, R.K. (1980) Strength and compressibility of weathered andesite lava. *J. S.A. Instn. Civ. Engrs.*, 22 (10), 489–499.

Blight, G.E. & Lyell, K. (1987) Lowering of the groundwater table by deep-rooted vegetation – the geotechnical effects of water table recovery. *9th European Conf. Soil Mech. & Found. Eng.*, Dublin, Ireland. Vol. 1, pp. 285–288.

Blight, G.E., Schwartz, K., Weber, H. & Wiid, B.L. (1991) Preheaving of expansive soils by flooding-failures and successes. *7th Int. Conf. Expansive Soils*, Dallas, USA. Vol. 1, pp. 131–135.

Brackley, I.J.A. (1975) *The Inter-relationship of the Factors Affecting Heave of an Expansive Unsaturated Soil*. PhD thesis, University of Natal, Durban, South Africa.

Brackley, I.J.A. (1980) Prediction of soil heave from suction measurements. *7th Reg. Conf. Africa Soil Mech. & Found. Eng., Accra*, Ghana. Vol. 1, pp. 159–166.

Brand, E.W. & Phillipson, H.B. (1985) *Sampling and Testing of Residual Soils*. Hong Kong, Scorpion Press.

Bycroft, G.N. (1956) Forced vibrations of a rigid circular plate on a semi-infinite elastic space on an elastic stratum. *Phil. Trans. Roy. Soc., London*, Series A, 248, 327–368.

Christian, J.T. & Carrier, D. (1978). Plane strain consolidation by finite elements. *J. Geotech. Eng. Div., ASCE*, Vol. 96, No. GE 4, pp. 124–137.

Finn, P.S., Nisbet, R.M., & Hawkins, P.G. (1984) Guidance on pressuremeter, flat dilatometer and cone penetration testing. In: *Site Investigation Practice: Assessing BS 5930, 20th Reg. Mtg. Eng. Group, Geological Society*. British Geol. Soc. Eng. Geol. Spec. Pub. 1. Guildford, UK, pp. 223–233.

Gibson, R.E. & Henkel, D.J. (1954) Influence of duration of tests at constant rate of strain on measured "drained" strength. *Géotechnique*, 7 (4), 6–15.

Hall, B.E., Legg, P.A. & Partridge, T.C. (1994) Characteristics of a deeply weathered residual norite. In: Wardle, B., Blight, G. & Fourie, A. (eds) *Geotechnology in the African Environment*. Rotterdam, Balkema. Vol. 1, pp. 41–48.

Hughes, J.M.O., Wroth, C.P. & Windle, D. (1977) Pressuremeter tests in sands. *Géotechnique*, 27 (4), 455–477.

Jennings, J.E. & Knight, K. (1975) A guide to construction on or with materials exhibiting additional settlement due to collapse of grain structure. *6th Reg. Conf. Africa Soil Mech. & Found. Eng., Durban, South Africa*. Vol. 1, pp. 99–105.

Jones, D.L. & van Alphen, G.H. (1980) Collapsing sands – a case history. *7th Reg. Conf. Africa Soil Mech. & Found Eng., Accra, Ghana*. Vol. 2, pp. 769–774.

Jones, G.A. & Rust, E. (1989) Foundations on residual soil using pressuremeter moduli. *12th Int. Conf. Soil Mech. & Found. Eng., Rio de Janeiro, Brazil*. Vol. 1, pp. 519–524.

Knight, K. (1961) *The Collapse of Structure of Sandy Sub-soils on Wetting*. PhD Thesis, University of the Witwatersrand, Johannesburg.

Knight, K. & Dehlen, G.L. (1963) The failure of a road constructed on a collapsing soil. *3rd Reg. Conf. Africa, Soil Mech. & Found Eng., Salisbury, Rhodesia*. Vol. 1, pp. 31–34.

Mair, R.J. & Wood, D.M. (1987) Pressuremeter Testing – Methods and Interpretation. In: *In-situ Testing*. CIRIA Ground Eng. Report. London, Butterworths.

McKnight, C.L. (1999) The stratigraphy and engineering geological characteristics of collapsible residual soils on the southern Mozambique coastal plain. In: Wardle, B., Blight, G. & Fourie, A. (eds) *Geotechnics for Developing Africa*. Rotterdam, Balkema. pp. 633–645.

Menard, L. (1965) The interpretation of pressuremeter test results. *Sols-Soils*, Vol. 1, pp. 26–31.

Moore, P.J. & Chandler, K.R. (1980) Foundation evaluation for a high rise building in Melbourne. *5th Southeast Asian Conf. Soil Eng., Taipei, Taiwan*. Vol. 1, pp. 245–259.

Partridge, T.C. (1989) The significance of origin for the identification of engineering problems in transported quaternary soils. In: P. de Mulder & A.D. Hageman, (eds) *Applied Quaternary Research*. Rotterdam, Balkema. pp. 119–128.

Pavlakis, M. (1983) *Prediction of Foundation Behaviour in Residual Soils from Pressuremeter Tests*. PhD Thesis, University of the Witwatersrand, Johannesburg, South Africa.

Pavlakis, M. (2005) The Menard pressuremeter in general geotechnical practice in South Africa. In: Gambin, M.P., Mangan, L. & Mestad, M. (eds) *50 Ans de Pressiometres*. Paris, Presses de l'ENPC/LCPC. pp. 100–118.

Pellissier, J.P. (1991) Piles in deep residual soils. *10th Reg. Conf. Africa Soil Mech. & Found. Eng., Maseru, Lesotho*. Vol. 1, pp. 31–40.

Pidgeon, J.T. (1980) The rational design of raft foundations for houses on heaving soil. *7th Reg. Conf. Africa Soil Mech. & Found. Eng., Accra, Ghana*. Vol. 1, pp. 291–300.

Russam, K. & Coleman, J.D. (1961) The effect of climatic factors on subgrade moisture conditions. *Géotechnique*, 11 (1), 22–28.

Schmertmann, J.H. (1955) The undisturbed consolidation behaviour of clay. *Trans. ASCE*, 120, pp. 1201–1227.

Schmertmann, J.H. (1970) Static cone tests to compute static settlement over sand. *J. Soil Mech. & Found. Div., ASCE*, 96 (SM3), 1011–1043.

Schreiner, H.D. (1987) *State of the Art Review on Expansive Soils*. Crowthorne, UK, Transport and Road Res. Lab.

Schwartz, K. & Yates, J.R.C. (1980) Engineering properties of aeolian Kalahari sands. *7th Reg. Conf. Africa Soil Mech. & Found. Eng., Accra*, Ghana. Vol. 1, pp. 67–74.

Sellgren, E. (1981) *Friction Piles in Non-Cohesive Soils, Evaluation From Pressuremeter Tests.* PhD Thesis, Chalmers University of Technology, Goteburg, Sweden.

Selvadurai, P., Bauer, G. & Nicholas, T. (1980) Screw plate testing of a soft clay. *Canadian Geotech. J.*, 17 (4), 465–472.

Skempton, A.W. & Bjerrum, L. (1957) A contribution to the settlement analysis of foundations on clay. *Géotechnique*, 7 (4), 168–178.

Smith, D.A. (1987a) Geotechnical application of screw plate tests, Perth, Western Australia. *8th Pan Amer. Conf. Soil Mech. & Found. Eng.*, Cartagena, Columbia. Vol. 2, pp. 153–164.

Smith, D.A. (1987b) Screw plate testing of very soft alluvial sediments, Perth, Western Australia. *8th Pan Amer. Conf. Soil Mech. & Found. Eng.*, Cartagena, Columbia. Vol. 2, pp. 165–176.

Strydom, J.H. (1999) Impact compaction trials: Assessment of depth and degree of improvement and methods of integrity testing. In: Wardle, B., Blight, G. & Fourie, A., (eds) *Geotechnics for Developing Africa.* Rotterdam, Balkema. pp. 603–611.

Tromp, B.E. (1985) *Design of Stiffened Raft Foundations for Houses on Collapsing Sands.* Johannesburg, South Africa, Schwartz, Tromp & Assocs.

Wagener, R.v.M. (1985) Personal communication with author.

Watt, I.B. & Brink, A.B.A. (1985) Movement of benchmarks at the Pienaars river survey base. In: Brink, A.B.A. (ed.), *Engineering Geology of Southern Africa.* Pretoria, South Africa, Building Publications. pp. 199–204.

Weston, D.J. (1980) Compaction of collapsing sand roadbeds. *7th Reg. Conf. Africa Soil Mech. & Found. Eng. Accra*, Ghana. Vol. 1, pp. 341–354.

Williams, A.A.B. (1975) The settlement of three embankments on ancient residual soils. *6th Reg. Conf. Africa Soil Mech. & Found. Eng.*, Durban, South Africa. Vol. 1, pp. 255–262.

Williams, A.A.B. (1980) Severe heaving of a block of flats near Kimberley. *7th Reg. Conf. Africa Soil Mech. & Found. Eng., Accra*, Ghana. Vol. 1, pp. 301–309.

Williams, A.A.B. (1991) The extraordinary phenomenon of chemical heaving and its effect on buildings and roads. *10th Reg. Conf. Africa Soil Mech. & Found. Eng.*, Maseru, Lesotho. Vol. 1, pp. 91–98.

Willmer, J.L., Futrell, G.E. & Langfelder, J. (1982) Settlement predictions in Piedmont residual soil. *Eng. & Constr. in Tropical and Residual Soils, ASCE, Geotech. Div. Spec. Conf.*, Honolulu, USA. pp. 629–646.

Zeevaert, L. (1980) Deep foundation design problems related to ground surface subsidence. *6th South East Asia Conf. Soil Eng.*, Taipeh, ROC. Vol. 2, pp. 71–110.

Chapter 7

Shear strength behaviour and the measurement of shear strength in residual soils

R.P. Brenner, V.K. Garga & G.E. Blight

7.1 BEHAVIOUR AND DIFFERENCES OF RESIDUAL SOILS FROM TRANSPORTED SOILS

The selection of appropriate strength parameters and the prediction of deformation and failure are important steps in the design of foundations and cut slopes in residual soils. In countries with widespread residual soils, strength testing practice, both in the laboratory and in the field mostly follows standard procedures, employing triaxial and shear box tests in the laboratory and some form of penetration test, vane or plate loading test in the field (Brand & Phillipson, 1985). The need for a clear understanding of the difference between transported and residual soils arises mainly with the preparation and handling of the specimens, and with the interpretation of the test results. A knowledge of the genesis of residual soils and of the factors affecting their shear strength will enable both engineers engaged in design and in material testing to appreciate the peculiarities of these materials in their response to deformation and shear, and will thus facilitate the selection of the most suitable design values for their work.

Residual soils develop a particular fabric, grain structure and particle size distribution *in situ*, which may make them fundamentally different from transported soils. The latter develop their fabric as a result of their mode of deposition and their stress history after deposition. In the following section, the special features which affect the stress-strain behaviour and the strength of residual soils will be summarized and discussed.

7.1.1 Factors affecting stress-strain and strength behaviour

Table 7.1 lists special features encountered with residual soils which are mainly responsible for the differences in stress-strain and behaviour in comparison with transported soils. (Also refer to Chapter 1 and Figure 1.1.)

Stress history

After deposition, transported soils are usually subjected to increasing effective stress due to increasing depth of burial (normal consolidation). This may change when part or all of the overburden is removed by erosion (resulting in an overconsolidated state). In the case of clays, which are deposited from suspension in still water, the stress history after deposition almost fully determines the void ratio and particle arrangement.

Table 7.1 Comparison of features that affect strength of residual and transported soils.

Factors affecting strength	Effect on residual soil	Effect on transported soil
Stress history	Usually not important	Very important, behaviour differs in normally and overconsolidated soils
Grain/particle strength	Very variable, varying mineralogy and many weak or porous grains are possible	Usually uniform; few weak grains because weak particles are broken down into smaller, strong particles during transport
Bonding	Can be an important component of strength mostly because residual bonds and/or cementing, can cause cohesion intercept and result in yield stress that can be destroyed by disturbance	Only occurs with geologically ancient deposits, can produce cohesion intercept and yield stress, can be destroyed by disturbance
Relict structure and discontinuities	Develop from pre-existing structural features in parent rock, including bedding, flow structures amygdales joints, slickensides, etc.	No relict structures, structure develops from deposition cycles and from stress history, formation of slickensided surfaces possible in clays
Anisotropy	Usually derived from relict rock structure, e.g., bedding or flow structures	Derived from deposition and stress history of soil, e.g., bedding, weak clay layers, etc.
Void ratio/density	Depends on state reached in weathering process, largely independent of stress history	Depends directly on stress history

Residual soils are formed by a weathering history and the particles evolve as a result of chemical processes (e.g., leaching, precipitation, etc.) (see Figure 1.1). Weathering is mainly a weakening process but may include unloading by erosion. Weathering may also cause some vertical and lateral unloading due to the loss of mineral matter from voids forming in the altering rock. This implies a progressive modification of the *in situ* stresses which modifies the effect of previous stresses on the structure of the progressively weathering material. It is therefore reasonable to consider the current structure of residual soils to be in equilibrium with and associated with their current state of stress. The effect of past stresses to which they have been subjected during their formation will usually be small (Vaughan, 1988), but may have affected their structure (e.g., via fractures or shearing in the original rock (see Plate C19).

Most saprolitic or lateritic residual soils behave as if over-consolidated. Hence the A-parameter during shear is usually positive, but small. In many cases, the A-value is zero or close to zero at failure. In other words pore pressures generated by undrained shear are usually relatively unimportant and may often be ignored in analysis. Figure 7.1 shows the variation of the A-parameter at failure with depth for unconsolidated undrained triaxial compression of a typical residual soil, the weathered andesite lava of Figure 1.13. Note that the maximum value of A is +0.5, the minimum is close to

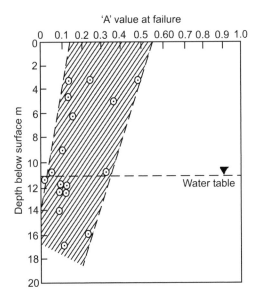

Figure 7.1 Profile of parameter A in a residual andesite lava.

zero and the mean value is less than +0.2. A tends to decrease with increasing depth, as the degree of weathering decreases.

Grain/particle strength

Weathering produces soil particles (mineral grains or agglomerations of grains) with variable degrees of weakening. The particles will, therefore, display a much wider variability in crushing strength than usually encountered with transported soils. Particle size distributions are less meaningful with residual soils than with transported soils because they depend on the degree to which the soil particles have been crushed during sample preparation.

Bonding

One of the characteristics of a residual soil is the existence of bonds between particles. These bonds represent a component of strength and stiffness that is independent of effective stress and void ratio or density. Bonding may also occur with certain transported soils (soft and stiff clays, silts, sands and even gravels) which are of great geologic age, i.e., when bonds have had sufficient time to develop. In engineering applications these bonds are usually not strong enough to consider their effects on strength or compressibility.

Possible causes for the development of bonds in residual soils are (Vaughan, 1988):

- cementation through the deposition of carbonates, hydroxides, organic matter, etc. at particle contacts in the progressively weathering soil (see Figure 1.10a and c).
- solution and re-precipitation of cementing agents, such as silicates or iron oxides.
- growth of bonds during the chemical alteration of minerals (see sections 1.5 and 1.7 for information on pedocretes).

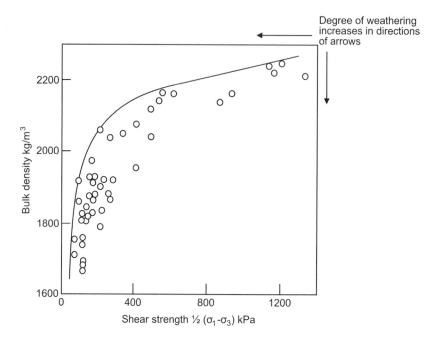

Figure 7.2 Relationship between unconsolidated-undrained shear strength under overburden stress and bulk density for saturated specimens of residual andesite.

The strength of bonds is variable because of different minerals and differences in weathering processes. But it should be kept in mind that even bonds so weak that a sample can scarcely be handled without destroying them, provide a component of strength and stiffness which can have a significant influence on the engineering behaviour of the soil *in situ*, especially at shallow depths and therefore low total stresses.

Figure 7.2 illustrates how the strength and therefore the bonds in a residual soil are progressively weakened as weathering proceeds, as indicated by reducing bulk density. It shows how the shear strength of a weathered andesite lava progressively weakens. The bonds in the partly weathered rock are of relatively high strength ($c_{uu} > 200$ kPa) but those in the completely weathered rock are relatively weak ($c_{uu} < 100$ kPa).

Interparticle bonds can be extremely sensitive to disturbance, which is important to consider in sampling, sample preparation and shear testing. If during the initial stages of shear testing the type of test or the stress path imposed is such that stresses become progressively non-uniform prior to failure, bonds may progressively be partly or fully destroyed, resulting in a decreased measured strength. Furthermore, a bonded structure may be partly destroyed during the processes of saturating and applying confining stresses to a test specimen, if applied pore pressure and stresses are not carefully and incrementally applied. Hence the measured strength may be considerably less than the undisturbed *in situ* strength.

Remoulding of a residual soil specimen (as in excavation and re-compaction) produces a "de-structured" soil in which most bonds have been destroyed, in most cases

irreversibly. Regeneration of broken bonds depends on their nature and may take from a few weeks to millennia, i.e., beyond the lifetime of engineering works, to occur, if it ever occurs. (See, for example, section 1.7).

Relict structures and discontinuities

The parent rock of a residual soil may have contained discontinuities of low shear strength, e.g., joints that had weathered and/or were filled with low strength gouge or clay. Where such seams were repeatedly and reversibly sheared *in situ* (e.g., by seasonal moisture changes) they may develop slickensided surfaces with a low residual strength ($\varphi < 10°$). These weak zones will, after decomposition of the rock, also exist in the residual soil (see Plates C19 and C20), and are usually difficult to discover by boring and drilling. Test specimens with such relict discontinuities will usually fail along these surfaces of weakness, even if the surfaces are not optimally aligned with the direction of maximum shear stress.

Many authors have concluded that the strength of saprolites in mass are governed almost entirely by their inherited fabric or structural features. The strength of the matrix material contained between features such as relict joints is governed by the degree of weathering and also by the degree of secondary cementation or lateritization (see Figure 1.10a). However, the overall strength of the soil mass is governed by the predominant orientation and frequency of structural features in relation to the direction of stress application, and to the strength characteristics of these features. Excellent examples of the truth of this statement have been given by Cowland & Carbray (1988), Irfan & Woods (1988) and Lo *et al.* (1988).

In certain cases, it may be possible to assess the strength of the soil mass by carefully mapping the saprolitic discontinuities and measuring the strength of the discontinuity surfaces either *in situ* or in the laboratory. This will be possible if the discontinuities are relatively infrequent, occur in a regular pattern and are clearly defined, as in the case of a jointed rock mass. In other cases, it may be possible to characterize the relict features as a regular soil anisotropy. For example, relict bedding in a residual shale can be idealized in terms of anisotropy of strength or permeability. In such a case it is possible to take a theoretical account of the anisotropy, as shown by Florkiewicz & Mroz (1989). In many cases, however, saprolitic discontinuities are hard to characterize and their effect on soil mass strength is difficult to assess directly.

Stress anisotropy

As a result of stress anisotropy in a soil, the response to a shear stress application depends on the direction of the resultant of applied and *in situ* stresses. In transported soils, stress anisotropy is directly associated with the mode of deposition and the stress history of the deposit. With residual soils, the anisotropic behaviour has usually mainly been inherited from the fabric of the parent rock, although relict *in situ* stresses may also play a role. This applies particularly to soils derived from metamorphic rocks and where mica is present. Plate-like clay minerals in a decomposing rock, for example, can become oriented during the shearing process to form a polished shear surface. Such surfaces may develop *in situ* (when not already present as relict joints), by strains accompanying soil genesis, but also by seasonal swelling and shrinkage (see Plates C19 and C20). Due to the randomness of these processes and the residual soil fabric, it is

obvious that in residual soils the stress anisotropy may vary *in situ* from point to point and hence from sample to sample.

In situ void ratio or density

Another important property governing the engineering characteristics of residual soils is soil structure expressed by the void ratio or the density of a test specimen. The void ratio in a residual soil is a function of the stage the weathering process has reached (see Figures 1.1 and 1.10a) and is usually not directly related to stress history. It may vary widely and is dependent on the characteristics of the parent rock.

In a weakly bonded soil, the void ratio has a strong influence on the shear strength, which increases with density (see Figure 7.2) (Howatt & Cater, 1982; Howatt 1988a). The void ratio also influences the deformation behaviour (see Figures 1.7 and 1.10a).

7.1.2 Effects of partial saturation

Because of climatic conditions (see Figure 1.3), groundwater tables in water deficient regions are often deep. Evapotranspiration often potentially exceeds infiltration (see Figure 6.38). This leads to deep desiccation of the soil profile. Therefore, residual soils frequently exist in an unsaturated state with continuous air in their voids. The pore air pressure will usually approximate to atmospheric pressure, but the pore water pressure will be sub-atmospheric, i.e., negative, due to capillary effects in the small pores of the soil. This negative pore water pressure or "suction", produces an additional component of effective stress, or in other words: the effective stress becomes greater than the total stress.

As introduced in section 1.13, the term "suction" is expressed by $(u_a - u_w)$

where u_a = pore air pressure, and

u_w = pore water pressure

The equation for the shear strength of a partly saturated soil can be written in terms of the two stress state variables $(\sigma - u_a)$ and $(u_a - u_w)$, where σ is the total stress, either as

$$\tau = c' + [(\sigma - u_a) + \chi(u_a - u_w)] \tan \varphi' \qquad (7.1a)$$

where χ is the dimensionless Bishop parameter (Bishop & Blight, 1963)

$$\text{or } \tau = c' + (u_a - u_w) \tan \varphi^b + (\sigma - u_a) \tan \varphi' \qquad (7.1b)$$

(Fredlund *et al.*, 1978)

where τ = shear strength,

c' = effective cohesion intercept,

σ = total normal stress,

φ' = effective angle of shear resistance,

φ^b = angle by which cohesion intercept increases with increasing suction.

Equation 7.1a can be written as:

$$\tau = c' + (\sigma - u_a) \tan \varphi' + (u_a - u_w) \chi \tan \varphi'$$

Comparing with equation 7.1b, it is seen that

$$(u_a - u_w)\chi \tan \varphi' = (u_a - u_w) \tan \varphi^b$$

In other words,

$$\tan \varphi^b = \chi \tan \varphi' \tag{7.1c}$$

Note that when $\chi = 1$, $\tan \varphi^b = \tan \varphi'$ which also follows from equation 7.1b. This shows that the φ^b approach is actually very clumsy. The χ parameter varies with both $(\sigma - u_a)$ and $(u_a - u_w)$ and, therefore, so does φ^b. It follows that the value of φ^b obtained from a set of measurements such as those portrayed by Figure 7.4a can only be an average value for a particular range of $(\sigma - u_a)$ and $(u_a - u_w)$, whereas the approach shown in Figure 7.3 gives specific values for χ (and therefore $\tan \varphi^b$) provided φ' is constant.

Equations 7.1a and b are an extension of the Mohr-Coulomb failure criterion. When a soil becomes saturated, the pore water pressure approaches the pore air pressure and equations 7.1a and b take the form commonly used for saturated soils. Hence there is little difference in the two forms of the shear strength equation and no advantage in using the one in preference to the other.

Figure 7.3 shows the relationship of the stress state variables $(\sigma - u_a)$ and $(u_a - u_w)$ according to equation 7.1a while Figure 7.4 shows the relationship according to equation 7.1b. Figure 7.3a shows a typical set of triaxial shear test results on an unsaturated soil in which u_a and u_w have been measured separately. The figure also shows the method used to determine the Bishop parameter χ. Figure 7.3b shows the same results plotted in $1/2(\sigma_1 + \sigma_3) - u_a$, $(u_a - u_w)$ and $1/2(\sigma_1 - \sigma_3)$ space. In this diagram, the angle of a line such as A'A to the direction of the $(u_a - u_w)$ axis would be the equivalent of Fredlund and Morgenstern's φ^b. Figure 7.3c shows typical relationships between the parameter χ in equation 7.1a and the degree of saturation S. The theoretical relationship was calculated for an idealized unsaturated soil consisting of an assemblage of spherical particles (see section 1.14.1). Figure 7.4 shows the Fredlund-Morgenstern interpretation via equation 7.1b.

The terms χ or φ^b must be determined experimentally or empirically (Bishop & Blight, 1963; Ho & Fredlund, 1982a; Khalili & Khabbaz, 1998). χ usually falls in a range between 0 and 1, but in rare instances, may exceed 1 (see equation 1.6, Figure 1.14, Blight, 1967b). The value of φ^b would correspondingly range in value from 0 to φ'.

In the case of an unconfined compression test, $\sigma_3 = u_a$, and equation 7.1a reduces to

$$\tau = c' + \chi(u_a - u_w) \tan \varphi', \text{ hence } \chi \tan \varphi' \text{ is given by}$$

$$\chi \tan \varphi' = \frac{\tau - c'}{u_a - u_w} = \tan \varphi^b \tag{7.1d}$$

and if $c' = 0$, then at failure:

$$\tan \varphi^b = \frac{\tau}{u_a - u_w} = \frac{1/2(\sigma_1 - \sigma_3)}{u_a - u_w}$$

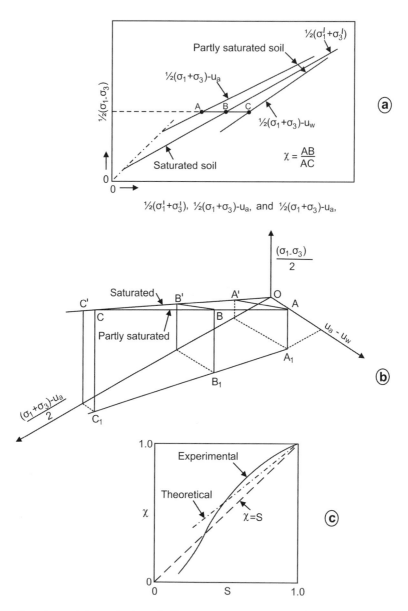

Figure 7.3 Experimental determination of effective stress parameter χ from results of triaxial shear tests (Bishop & Blight, 1963).

Figure 7.4b shows experimental relationships between $\frac{1}{2}(\sigma_1 - \sigma_3)$ and $(u_a - u_w)$ for a series of unconfined compression tests on compacted specimens of a clayey sand, as well as the corresponding relationship (from equation 7.1d) between χ and φ^b. This illustrates that the usually possible range of values for χ is 0 to 1 and for φ^b, is 0 to 45°.

The evaluation of soil water suction as a contribution to shear strength becomes particularly important with slope stability problems in residual soils. For example, a

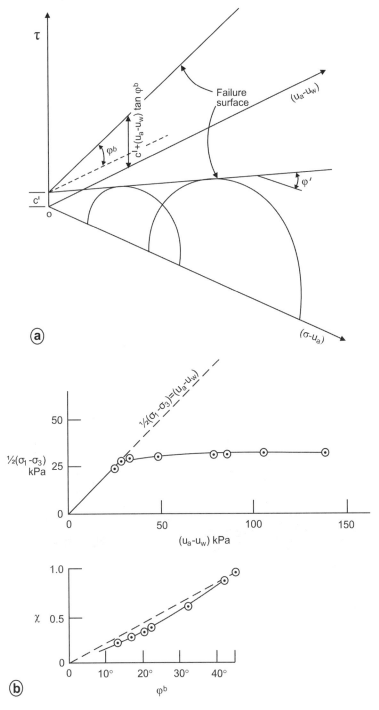

Figure 7.4 (a) Morgenstern & Fredlund's interpretation of shear strength in terms of the stress state variables for an unsaturated soil, $(\sigma - u_a)$ and $(u_a - u_w)$. (b) Relationship between $\frac{1}{2}(\sigma_1 - \sigma_3)$ and $(u_a - u_w)$ for an unsaturated clayey sand in unconfined compression and corresponding relationship between Bishop parameter χ and Fredlund parameter φ^b.

matrix suction of, say, 100 kPa can increase the apparent cohesion of the soil by about 36 kPa.

7.1.3 Measuring the shear strength of residual soils

Because of the many complexities of shear strength behaviour in residual soils, two possible approaches may be adopted in measuring their strength:

1 Suitably large scale shear tests (e.g., Blight, 1984; Chu et al., 1988; Chang & Goh, 1988; Premchitz et al., 1988) can be performed. Provided the scale of the test approaches that of the prototype structure and the instrumentation is well designed and robust, such tests can yield reliable and valuable information. However, large scale field tests suffer the disadvantage of being costly and time-consuming. Because of the cost, it is seldom possible to do more than a minimum number of tests, and knowledge of soil variability suffers.
2 A large number of small scale in situ or laboratory tests can be performed. Suitable in situ tests could include semi-rational tests such as the vane and pressuremeter and empirical or semi-empirical tests such as the standard penetration and cone penetrometer tests. Suitable laboratory tests include the unconsolidated undrained triaxial compression test and shear box tests. Possible refinements to in situ tests are vane tests using vanes of different shapes to assess directional strength and cone penetrometer tests with measured pore water pressures and sleeve friction. In the laboratory, shear box tests with the direction of shearing in specific orientations can be used to explore the effects of strength or stress anisotropy.

This approach has the advantage of enabling soil variability to be explored both laterally and with depth, but with lesser reliability for the final choice of shear strength values for design or analysis.

A number of workers, e.g., Burland et al. (1966) and Garga (1988) have shown that for stiff materials containing discontinuities, small scale strength tests may greatly over-estimate soil mass strength. However, the lower limit to a statistical population of small scale shear strengths approaches the strength of the soil in mass.

This is simply because the lowest measured small-scale strengths correspond to the strengths of the discontinuities that govern the strength of the soil in mass. Figure 7.5 (Blight, 1969) shows a comparison of strengths measured by various small-scale methods on a lateritic residual weathered shale, with the strength-in-mass back-figured from a sliding failure through the foundation of a waste rock dump founded on this material. The comparison illustrates the above statement. The scatter evident in Figure 7.5 is quite characteristic of strength measurements in fissured residual soils. The difference between the undisturbed and the remoulded vane shear strengths is particularly marked. The undisturbed values represent the strength of material between saprolitic discontinuities while the remoulded values represent the strength on artificially produced fissure surfaces. Comparing these values with the calculated strength of the soil in mass, makes it clear that the strength in mass is almost entirely governed by the strength along discontinuities. This strength is represented by the lower limit to the strength measured in small-scale tests. Various other examples of this kind will be given in what follows.

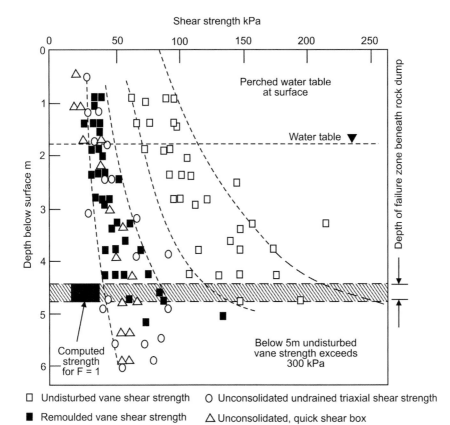

Figure 7.5 Comparison of various small strength measurements in a soil residual from shale with strength-in-mass calculated from failure of a waste rock dump.

The effect of the spacing of discontinuities, joints or fissures, on measured shear strength is further illustrated by Figures 7.6a and 7.6b which both show that the measured strength for a specimen of stiff fissured soil such as London clay becomes less and less realistic as its size decreases. Test specimens must have a minimum dimension that is 2 or more times the spacing of the discontinuities if measured strengths are to be realistic. These observations are supported, specifically for residual soils, by Figure 7.7, which shows the considerable effect of specimen size on measured strengths for a soil residual from a vesicular basalt in Brazil (Garga, 1988). It is obvious from Figures 7.6 and 7.7 that the strength of a stiff fissured soil may be over-estimated by a factor of up to 5 if too small a specimen size is chosen. It should also be noted from Figure 7.7 that the size effect is more pronounced at $\sigma_n = 50$ kPa (over 6 for an "intact" lump and 3.5 for fissured soil) and reduces progressively as σ_n is increased to 200 kPa and then 350 kPa (the strength ratio for an intact lump reduces to 3 and then 2.5 and to 1.8 and 1.5 for fissured soil). This shows both that the defects in the samples are in the nature of voids or partings and that the "intact" lump was not intact, but also suffered from similar defects to the "non-intact" specimens.

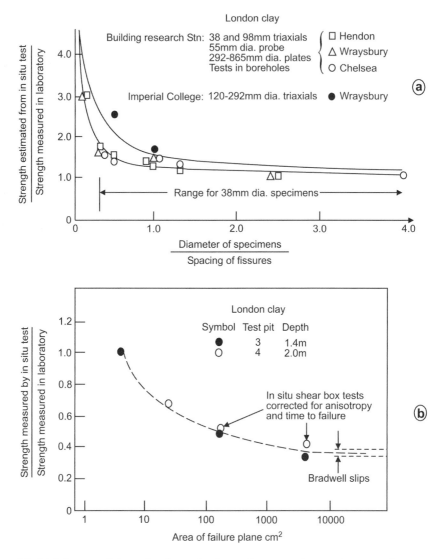

Figure 7.6 (a) Influence of the ratio of sample size to fissure spacing on the strength measured in laboratory tests on London clay. (b) Influence of area of failure plane on strength of London clay (Lo, 1970 and Marsland, 1972).

Garga's conclusions on the effects of discontinuities and fissures on the strength of a residual soil in mass are worth quoting, and now follow:

"1   Discontinuities and fissures significantly affect the mass strength of residual soils. This behaviour is similar to that previously reported for stiff fissured clays of sedimentary and glacial origins.

2   The drained strength of fissured dense soil from 500 mm direct shear samples has been found to be 1.5–3 times [less than]* the strength from 36 mm diameter

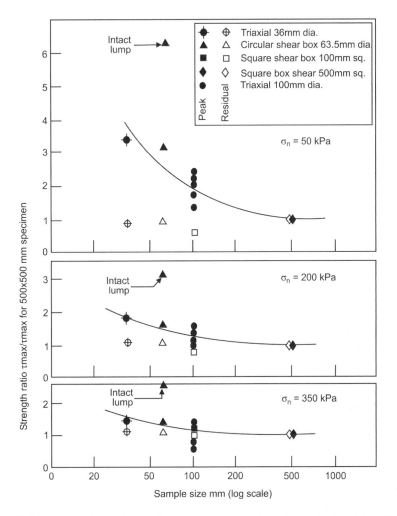

Figure 7.7 Effect of specimen size on shear strength of weathered vesicular basalt lava (Brazil).

triaxial samples in the normal stress range of 50–350 kPa. In contrast, the effect of sample size has been found to be insignificant for the more uniform vesicular residual soil.

3   The frictional component of strength in terms of effective stress shows little variation with sample size. For the dense basalt residual soil, the reduction in strength with size may be attributed essentially to a reduction in cohesion.

4   The limited data suggest that it may be possible to conservatively estimate the field strength of fissured residual soils by ignoring the value of the cohesion intercept from the strength measured on small-sized samples."

*(It should be noted that there appears to be an error in conclusion 2. where marked by [ ]. As the results clearly show that the 500 mm samples gave the lowest strength, the author must have meant to say "... to be 1.5–3 times [less than] the ...).

7.2 LABORATORY STRENGTH TESTING

7.2.1 Types of laboratory shear strength tests

There are two generic types of testing methods commonly used for the shear strength testing of soils in the laboratory, namely the direct shear test and triaxial compression

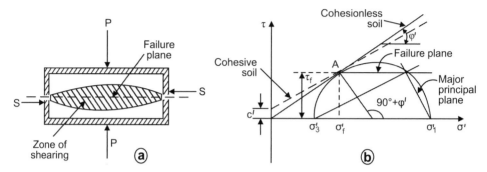

Figure 7.8 Stress system applied in direct shear test: (a) forces acting on specimen. (b) failure envelope through point A and tangent to relevant Mohr's circle.

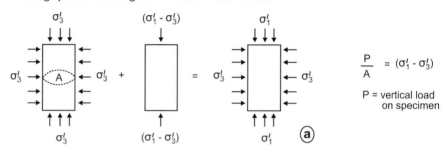

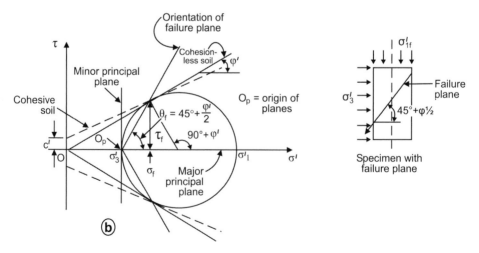

Figure 7.9 Stress system applied in triaxial compression test: (a) stresses on specimen. (b) Mohr's circle and orientation of principal and failure planes.

or extension tests. The systems of stresses applied in these two types of tests are shown schematically in Figures 7.8 and 7.9, respectively. Both tests have their advantages and disadvantages, but certain field conditions may be simulated better by one type than by the other. The main features of these two types of test are summarized in Table 7.2.

The triaxial test is, in theory, superior to the direct shear test. The great advantages of a direct shear test, however, are the speed and simplicity of carrying out a test. For a block sample of a residual soil (which is often coarse-grained or may contain coarse

Table 7.2 Comparison of direct shear and triaxial tests.

Direct shear test	Triaxial test
Advantages • Relatively simple and quick to perform • Enables relatively large strains to be applied and thus the determination of residual strength • Less time is required for specimen consolidation and drainage because drainage path length is small (half specimen thickness) • Enables shearing along a predetermined direction (e.g., plane of weakness, such as relict bedding)	Advantages • Enables the control of drainage and the measurement of pore pressures. Both drained and undrained shearing possible, with or without measured pore pressure • Stress conditions in the sample remain more or less constant and are more uniform than in direct shear test. They are controllable during the test and their magnitude is known with fair accuracy • Volume changes during shearing can be determined accurately • Both triaxial compression (σ_1 vertical) and triaxial extension (σ_1 horizontal) tests can be done
Disadvantages/Limitations • Drainage conditions during test, especially for less pervious soils, are difficult to control. Essentially, only drained tests are possible. Shearing takes place within a zone (Figure 7.8a) • Pore pressures cannot be measured • Stress conditions during the test are indeterminate and a stress path cannot be established. The stresses within the soil specimen are non-uniform. Only one point can be plotted in a diagram of shear stress τ versus normal stress σ, representing the average shear stress on the horizontal failure plane. Mohr's stress circle can only be drawn by assuming that the horizontal plane through the shear box is the theoretical failure plane. During straining the direction of principal stresses rotates • Shear stress over failure surface is not uniform and failure may develop progressively • Saturation of fine-grained specimens (e.g., by back-pressuring) is not possible, but suction can be set to zero by inundating specimen • The area of the shearing surface changes continuously. Change of area must be corrected for	Disadvantages/Limitations • Influence of value of intermediate principal stress, σ_2, cannot be evaluated. In certain practical problems which approximate the conditions of plane strain, σ_2 may be higher than σ_3. This will influence c' and φ' • Principal stress directions remain fixed, conditions where the principal stresses change continuously cannot easily be simulated • Influence of end restraint (end caps) causes non-uniform stresses, pore pressures and strains in the test specimens and barrel-shape deformation, all of which must be considered and corrected for where possible

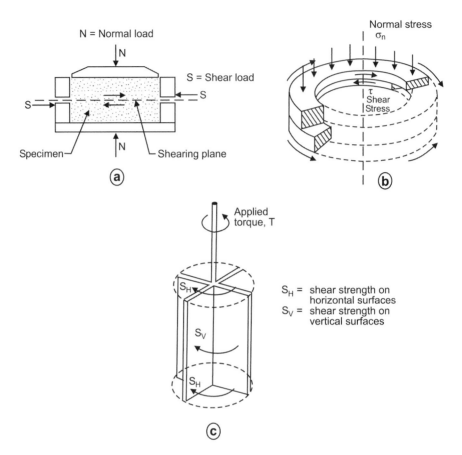

Figure 7.10 Principles of devices to conduct shear tests in the laboratory: (a) shear box, (b) ring shear apparatus, (c) laboratory or field vane shear apparatus.

particles and aggregations of minerals), the direct shear test is often more advantageous than the triaxial test as it may be easier to trim a square specimen than a cylindrical one, and with available shear box sizes of up to 500×500 mm, a larger specimen will give a better representation of *in situ* conditions than a triaxial specimen of maximum diameter, say, 100 mm (see Figures 7.6 and 7.7).

Figure 7.10 illustrates the principles of various devices to conduct direct shear tests in the laboratory, the most common and simplest one being the conventional square shear box. The ring shear apparatus (Bishop *et al.*, 1971) was developed to overcome certain disadvantages of the conventional shear box for the measurement of residual shear strength at very large shear strains. It is usually employed with reconstituted specimens, but can also be used to test undisturbed specimens.

In the vane shear test, direct shearing takes place between a cylindrical volume of soil and the surrounding material. Both field and laboratory vane shear devices are available. The vane can measure either the drained or undrained shear strength by suitably adjusting the rate of loading.

7.2.2 Shear box test

The results that can be obtained from a shear box test are:

- the angle of shearing resistance (peak and residual),
- the cohesion intercept (peak and residual),
- the volume change response of the soil to shearing (dilatant or contractant).

A large number of test variables must be specified when planning a laboratory testing programme. However, in routine testing for engineering projects the available equipment usually limits the choice of the test variables. The following variables may have to be decided upon before starting a test series:

- minimum size of shear box and thickness of specimen,
- status of consolidation and drainage, saturation condition during test,
- controlled strain or controlled stress test,
- rate of straining or stressing,
- normal stresses to be applied,
- maximum horizontal displacement to which specimen should be taken.

Box size and shape and specimen thickness

Shear boxes are usually square (but occasionally circular boxes are used). With square boxes it is much easier to account for the reduction in area as shearing proceeds. Typical sizes for the square box are: 60, 100 and, more rarely 300 mm or more. For circular shear boxes common sizes are 50 and 75 mm diameter.

The maximum particle size of the soil dictates the minimum thickness of the test specimen (Cheung *et al.*, 1988). According to ASTM D3080 the following guidelines apply:

- the specimen thickness should be at least six times the maximum grain size of the soil, and not less than 12.5 mm,
- the specimen diameter (or width) should be at least twice the thickness.

An alternative specification quoted by Cheung *et al.* (1988) recommends:

- the specimen thickness should be at least four to eight times the maximum grain size of the soil,
- the specimen size (square) or diameter should be at least eight to twelve times the maximum grain size.

Cheung *et al.* (1988) found that a 100 mm square shear box with 44 mm thick samples was adequate for testing residual granitic soils with a maximum grain diameter of up to approximately 8 mm. When smaller-sized shear boxes were used irregular stress-strain curves and higher shear strengths were obtained, probably because of excessive single particle load carrying or crushing and gouging displacement within the confines of the box. (It should be noted that shear boxes for 100 mm square by 20 mm thick specimens are available as standard equipment. The specimen thickness

of 44 mm was achieved by Cheung *et al.* by removing the two porous stones and the lower loading platen).

As mentioned above, when residual soils contain discontinuities and fissures as a result of relict structure, their large-scale strength will inevitably be significantly affected. Garga (1988), for example, found that the drained strength of fissured dense soil (residual basalt) from 500 × 500 mm and 290 mm high shear box specimens was 1.5 to 3 times less than the strength from 36 mm diameter triaxial specimens in the normal stress range of 50 to 350 kPa (see Figure 7.7). With relatively uniform specimens the size of the shear box was found to have less effect.

Status of consolidation, drainage and saturation conditions

Shear box specimens can be sheared as nominally unconsolidated, undrained, or consolidated, undrained or drained specimens. Thus, the following test categories are theoretically possible, but not necessarily attainable because of the lack of control of specimen drainage:

- unconsolidated, undrained (UU),
- consolidated, undrained (CU),
- consolidated drained (CD).

With specimens of standard thickness, say 20 mm in the 100 mm square shear box, the drainage path is much shorter than in the triaxial test. This allows excess pore pressures to be dissipated fairly rapidly, but requires very high rates of shear to achieve undrained shear. High rates of shear straining may give falsely high strengths because of viscosity effects. Hence only drained shear in shear boxes is recommended, as only drained shear can be reliably achieved.

For pervious soils, the CD test is the most appropriate. The result will give drained strength parameters, c' and φ'. For impervious cohesive soils (clays and clayey silts) the UU or "quick" test or the CU test are often attempted, and CD tests are also possible and preferable. Pore pressures cannot be measured or drainage controlled, nor can specimens be fully saturated. Hence interpretations of test results other than CD tests may be unreliable.

Tests for which drainage is allowed should be performed with the specimen under normal load and fully immersed in water for at least 16 hours to eliminate the effects of capillary stresses.

Controlled strain or controlled stress tests

The shear stress can be applied either in increments, measuring the resulting displacement (i.e., stress-control), or at a constant displacement rate, measuring the resulting stress (i.e., strain-control). Stress-controlled tests are not common, but are convenient if tests are to be run at a very low loads (a low applied stress can easily be applied and kept constant) and when the creep behaviour of the soil is of interest. Stress-controlled tests employ step-wise load increments and cannot determine the peak shear stress accurately. They are unsuitable for residual strength measurements, as it is difficult to decide what rate of movement constitutes failure.

The strain-controlled test is easier to perform and allows the ultimate, large displacement or residual shear strength to be determined. For these reasons, direct shear tests are usually conducted in displacement or strain control.

Rate of shearing

Creep considerations aside, the shear strain or displacement rate to be applied should depend on the permeability or coefficient of consolidation of the specimen. It is usually possible to select a shear strain rate such that deviation from a fully drained condition is not significant. Based on investigations by Gibson & Henkel (1954), Head (1982) recommends a time to failure, t_f, for drained direct shear tests of:

$$t_f = 12.7t_{100} \qquad (7.2a)$$

where t_{100} is the time to 100% of primary consolidation. The value of t_{100} can be obtained by extrapolating the linear portion of the square root of time plot of the consolidation phase of the test to its intercept with the time axis at 100% consolidation. Equation 7.2a is based on attaining 95% pore pressure dissipation at the centre of the specimen. ASTM D3080 recommends:

$$t_f = 50t_{50} \qquad (7.2b)$$

where t_{50} is the time required for the specimen to reach 50% primary consolidation. This equation gives essentially the same time to failure as equation 7.2a (Cheung et al., 1988).

According to Blight (1963b), (see Figures 7.24 and 7.25) the time to failure for 95% consolidation on the shearing plane is given by:

$$t_f = \frac{1.6H^2}{c_v} \qquad (6.8a)$$

where H is half the specimen thickness.

When t_f has been determined, the maximum permissible rate of shearing in a drained direct shear test can be estimated from:

$$\text{Rate of shearing} < \frac{\delta_f}{t_f} \qquad (7.3)$$

where δ_f is the horizontal displacement of the shear box at peak strength (failure). This value is not usually known and has to be established by a trial test.

Normal loads or stresses to be applied

The normal pressures applied to the test specimens should generally straddle the maximum stress which is likely to occur *in situ* for the design being investigated. Tests under at least four different values of normal stress should be used to define the strength envelope.

With cohesionless soils, the strength envelope usually passes through the origin, but with heavily over consolidated soils or soils having a bonded structure, there may be

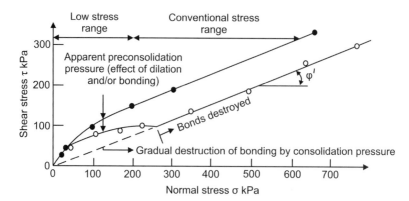

Figure 7.11　Effect of bonding on the cohesion intercept of a drained strength failure envelope (After Vargas, 1974 and Rodriguez, 2005).

a cohesion intercept. If this component of strength is of importance in an engineering application of the strength test result, tests have to be carried out with low normal stresses on carefully handled undisturbed specimens. In the normal stress range typical values of applied stresses are between about 200 and 800 kPa. Figure 7.11 shows the effect of soil bonds on the shear strengths of two residual clays from Brazil (Vargas, 1974 and Rodriguez, 2005) in producing an enhanced strength at low normal stresses. Note from the test by Rodriguez that an enhanced strength at low stresses does not necessarily mean that the soil has a true cohesion.

Density of compacted specimens

If tests are going to be carried out on compacted specimens, the density for testing should be defined. The angle of shearing resistance can then be evaluated as a function of the density (or void ratio in the case of cohesionless soils).

Maximum shear displacement

The strain-controlled direct shear test is particularly useful when the relevant engineering problem requires a knowledge of the residual strength. The most suitable equipment to carry out such tests is the ring shear apparatus (Bishop *et al.*, 1971), which is now commonly available. Most shear boxes can be used for multiple reversed shear cycles for studying post-peak behaviour. Such devices enable horizontal displacements of any length, to reach the residual shear strength. The shear displacement required to reach the residual strength may be as high as 300 mm, making many reversals necessary.

Tests that do not require the determination of the residual strength may be terminated after the peak strength has been passed but at not less than about 15 mm of shear displacement. With soils not showing a peak strength (which may occur with weaker specimens) the tests should be carried to a displacement of about 20 mm.

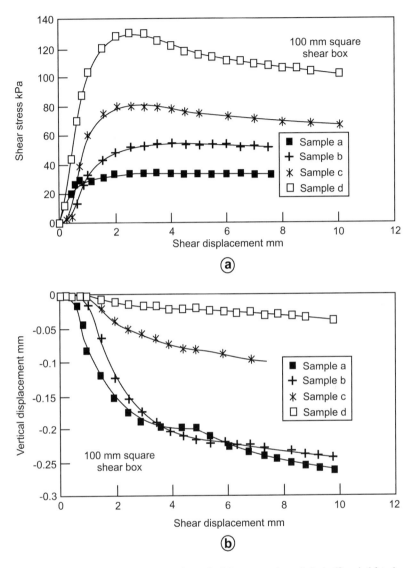

Figure 7.12 Shear box tests on soil residual from weathered shale (South Africa).

Examples of direct shear results for residual soils

Figure 7.12 shows the results of a set of slow consolidated drained shear box tests on a soil residual from weathered shale. The upper diagram (a) shows the development of shearing resistance with increasing shear displacement, while the lower diagram (b) shows the compression or contraction of the specimens as they were sheared. Figure 7.13 shows 130 mm diameter ring shear tests on the same soil. Whereas the shear box tests were made on trimmed undisturbed specimens, the ring shear tests

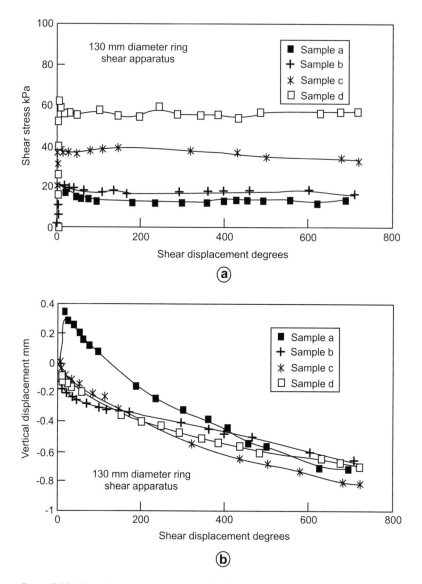

Figure 7.13 Ring shear tests on soil residual from weathered shale (South Africa).

were on semi-disturbed specimens, the semi-disturbance being caused by trimming small segments of soil to form the ring specimens.

Whereas the shear box tests had to be terminated at a shear displacement of 10 mm, the ring shear tests were taken through two complete revolutions (720°), which corresponds to a shear displacement of over 800 mm. (Note that in Figure 7.12, the shear stress peaked at a displacement of 2–3 mm. This distance is almost invisible in Figure 7.13 which obscures the peaking characteristic.). After the first stage of testing on

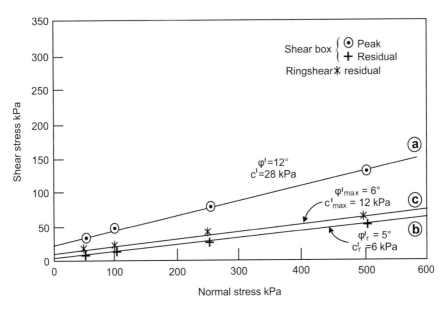

Figure 7.14 Failure envelopes for shear tests on soil residual from weathered shale (South Africa): (a) 100 mm square shear box – peak strength. (b) 100 mm square shear box – residual strength (5 reversals). (c) 130 mm diameter ring shear apparatus – residual strength.

100 mm square specimens, shown in Figure 7.12, the specimens were subjected to 5 reverse and forward shearing movements of 10 mm each so as to reach the residual shear strength (a total displacement on the shear plane of 110 mm).

The failure envelopes corresponding to Figures 7.12 and 7.13 are shown in Figure 7.14. In this case the ring shear tests gave a set of slightly larger residual shear strength parameters than did the reversed shear box tests.

Figure 7.15 shows a comparison of tests on semi-undisturbed and undisturbed segmental ring shear specimens. Here, the "undisturbed" specimens were prepared by painstakingly cutting small blocks of undisturbed clay to fit as exactly as possible into the ring shear apparatus whereas less care was taken with the "semi-undisturbed" specimen. Although there were significant differences in the shear stress versus shear displacement and vertical displacement versus shear displacement curves, the peak and residual shear strength envelopes were very close to each other, with the "undisturbed" specimens giving a slightly higher strength (Figure 7.16).

7.2.3 Triaxial test

The triaxial testing equipment has a considerable versatility and permits of a large variety of test procedures to determine triaxial shear strength, stiffness and characteristic stress ratios (e.g., K_0) of a soil. In addition, the test can be used to measure consolidation and permeability characteristics. The state-of-the-art in triaxial testing of cylindrical soil specimens was established by Bishop & Henkel (1962).

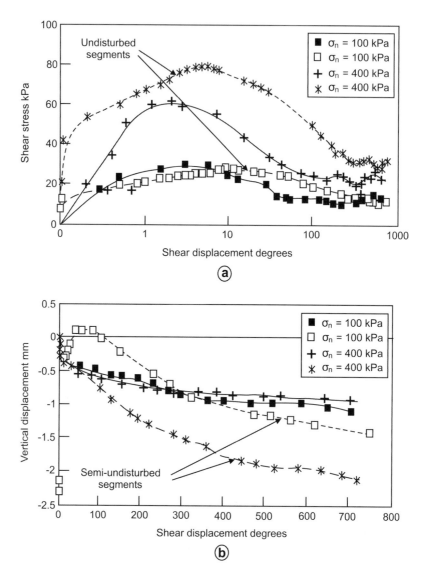

Figure 7.15 Ring shear tests on clay residual from smectitic mud-rock (South Africa). Comparison of "undisturbed" segmental and semi-undisturbed segmental specimens.

In practice, the following triaxial tests are routinely carried out, although other types of tests are also possible:

- unconsolidated undrained (UU) test with or without pore pressure measurement,
- isotropically or anisotropically consolidated undrained compression or extension (CIU or CAU) test with or without pore pressure measurement,
- isotropically or anisotropically consolidated drained compression or extension (CID of CAD) test.

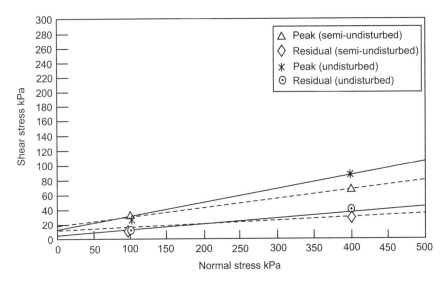

Figure 7.16 Comparison of results of ring shear tests on "semi-disturbed" and "undisturbed" specimens of clay residual from smectitic mud-rock (South Africa).

A diagrammatic layout of the triaxial cell is shown in Figure 7.17. The specimen is sealed in a thin latex rubber membrane and subjected to an all-round fluid pressure. A load is applied axially, through a piston acting on a top cap and controlling the magnitude of the deviator stress. Figure 7.17 shows a triaxial cell suitable for testing either saturated or unsaturated soils. The load is preferably measured by a load cell internal to the triaxial pressure cell.

In a compression test the applied axial load represents $A(\sigma_1 - \sigma_3)$, where A is the specimen cross-sectional area. The net axial stress is the major principal stress σ_1 {$(\sigma_1 = (\sigma_1 - \sigma_3) + \sigma_3)$}. The intermediate and minor principal stresses, σ_2 and σ_3 are equal to σ_3 and correspond to the cell or confining pressure. In an extension test, the confining pressure equals σ_1 and the deviator stress is negative (a tension is applied to the loading ram). The axial stress is $\sigma_3 = $ {$(\sigma_1 - \sigma_3) + \sigma_1$}. Connections to the ends of the specimen enable the drainage or injection of water from the base, or of air from the top cap. Alternatively, the measurement of pore water and air pressure under conditions of no drainage, or the control of pore water and/or air pressure is possible. A standard test is usually carried out in two separate stages, namely the application of the confining pressure σ_3 or σ_1 followed by the application of a deviator stress $(\sigma_1 - \sigma_3)$, either compression or tension. (For extension tests, a tension link is provided between the specimen top cap and the loading ram and between the top of the loading ram and the loading frame. The base of the cell must be clamped to the base of the loading frame to resist the tensile force applied to the loading ram.)

It should be pointed out that the stress systems applied to the test specimen in these tests do not necessarily match the stresses acting at a point in the soil for which the specimen should be representative. Also the stress paths for loading (or unloading) the specimen *in situ* and in the laboratory are often not the same. It follows from this that the application of triaxial test results to practical problems requires considerable

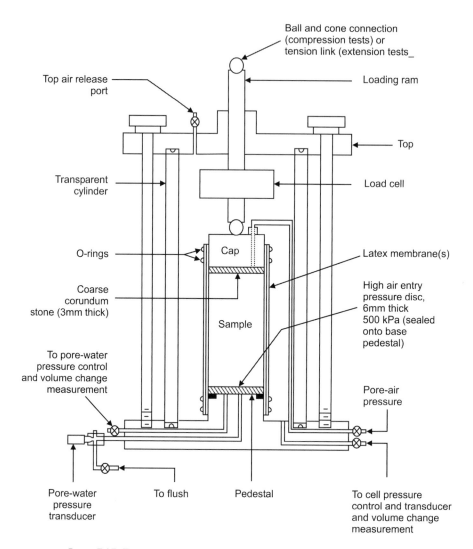

Figure 7.17 Triaxial cell equipped for testing unsaturated soil specimens.

interpretation. An evaluation of triaxial test results should therefore always be based on a knowledge of how and by what means the data have been obtained, as well as details of the problem to which the results will be applied.

Triaxial test variables

The results that can be obtained from a triaxial test, depending on the type of test and available equipment are:

- the strength envelope with peak angle of shearing resistance and cohesion intercept,
- the pore pressure response to shearing (in undrained tests),

- the volume change response to shearing (in drained tests on saturated soils and drained and undrained tests on unsaturated soils,
- tangent and secant moduli (or corresponding unloading and re-loading moduli)
- consolidation characteristics,
- permeability under different confining pressures and flow gradients,
- tests on unsaturated soils are usually drained with respect to the pore air and undrained with respect to pore water, i.e., constant u_a, constant water content tests.

Sample size

For testing residual soils, the specimen diameter should not be less than 76 mm. Specimens with smaller diameters are not considered representative, because of the scale effect relating to fissures and joints in the soil. In addition, the specimen diameter should not be less than 8 times the maximum particle size.

The ratio of specimen length to diameter must be at least 2 to 1, but not greater than 3 to 1 to minimize both effects of end restraint and tendencies to buckle under compressive stress.

Consolidation prior to shear

The specimen is either consolidated under a specified stress system (σ_1, σ_3) prior to shear, or no consolidation is allowed (consolidated (C) and unconsolidated (U) tests, respectively). In saturated soils (clays, silts), for a series of samples taken from the same depth, the compressive strength from unconsolidated, undrained (or UU) tests is found to be independent of the cell pressure σ_3, (with the exception of fissured clays and compact silts at low cell pressure). The strength envelope in terms of total stresses is approximately horizontal, i.e., $\varphi_{uu} = 0$. The undrained strength is then the apparent cohesion c_{uu}:

$$c_{uu} = \tfrac{1}{2}(\sigma_1 - \sigma_3)_{max} \tag{7.4}$$

where $(\sigma_1 - \sigma_3)_{max}$ is the deviator stress at failure. For normally consolidated soils there is a unique relationship between strength and water content at failure. The unconsolidated undrained (UU) test is used for end-of-construction stability analyses.

For residual soils, which are often both fissured and unsaturated, the undrained strength will increase with increasing confining pressure and the strength envelope may be a straight line, but is often curved. However, as the confining pressure increases, the air in the voids becomes compressed and passes into solution. Finally the stresses may be large enough to cause full saturation and φ_u will approach zero. Figure 7.18 shows a set of UU test results for an unsaturated residual andesite lava. Note that whereas the UU strength envelope in terms of total stresses is curved, the corresponding effective stress envelope is a straight line (Blight, 1963a).

Consolidation stress system

The confining stress system during consolidation can either be isotropic $(\sigma_1 = \sigma_2 = \sigma_3)$ or anisotropic $(\sigma_2 = \sigma_3$ and $\sigma_1 > \sigma_3)$. In the anisotropic case the additional axial stress

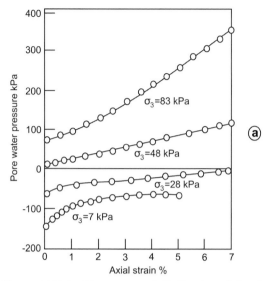

Changes in pore water pressure and effective stress during shear

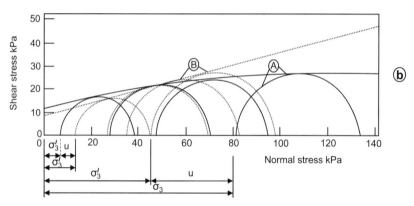

Mohr circles at failure drawn in terms of total and effective stresses

A - Total stress circles and failure envelope
    (φ approaches zero as soil is saturated by compression)
B - Effective stress circles and failure envelope.

Figure 7.18 Typical results of UU triaxial shear tests on an unsaturated residual andesite soil (effective stress has been taken as $\sigma' = \sigma - u$).

$(\sigma_1 - \sigma_3)$ is usually applied through a deadload hanger system. Since in general the stress conditions in the ground are not isotropic, i.e., $\sigma_v \neq \sigma_h$, consolidation under an anisotropic stress system gives a more realistic starting point for a triaxial test than isotropic consolidation. The most frequently used anisotropic stress system is σ_1' with $\sigma_3' = K_0\sigma_1'$, so-called K_0-consolidation. However, for many practical applications

the stresses at failure for isotropically and anisotropically consolidated tests are not much different and specimens are therefore usually consolidated isotropically, which is simpler to do.

Loading (deviator) stress system

Theoretically, it is possible to load the specimen to failure along any stress path, both in the consolidation and the shearing stages. A stress path is a curve (usually) in $1/2(\sigma_1 - \sigma_3)$ versus $1/2(\sigma'_1 + \sigma'_3)$ space representing the successive stress states the specimen is subjected to during loading. (Note that $1/2\sigma_1 - \sigma_3)$ is an effective stress because the same pore pressure is subtracted from σ_1 and σ_3 to arrive at σ'_1 and σ'_3.) Four frequently used stress paths are:

- compression loading   (σ'_1 increasing, σ'_3 constant)
- compression unloading (σ'_1 constant, σ'_3 decreasing)
- extension loading   (σ'_1 constant, σ'_3 increasing)
- extension unloading   (σ'_1 decreasing, σ'_3 constant)

Other stress paths sometimes employed are:

- constant mean principal stress $p' = 1/3(\sigma'_1 + 2\sigma'_3)$
- constant stress ratio $3/2(\sigma'_1 - \sigma'_3)/(\sigma'_1 + 2\sigma'_3) = q'/p'$
  where $q' = 1/2(\sigma'_1 - \sigma'_3)$.

    Undrained strength, effective angle of shearing resistance φ' and stiffness, all vary between compression and extension tests, as shown in Figure 7.19 (Baldi *et al.*, 1988). The highest values are usually obtained with triaxial compression.

    As examples of loading and unloading tests, Blight (1963c) conducted triaxial compression and extension tests on undistorted specimens of a weakly cemented sand residual from a micaceous schist. The strength parameters were $c' = 117$ and 93 kPa and $\varphi' = 39°$ and $31°$ for the compression and extension tests respectively. These results, illustrated in Figure 7.20, show the effect of the sub-horizontal schistose laminations in the soil. In extension, failure took place parallel to the laminations, whereas in compression the failure surfaces cut across the laminations. Note that in these tests the soil was unsaturated and u_a and u_w were measured separately (also see Figure 7.3a). In each case, the strength lines are compared with those for identical tests done on saturated soil and therefore representing true effective stresses.

    Bishop & Wesley (1975) developed a hydraulic triaxial cell which, in principle, is capable of any stress path test. This equipment is available commercially, but it has only been designed for specimens with a diameter of 38 mm. A conventional triaxial cell can be used for stress path testing if the deviator stress is applied by a dead load supported by a hanger on the loading ram, or if a displacement-controlled loading press is modified to apply a controlled deviator stress by use of a double acting pneumatic actuator attached to the reaction frame. A combination of the two methods has proved to be practical (Baldi *et al.*, 1988).

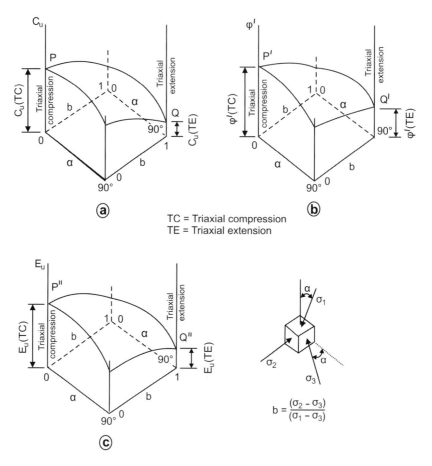

Figure 7.19 Hypothetical surfaces representing strength and stiffness parameters vs α and b axis (Baldi et al., 1988): (a) undrained strength, (b) drained strength, (c) secant stiffness at 0.1% strain. σ_1 and σ_3 vary in a single plane only and the triaxial test can furnish results only at points P and Q of the surface.

Saturation conditions and back pressure application (for CU and CD tests)

Residual soils are usually unsaturated when sampled, and saturation by back pressure is considered routine in many applications.

Saturating a residual soil specimen represents a condition which is more severe than the natural state usually experiences. In residual soil areas the ground water table is often deep and foundations well above it. In a sequence of usually wet years it may happen that the ground water table rises by several metres. Saturation therefore represents the least favourable condition of the residual soil. Saturation has, in general, little effect on the friction angle φ' and will reduce capillary suctions in the soil. Only in cases where saturation causes a weakening of existing cementation bonds in the soil will the cohesion in terms of effective stresses be reduced.

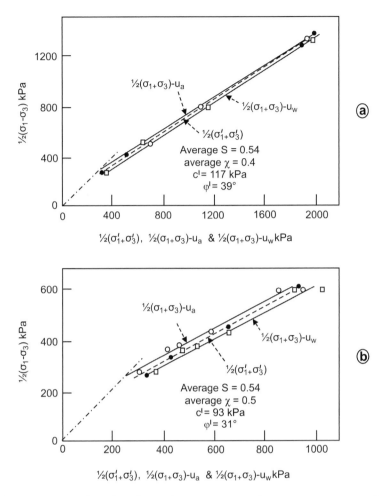

Figure 7.20 Comparison of results of triaxial compression and extension tests on a decomposed residual micaceous schist. Also note the comparison between tests on saturated and unsaturated specimens (σ_1' and σ_3' denote saturated specimens). (a) σ_1 normal to schistose laminations (b) σ_1 parallel to schistose laminations.

The extent to which saturation of the test specimen has been achieved can be checked by measuring Skempton's pore pressure parameter $B = \Delta u / \Delta \sigma_3$. This parameter is, however, dependent on the porosity of the specimen and on the compressibility of both soil structure and pore fluid (Black & Lee, 1973). This implies that for the same B-value, the degree of saturation is not the same for soils with different stiffnesses. Higher saturation will exist in the stiffer soil.

Saturation by applying a back pressure involves increasing the pore water pressure of the specimen and thereby compressing and dissolving the air in the pores according to Boyle's and Henry's laws respectively. Lowe & Johnson (1960) have shown that the theoretical back pressure Δu_{bp} required to bring a specimen from an initial degree of

saturation, S_0, to a final degree of saturation, S, under conditions of constant overall soil volume and increasing water content is:

$$\Delta u_{bp} = u_{a0}[S - S_0(1 - H)]/[1 - S(1 - H)] \tag{7.5}$$

where u_{a0} = initial pressure of air in voids (usually atmospheric = 100 kPa absolute),
H = Henry's coefficient of solubility in volumetric terms (0.02 volume of air per volume of water, approximately, at 20°C).

For complete final saturation $(S = 1)$

$$\Delta u_{bp100} = 49u_{a0}(1 - S_0) \text{ with } u_{a0} = 100 \text{ kPa (absolute)}$$

To saturate from $S_0 = 70\%$, the theoretical back pressure for $S = 100\%$ would be about 1470 kPa. (Also see Section 4.15).

Single-stage and multi-stage tests

In conventional triaxial testing with specimens consolidated under a selected stress system, a new specimen is set up for every test. The specimens belonging to one test series should be as near identical as possible. This requirement is difficult to fulfil with most residual soils because of their great variability. In the multi-stage triaxial test, the shear strength parameters, c' and φ', are determined by testing a single specimen. When failure has started to develop in the first stage, the deviator stress $(\sigma_1 - \sigma_3)$ is released and the lateral stress σ_3 is increased to a new value. After re-consolidation, the deviator stress is again increased until failure recommences at the new effective confining pressure. This procedure is repeated until sufficient failure points (usually three) have been obtained or until the deformation of the specimen is too large for reliable results to be expected (Figure 7.21). Since only one specimen has to be set up, the multi-stage test can save testing time and thus may also be economically attractive. (However it is doubtful if small savings in the cost of testing at the expense of the quality of the test results can be regarded as justifiable).

Lumb (1964) applied the multi-stage test to drained tests on undisturbed unsaturated residual soils from Hong Kong. The results for c' and φ' were practically indistinguishable from the single-stage tests. He found that the maximum axial strain necessary to obtain at least three failure points was not a serious problem with undisturbed specimens. The largest total strain did not exceed 23%. However, if the soil contains relict bonds such as those illustrated in Figure 7.11, multi-stage testing should not be used as the effect of failure in stage 1 on subsequent stages is not known and cannot be determined.

Ho & Fredlund (1982b) developed a multi-stage testing procedure for unsaturated soils to measure the increase in shear strength due to soil suction (i.e., φ^b). The quantities c', φ' and φ^b can be obtained from a graphical procedure. Their procedure is set out in Figure 7.22. $(\sigma_3 - u_a)$ is kept constant for all 3 stages, as is u_w, but $(u_a - u_w)$ is increased by increasing u_a.

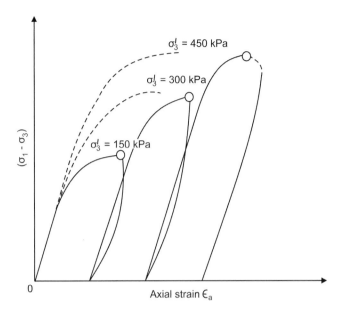

Figure 7.21 Principle of multi-stage testing: Stress-strain curves for a multi-stage test.

Controlled strain or controlled stress test

The most convenient way to shear a specimen is by applying a constant rate of axial strain. The controlled-stress test is needed to simulate certain types of field loading. Lundgren *et al.* (1968) has discussed the advantages and disadvantages of stress-controlled shearing:

Advantages:

- Load increments may be selected in both magnitude and duration, so that complete pore pressure equalization is obtained.
- The deformation versus time relationship may be observed during each load increment.
- For structurally sensitive soils an indication of the yield stress may be obtained.

Disadvantages:

- Failure may be abrupt and result in a complete collapse of the specimen. Determination of the ultimate strength is difficult and determining residual strength is not easily possible.
- In drained tests, application of the failure load increment (the increment resulting in failure) will usually cause failure under undrained or only partially drained conditions.
- In undrained tests, the pore pressure induced by the failure load increment cannot be measured accurately.

Stage	σ_3	u_a	u_w	σ_1(peak)	σ_3-u_a	σ_1-u_a	$\dfrac{\sigma_1-\sigma_3}{2}$	u_a-u_w	$\dfrac{\sigma_1+\sigma_3}{2}-u_a$
I	241.3	103.4	68.9	725.4	137.9	622.0	242.1	34.5	380.0
II	344.8	206.9	68.9	952.3	137.9	745.4	303.8	136.0	441.7
III	448.2	310.3	68.9	1143.0	137.9	832.7	347.4	241.4	485.3

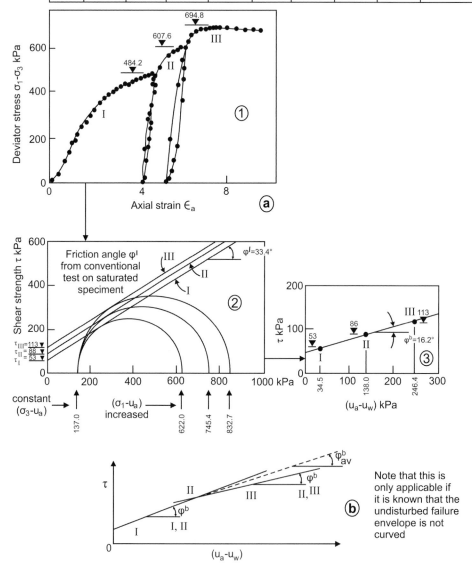

Figure 7.22 (a) Determination of the strength parameters of an unsaturated specimen by Fredlund's method and multi-stage testing. (b) Determination of average φ^b when strength in stage III is reduced due to disturbance.

Measurement of pore water pressure during shearing

Undrained tests can be conducted with or without measuring pore pressures. If only the value of the undrained strength is of interest the measurement of pore pressure is not required. Pore pressures, on the other hand, must be measured if the strength parameters in terms of effective stress as well as the contractivity/dilativity behaviour are of importance in treating the engineering problem for which the test is conducted.

In the case of unsaturated soils, with both air and water present in the voids, the measurement of pore water pressure requires modification of the test equipment and additional considerations due to the effects of surface tension or capillarity. In clayey soils, especially when compacted, the difference between u_a and u_w may amount to several hundreds of kPa (Bishop & Blight, 1963).

The method used to measure pore water pressure consists of bringing a rigid fine-pored filter (normally of ceramic) saturated with water in contact with the soil specimen (see Figure 7.17). The pore water pressure in the specimen is then defined as the pressure required to prevent flow of water through the porous filter. In an unsaturated soil the filter must have a sufficiently high air entry pressure to prevent air from the specimen displacing water from the pores in the filter. The air entry pressure is defined as the air pressure required to displace water from the pores of a saturated porous element. If the air entry pressure of the porous stone at the base of the specimen is too low, air will enter the porous stone and water will be drawn from it into the specimen. The pressure measured on the remote side of the porous element by the pressure sensing device will then be the pore air pressure and not the pore water pressure. The recommended minimum air entry pressure of porous ceramics for measuring pore water pressures in triaxial tests is 500 kPa.

Cell and consolidation pressures to be applied

Cell pressures in CU and CD tests are usually in the range from 0 to about 1200 kPa. With a saturating back pressure in the order of 300 to 500 kPa, this results in maximum effective consolidation pressures of 700 to 900 kPa. At high effective stresses the angle of shearing resistance is almost insensitive to saturation even though the soil specimen was originally unsaturated. Also effects of disturbance are minimized. On the other hand, high stresses in combination with back pressuring may lead to a destruction of weak bonds and/or particle crushing which in turn decreases the angle of shearing resistance. At low stresses, the strength envelope is often curved and can only be defined by using consolidation stresses of 10 kPa or less (see Figure 7.11). Such low pressures may be of engineering importance, for example in the analysis of shallow slope failures, where the overburden stress may only be of the order of 10 to 20 kPa.

The consolidation stresses applied in a test series are usually equally spaced (e.g., 20, 40, 60 and 80 kPa). Four tests or more are required to define the strength envelope adequately.

Rate of strain

In undrained tests without pore pressure measurement, the undrained shear strength, c_{uu} decreases with increasing time to failure as a result of soil creep (Casagrande & Wilson, 1951). For clays, the decrease may be 5% for a 10 fold increase in the time

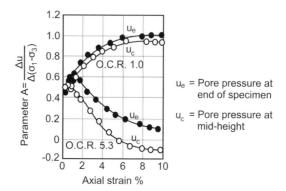

Figure 7.23 Typical pore pressure difference between ends and centre of triaxial shear specimens during rapid shearing (Blight, 1963b).

to failure. A commonly used testing rate for routine UU tests is 1 to 2 per cent axial strain per minute.

For undrained tests (UU and CU) with pore pressure measurement, the rate of deformation must be slow enough for non-uniformities in the pore pressure distribution to equalize and in drained tests time must be allowed for complete drainage to occur. This results in similar times to failure in both drained and undrained tests.

In undrained tests the non-uniformity in pore pressure in the test specimen results from the non-uniformity in stress and strain due to end restraint (Blight, 1963b). The pore pressure at the ends is usually higher than in the centre of the specimen. Thus, when measuring the pore pressure at the base of the specimen and using a rate that is too fast for equalization, the recorded value will be too high. This has an effect on the position of the failure envelope; the apparent cohesion intercept being too large (Bishop *et al.*, 1960). The time for equalization depends on the size of the specimen and the coefficient of consolidation and the drainage conditions.

Figure 7.23 shows pore pressure changes with axial strain recorded at the ends (u_e) and centre (u_c) of two triaxial compression specimens strained at a rapid rate. In both the normally and overconsolidated specimens the pore pressure recorded at the end of the specimen was greater than that recorded at the centre, and hence the effective stress would have apparently been less than the actual average value.

Figure 7.24 shows experimental equalization curves constructed from tests on two clays which were run at a range of rates of strain. The "theoretical curves" are those calculated from the theory of Gibson & Henkel (1954). The experimental curves indicate that actual equalization of pore pressure takes place more rapidly than the theory suggests. This is despite the fact that the theoretical analysis assumes that the drains (and in particular, paper side drains) are infinitely permeable which they are not.

It is apparent from Figure 7.23 that at a particular degree of pore pressure equalization the error in the value of σ'_3 due to unequalized pore pressure will depend on the stress history of the soil. With heavily over-consolidated soils an appreciable error in the value of σ'_3 may occur even though the degree of equalization appears satisfactory. With normally consolidated soils, on the other hand, errors in σ'_3 are likely to be small,

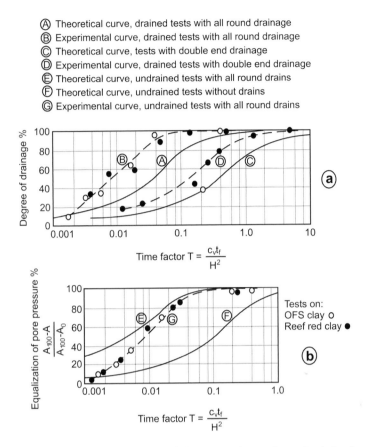

Ⓐ Theoretical curve, drained tests with all round drainage
Ⓑ Experimental curve, drained tests with all round drainage
Ⓒ Theoretical curve, tests with double end drainage
Ⓓ Experimental curve, drained tests with double end drainage
Ⓔ Theoretical curve, undrained tests with all round drains
Ⓕ Theoretical curve, undrained tests without drains
Ⓖ Experimental curve, undrained tests with all round drains

Figure 7.24 (a) The relation between degree of drainage and time factor in drained tests. (b) The relation between equalization of pore pressure and time factor in undrained tests (Blight, 1963b).

even with testing rates which are too rapid for proper equalization of pore pressure. In deciding on a test duration for an undrained test, the choice of a desirable degree of equalization must be considered in relation to the error in σ_3' likely to occur at this degree of equalization. Lower degrees of equalization can be tolerated with soils that are normally or lightly overconsolidated. Higher degrees of equalization should be aimed at in tests on heavily overconsolidated soils (and soils which fail by shearing along a narrow zone).

For tests without drains:

$$t_f = \frac{1.6H^2}{c_v} \tag{6.8a}$$

where 1.6 is the time factor corresponding to 95% equalization in tests without drains and H is half the specimen height.

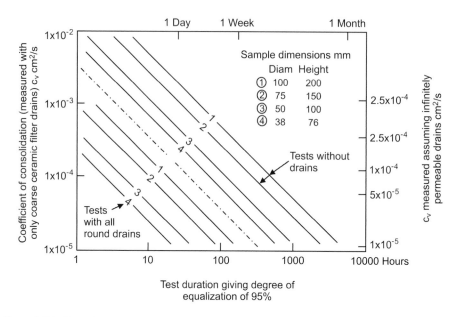

Figure 7.25 Chart for finding test duration giving 95% pore pressure equalization (Blight, 1963b).

Similarly, for test with all-round drains:

$$t_f = \frac{0.07H^2}{c_v} \tag{6.8b}$$

The meaning of the term "test duration" will depend on the object of the test. If the object is to measure the peak shear strength parameters of a soil only, the duration of the test may be taken as the time to failure. If complete and accurate information on the whole stress path is required, the duration will be the period between the start of the test and the first significant stress and pore pressure measurement.

In Figure 7.25 equations 6.8a and 6.8b have been presented in the form of a chart which enables a test duration giving a degree of equalization of 95% to be read off knowing c_v, the coefficient of consolidation of the soil (Blight, 1963b).

Triaxial testing of stiff fissured clays

Clays formed by the *in situ* weathering of shales, lavas and dolerites usually contain relict joints that are commonly inclined at angles of between 45° and 60° to the horizontal. Because of the unfavourable orientation of the joints and fissures, failure in triaxial tests on these soils often occurs by sliding along a single inclined plane. Failure planes become visible at an axial strain of 1 to 3% and deformation usually appears to be confined to this surface with the two sections of the specimen above and below the failure surface acting as relatively rigid blocks. The axial stress on the specimen seldom falls after the appearance of the failure plane even though the contact area across the plane is continually decreasing.

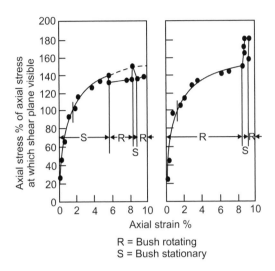

Figure 7.26 Stress-strain curves for triaxial tests on specimens of a typical indurated fissured clay.

Figure 7.26 shows the stress-strain curves for two typical tests on specimens of a stiff fissured clay. The specimens were enclosed by a single latex rubber membrane with a thickness of 0.15 mm and were tested in undrained compression at a strain rate of 0.05% per hour. The effective confining stress in these tests was 105 kPa. The loading rams of the triaxial cells were guided by lubricated rotating bushes so that the effects of ram friction could be assessed.

In Figure 7.26 the axial stress at which the shear planes first became visible has been allocated a relative value of 100%. The axial stresses increased considerably after the formation of the failure planes whether the ram bushes were rotating (R) or stationary (S).

It appears to be generally accepted (e.g., Andresen & Simons (1960), Warlam (1960), Bishop et al. (1965)) that the use of a rotating bush virtually eliminates ram friction. If this is the case, it will be seen from the difference between the stress-strain curves with the bushes stationary and rotating that even at relatively large axial strains of 5 to 10% ram friction accounted for only 5 to 20% of the measured deviator stress. The remaining increase in deviator stress after the shear planes had formed can be ascribed to restraint developed by tension in the rubber membrane pulling against the ends of the sliding blocks of soil.

Bishop & Henkel (1962) considered that the membrane correction would not exceed 14 kPa at axial strains of 4 to 5%. However, an analysis of the mechanics of the problem shows that, if the membrane does not slip over the surface of the specimen, local strains in the rubber may be as much as 30 times the nominal axial strain of the soil specimen. This together with the evidence of Figure 7.26 indicates that membrane corrections for stiff fissured clay specimens might far exceed 14 kPa.

An investigation into membrane restraint was carried out with a 38 mm diameter by 76 mm high rigid dummy specimen made of perspex. The specimen (see sketch in

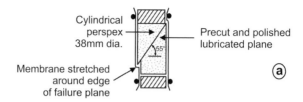

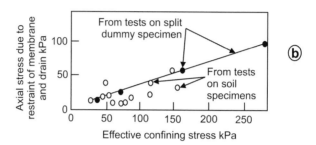

Figure 7.27 Membrane and drain restraint in triaxial tests on specimens failing on a single plane: (a) Dummy specimen used in investigation. (b) Measured restraint in tests on real and dummy specimens.

Figure 7.27a) was cut along a diagonal plane inclined at 55° to the ends of the cylinder. The two faces of the pre-cut "failure plane" were polished and the outer surface of the cylinder was roughened using emery cloth. After lubricating the failure surface with silicone grease, the dummy specimen was tested at a range of confining stresses using a cell with a rotating bush. The coefficient of friction along the lubricated failure plane was only 0.02 (Blight, 1963b) and hence the measured axial stress in these tests was almost entirely due to membrane restraint.

Figure 7.27 (derived from triaxial tests on the split dummy specimen as well as on indurated clay specimens) shows an experimental relation between effective confining stress and the increase in axial stress over a strain interval of 1% from the strain at which a shear plane first became visible. The increase in the resistance of the clay specimens represents the effect of membrane and paper side drain restraint and is compared with results from tests on the perspex dummy specimen. It appears from this comparison that the dummy specimen gave a fairly realistic representation of the behaviour of specimens of indurated fissured clay).

Selected measurements from the investigation are listed in Table 7.3.

The axial strain referred to in Table 7.3 is the strain that follows the first formation of a failure surface. To apply a correction for membrane restraint at any subsequent strain, this failure strain must be identified. In practice the simplest procedure is to watch the triaxial specimen closely and note the deviator stress at which a failure surface first appears. This stress is then taken as the failure stress for the specimen.

Table 7.3 Investigation into effects of membrane restraint on triaxial specimens failing on a single plane. Tests on a perspex dummy specimen.

Normal axial strain per cent	Confining stress kPa	Membrane and drain restraint with specimen enclosed by	
		I membrane 0.15 mm thick kPa	I membrane + I wet filter paper drain kPa
I	70	18	27
I	280	53	91
5	70	35	56
5	280	102	130

Examples of triaxial test results

Figures 7.28 to 7.30 are examples of triaxial shear test results.

Figure 7.28 shows typical consolidated undrained (CU) triaxial test results for a residual andesite lava. Figure 7.28a shows typical stress paths and a failure envelope for the soil with, inset, values of the pore pressure parameter A at failure. Figure 7.28b shows failure stress points for specimens of the same residual lava taken from 5 different sites within a distance of about 10 km of each other. This illustrates the inherent variability of residual soils, all of which have the same origin, but which arise from a number of different lava flows and in which the degree of weathering and age also vary (Blight, 1996).

Figure 7.29 (Bishop & Blight, 1963) shows a set of pore pressure measurements made in consolidated undrained triaxial compression tests on specimens of unsaturated soil at increasing values of $(\sigma_3 - u_a)$. The pore air (u_a) and water (u_w) pressures have been measured separately in a triaxial cell equipped like the one illustrated in Figure 7.17. It will be noted that as $(\sigma_3 - u_a)$ is increased, the suction $(u_a - u_w)$ decreases. Also that in each test $(u_a - u_w)$ decreased initially, as the soil tended to compress and then increased as the soil started to dilate. In the test at a value for $(\sigma_3 - u_a)$ of 204 kPa, the soil became saturated by compression in the initial stages of the test and then desaturated as the soil dilated at larger strains, even though the dilation was not shown by an increase in volumetric strain.

Figure 7.30 (Bishop & Blight, 1963) shows two sets of experimental results like those shown diagrammatically in Figure 7.3a. On each of the diagrams, the degree of equalization of pore pressure established by comparing simultaneous measurements of u_a at the base and mid-height of the specimen are recorded. Note that the position of the strength line for saturated soil lies between the characteristic lines in terms of u_a and u_w for the unsaturated soil.

7.2.4 Determination of K_0 from the triaxial test

The coefficient of earth pressure at rest, K_0, is defined by

$$K_0 = \frac{\sigma_h'}{\sigma_v'} \quad \text{for } \varepsilon_h = 0 \tag{6.9b}$$

where the subscripts v and h refer to the vertical and horizontal directions, respectively.

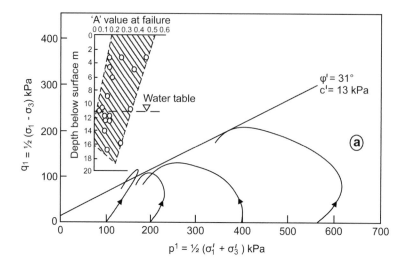

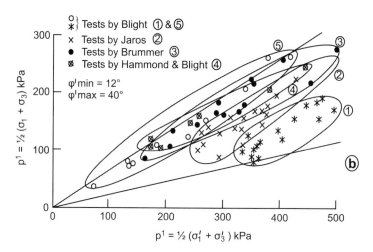

Figure 7.28 (a) Typical stress paths for consolidated undrained triaxial shear of residual andesite lava. (b) Summary of triaxial shear test results on undisturbed weathered andesite lava from five different sites in the Johannesburg graben. Tests were unconsolidated undrained (with measured pore pressures) on 75 mm dia. specimens.

The usual object of K_0 testing is to evaluate K_0 for a range of values of σ'_v and for both normally consolidated and overconsolidated conditions. Two basic ways have been used to evaluate K_0 directly in the triaxial test:

(i)   For zero lateral yield to occur, the change in volume of a specimen subjected to vertical compression must equal the horizontal cross-sectional area, A, multiplied by the vertical compression δ_v, i.e., $\Delta v = A \delta_v$, $\varepsilon_h = 0$ (see inset on Figure 7.31).

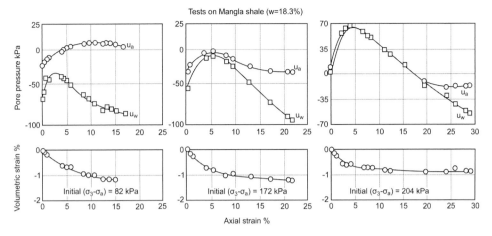

Figure 7.29 Pore pressure changes during undrained shear of a partly saturated compacted soil residual from weathered shale.

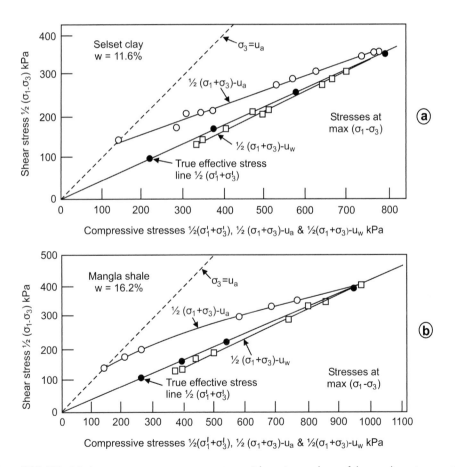

Figure 7.30 Triaxial shear at constant water content with various values of $(\sigma_3 - u_a)$ on two partly saturated soils. Degree of pore pressure equalization: (a) Selset clay 95%. (b) Mangla shale 90%.

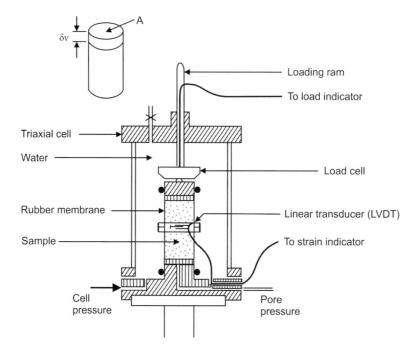

Figure 7.31 Triaxial test with a linear transducer (LVDT) to record or control radial strain.

Starting with a fully consolidated specimen, the vertical stress σ'_v is increased at a rate slow enough to maintain full consolidation. δ_v and Δv are observed and σ'_h is adjusted continuously to maintain the condition $\Delta v = A\delta_v$.

This is a very simple technique that requires no additional apparatus, and appears to give reasonably accurate results.

(ii)　The horizontal strain ε_h may be monitored directly and kept to zero by continually adjusting σ'_h as σ'_v is varied. Various devices have been used to monitor horizontal strains. The most commonly known are the Bishop lateral strain indicator (Bishop & Henkel, 1962), the strain-gauged lateral strain indicator made of a brass half-loop, or the version of the Bishop lateral strain indicator fitted with a LVDT (linear voltage differential transformer) indicated in Figure 7.31.

Figure 7.32 shows a set of K_0 test results obtained for a soil residual from a smectitic mud-rock. (See also Figures 7.15 and 7.16). Figure 7.32a shows the measured relationship between the applied σ'_v and the corresponding σ'_h adjusted to maintain zero lateral strain. Note that the relationship is reasonably linear for loading and that K_0 has a value of about 0.63. As the soil is unloaded, K_0 increases progressively, reaching a value of 1 when σ'_v reaches 400 kPa and 3 when $\sigma'_v = 50$ kPa.

Figure 7.32b shows the corresponding void ratio – σ'_v relationship. Note that during initial consolidation, the specimen swelled from a void ratio of 1.39 to almost 1.47 (6%). Hence the soil was both desiccated and overconsolidated at the start of the test.

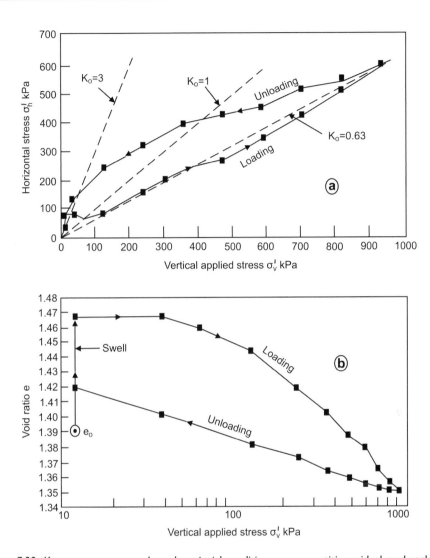

Figure 7.32 K_0 measurements made under triaxial conditions on a smectitic residual mud-rock clay.

K_0 can also be measured indirectly, as indicated by equation 6.1 and illustrated for a residual andesite lava profile by Figure 6.4.

7.3 FIELD STRENGTH TESTING

7.3.1 General

Field tests are advantageous for residual soils for the following reasons:

- The disturbance caused by sampling, stress release, transportation, storage, etc. is mostly eliminated.

Table 7.4 In situ strength tests.

Test	Remarks and limitations
Field direct shear test	Usually on the surface exposed in shallow pits, time-consuming and expensive. Rational interpretation
Field vane shear test	Unsuitable for soils containing large particles. Rational interpretation
Borehole shear test	Limited area of contact, multistage test, only for shallow depths. Rational interpretation
Standard penetration test	Mainly used for granular soils and stiff clays, requires pre-drilled hole. Empirical interpretation
Cone penetration test	For soft/loose to medium stiff/dense, predominantly fine-grained soils. Semi-empirical interpretation
*Pressuremeter (prebored, or self-bored)	For all types of soils and soft rock; requires a high quality borehole. Difficult to use in stoney soils. Semi-empirical interpretation
*Plate bearing test or screw plate bearing test	Usually close to the ground surface, used mainly for settlement evaluation rather than bearing capacity, expensive. Semi-empirical interpretation

*See sections 6.4.1 to 6.4.4 for details.

- The test specimen size can be increased and becomes more representative of the soil mass. There are many tests available to measure strength *in situ*, either directly, or by an indirect measure, e.g., of penetration resistance. Only few of these have found wide-spread use in geotechnical and site investigation practice. Table 7.4 lists some of these tests and their limitations.
- Most residual soils behave as if overconsolidated. Pore pressure changes induced by shearing are therefore small and have little effect on the shear strength. Their effect can thus usually be neglected. (See, e.g., Figure 7.1.)

When planning a site investigation program that involves the determination of shear strength parameters, one usually has to use the locally available testing tools. These have to be used in an optimal way within the limits of the allocated budget and time, and in combination with laboratory testing. Advantages and disadvantages of the available methods must be assessed, keeping in mind which parameters are actually needed for the design and how the reliability of these parameters may actually influence the design. It is therefore essential that the geotechnical engineer have a sound understanding of the various field testing methods, their capabilities and limitations, the test variables, and of the factors which influence the test, when specifying an investigation program and later when analysing the results.

In this section, the procedures for the field direct shear, the field vane shear, the borehole shear, the Standard Penetration Test (SPT) and the Cone Penetrometer Test (CPT) will be described. Pressuremeter and plate bearing tests are mainly used to measure stress-deformation properties of the *in situ* soil. Only when they are carried to failure (which may be difficult with many residual soils) can the results be used to calculate strength parameters. These tests are therefore also discussed in Chapter 6 in terms of compressibility and settlement. The SPT and CPT are indirect tests, i.e., they do not measure the strength directly.

7.3.2 Field direct shear test

In situ direct shear tests are not frequently employed, because of their relatively high cost. Most applications reported in the literature concern rock materials, because these are usually heterogeneous and stratified and require larger test specimens to produce meaningful results. This is also often true for residual soils for which discontinuities and relict joints have an important influence (see Plate C19). The field direct shear test is particularly suitable for simulating the stress conditions that exist on a potential failure plane in a slope. It also enables shearing under the relatively low normal stress that occurs with shallow failure surfaces. Hence, field direct shear tests in residual soils are mainly employed in connection with important slope stability problems.

The main purpose of the test is to obtain the values of peak and residual strength for either the intact material or for discontinuity surfaces, including relict joints. The test is generally carried out at the bottom of shallow trenches or pits and less commonly in shafts.

Most of the tests are set up so that the shear plane will be horizontal. Ideally, the shear plane should be parallel with major discontinuity sets (e.g., relict joints or slickensides) or coincide with a single major discontinuity (e.g., the exposed shear surface of a slip).

The size of the specimen should be at least ten times the largest particle size. Typical specimen sizes are 300×300 mm and 500×500 mm for soils and weak rock.

Excavation of the test pit and of the soil pedestal (test specimen) must be done with utmost care to avoid disturbance of discontinuities in the specimen. For trimming the specimen, handsawing and cutting should be used. Once the test pedestal has been shaped, it must be protected using plastic sheets to minimize changes of water content. The final trimming must be done with a minimum of delay to avoid changes in water content. Special precautions are required for tests below the water table to avoid the effects of water pressure and seepage. If it is intended to shear the specimen along a specified discontinuity, the spatial orientation of this discontinuity should be carefully identified in terms of strike direction and dip before starting to trim the specimen.

The equipment for applying the normal load consists of weights, kentledge, hydraulic jacks, flat jacks acting against the roof of a tunnel or adit or against an anchor system. It is important that the reaction system ensures uniform transfer of the normal load to the sample and minimum resistance to shear displacements. During the test the alignment of the normal force must be maintained, as the shear displacement increases.

The system for applying the shear force must provide a uniform load over the plane of shearing. Reactions are often provided by the excavation side walls or the blade of a bulldozer. In certain cases, a bulldozer blade can also be used to apply the shearing load. It is important to allow for sufficient travel in the shear force application and deformation measurement systems so that the test can be carried out without a need to reset the deformation gauges. LVDTs are better than dial gauges because they are available with longer travels and can be logged electronically, remote from the test pit. Applied loads should be measured to an accuracy of $\pm 2\%$ of the expected maximum value. Movement reference points must be well fixed and sufficiently remote from the test to ensure they are not influenced by the test forces. The effects of temperature variations on displacement should be evaluated by measuring temperatures within

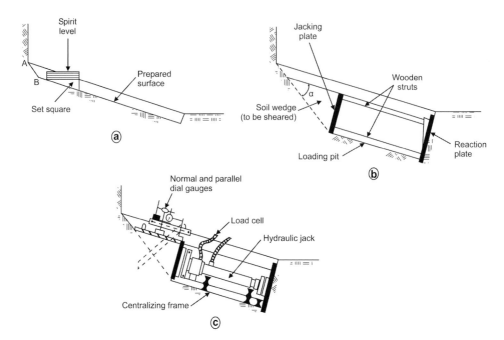

Figure 7.33 Stages in performing an *in situ* soil wedge test: (a) Preparation of tests area (in test pit). (b) Cutting sides of soil wedge. (c) Loading (Mirata, 1974).

the test set-up and correcting where necessary. Thermocouples can easily be used to monitor temperatures at several points of the test set-up.

Examples of in situ direct shear tests

Field direct shear tests carried out on residual soils or soft rocks have been described by James (1969), Mirata (1974), Brenner *et al.* (1978), Brand *et al.* (1983) and Brand (1988), and others with various shearing devices.

James (1969) reported on tests performed on sub-horizontal mudstone bands at a dam site. The test blocks were 0.6 m square. The nominal load was applied through an anchor system with the anchors taken to a depth of about 3 to 3.5 m which allowed a normal pressure of 600 to 800 kPa. Horizontal (shear) loads were applied by jacking against a concrete block built on one side of the excavation. The shearing rate was 0.5 mm/min. The test was carried out in stages. After shearing under the first normal load, the load was released when failure became imminent. Then the next normal load was applied and time allocated for consolidation. After completion of three stages, the block was restored to its original position under a nominal vertical load. Shearing was then repeated under a nominal vertical load until the residual shear strength was reached.

Mirata (1974) introduced a shear test he called the "*in situ* wedge shear test". A wedge of soil is sheared along its base by means of a single hydraulic jack (Figure 7.33). The test has been applied in unsaturated stiff fissured clays for the solution of slope

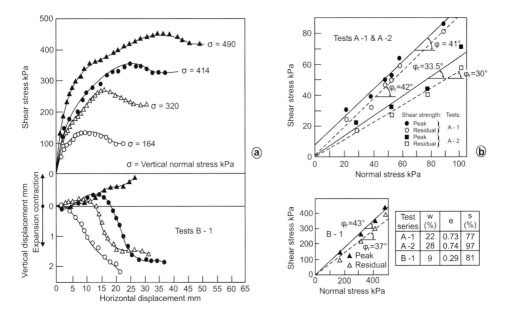

Figure 7.34 Results of field direct shear tests on weathered granite using the Brenner *et al.* (1978) apparatus: (a) Shear stress and vertical displacement. (b) Strength envelopes (peak and residual).

stability problems. Its principle is to alter the inclination of the failure plane with respect to the direction of loading. In this way the ratio of shear strength to normal stress can be varied over approximately the same range of normal stress as encountered in slope stability problems. The undrained strength parameters thus obtained are used in conjunction with a total stress type of stability analysis. The test procedure is also applicable to saturated soils above the water table.

The field shear test procedure developed by Brenner *et al.* (1978) was designed for use in residual soil slopes. The equipment is highly portable and consists of a 305 × 305 mm and 150 mm deep shear box which derives its normal and shear reaction from a light steel frame loaded with sand bags. The shear force is applied by means of a hand-driven screw jack and proving ring assembly, which was constructed by modifying a field CBR apparatus. For the normal stress, up to 10 kPa, a remote controlled hydraulic jack was employed which could be aligned by means of ball and socket bearings. The normal stress was measured by means of a load cell.

Typical test results obtained with this equipment are shown in Figure 7.34. The tests were carried out at two locations cut in a slope of residual granite soil with different grades of weathering. Location A was in a clayey sand (completely weathered granite), while location B was in a gravelly silty sand (highly weathered granite). At location A two test series were carried out, i.e., one at field water content (A-1) and the other soaked (A-2). Soaking was accomplished by lining the test pit with plastic sheeting and submerging the specimen block (which was ready to be sheared) in water for about 12 hours. At location B tests (B-1) were conducted at field water content.

A field shear box described by Brand *et al.* (1983) and Brand (1988) was developed by the Geotechnical Control Office (GCO), Hong Kong, for use in residual soils derived from granite and volcanic rocks. Each component can be carried by one person and the apparatus is assembled at the test site.

Typical test results obtained with the GCO shear box are shown in Figures 7.35 and 7.36. These tests were carried out on completely weathered granite. One series of four tests was carried out under natural water content conditions and the other under soaked conditions. It can be seen that peak shearing resistance is reached at very low shear displacements (1 to 2 mm). Note the very high ratios of τ/σ which are indicative of a cohesive material, as confirmed by Figure 7.36. The displacement to the right of the soaked strength envelope relative to the unsoaked shows that the suction at natural water content amounted to 13 kPa, or an increase of apparent cohesion of 9 kPa.

7.3.3 Vane shear test

The test is usually used only in fine grained soils which are soft enough to permit penetration and rotation of the vane blades. Hence, the usual range of application comprises low strength clayey soils, i.e., soft to medium clays which are free of stones and pedogenic nodules. However, modified vanes have been used successfully in residual soils with peak strengths up to 300 kPa (see Figure 7.5). Sowers (1985) stated that the vane shear test is not representative of the controlling weakness of the mass. This is well illustrated by the results shown in Figure 7.5 where the undisturbed vane shear strengths far exceeded the strength in mass of the soil, represented by the strength back-calculated from a slide failure. However, as Figure 7.5 also shows, the remoulded vane shear strength approaches the strength of the soil in mass. Possible applications are to clays and silts originating from deeply weathered lava, mud rocks and shales which do not contain gravelly particles (Blight, 1985).

Principle of the vane test

The vane usually consists of four thin rectangular spring steel plates attached by brazing at right angles to a torque rod in a cruciform. The vane is pushed vertically into the soil. A torque is applied to the rod by means of a torque measuring head or torque wrench, causing the blades to rotate and thus producing a shear failure along a cylindrical surface. The shear strength is calculated from the maximum measured torque required to shear the clay along the cylindrical surface.

The vane shape commonly regarded as "standard" has blades with a height to width ratio, $H/D = 2$. The actual values of H and D depend on the strength of the soil to be tested. For example, a vane with $H = 100$ mm and $D = 50$ mm can be used for strengths between about 50 and 70 kPa. According to Andresen (1981), this is the smallest size suitable for accurate determination of the shear strength of soft clays. However, smaller vanes (as small as $H = 38$ mm) have been used successfully in stiff residual soils (e.g., Blight *et al.*, 1970).

Vane tests may be carried out at the bottom of a pre-bored hole or an excavation or by pushing the vane into the ground from the surface to the required depth. The vane can also be pushed into the side of an excavation. Pushing a vane in from the ground surface is rarely possible in residual soils as their strength is usually far too

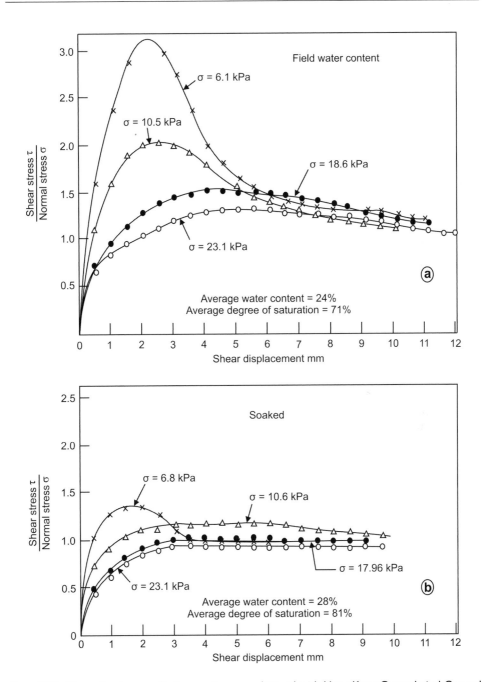

Figure 7.35 Normalized stress-displacement curves obtained with Hong Kong Geotechnical Control Office direct shear machine on residual granite: (a) At natural water content. (b) Under soaked conditions (Brand et al., 1983).

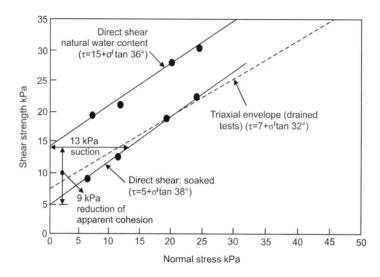

Figure 7.36 Strength envelopes for direct shear tests shown in Figure 7.35 compared with strength envelope obtained from drained triaxial tests on the same material. Note the drop in cohesion caused by soaking and the closeness of the envelopes obtained for the soaked material and with the drained triaxial test (Brand *et al.*, 1983).

high. Figure 7.37 is a diagram of the usual vane shear arrangement for use in residual soils.

Figure 7.5 showed the results of vane shear tests in a residual shale profile, while Figure 7.38 shows a vane strength profile measured in a residual andesite lava. Note that in this profile, as with that referred to in Figure 7.5, the remoulded vane strength agrees reasonably well with the strength envelope established by means of laboratory triaxial tests, whereas the measured undisturbed strength is too large. This illustrates the effects of soil discontinuities, with the undisturbed vane strength representing the strength of the intact soil and the remoulded strength representing that of a fissured material.

Effect of vane insertion

Drilling a hole for the vane test causes disturbance by stress release below the bottom of the hole. Tests at the Norwegian Geotechnical Institute in soft transported clays indicated that a vane should be advanced below the bottom of the hole to at least six times the borehole diameter (Andresen, 1981), e.g., 1 m in the case of a 150 mm diameter hole. However, in stiffer materials, the depth of influence of the hole is likely to be a lot less, and a distance of 0.5 m is commonly used.

Mode of failure

Finite element analyses have indicated that the shear stress distribution on the vertical sides of the cylindrical failure surface are reasonably close to the conventional

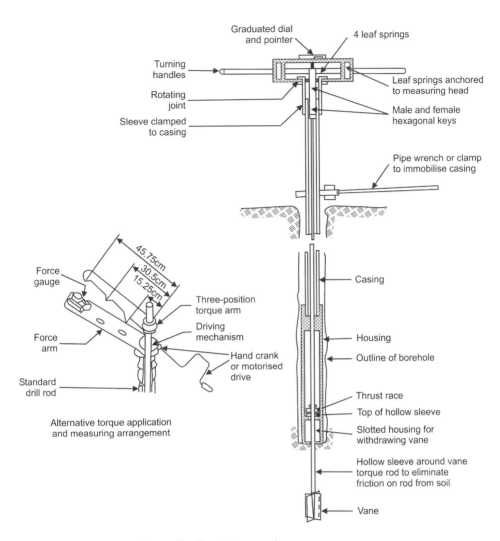

Figure 7.37 The *in situ* vane shear test apparatus.

assumption of being uniform (Wroth, 1984). On the horizontal surfaces of the failure cylinder, the shear stress distribution is highly non-uniform with high peaks at the edges. However, with $H/D = 2$ the major contribution to the measured shear torque arises from the cylindrical surface (86%) with the contribution by the ends being only 14%. The failure mode is most likely that of simple shear up to peak shear stress, whereas the direct shear mode develops in the post-peak phase when the cylindrical failure surface is formed (Chandler, 1988). The shape of the failure surface in stiff residual soils has been checked by sampling the failure zone by means of a thin-walled tube sampler and exposing the failure zone. The failure surface is indeed a right cylinder and very little disturbance is visible in the soil forming the outside of the surface.

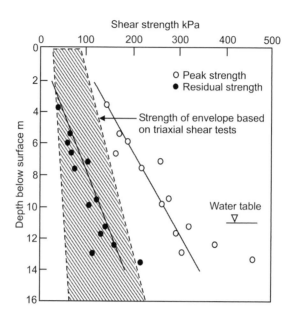

Figure 7.38 In situ strength profile for residual andesite lava.

Shearing under undrained conditions

In order to enable a definitive interpretation of vane test results, shearing must take place either under completely undrained or completely drained conditions. Undrained conditions can be assured if, for practical purposes, the average degree of consolidation in the failure zone U is less than 10%. The corresponding time factor has been established by Blight (1968) based on an approximate theory supported by experimental data. It takes the form of an experimental drainage versus time factor curve in which the time factor

$$T = \frac{c_v t_f}{D^2} \tag{7.7}$$

is related to U (see Figure 7.39).
where c_v is the coefficient of consolidation,
   t_f the time to failure,
   D the vane diameter.

For a commonly used vane size of D = 65 mm, t_f is typically about 1 minute. From the U vs T drainage curve, T is less than 0.05 when U < 10%. Hence, c_v must be less than about 110 m²/y or 3.5 × 10⁻² cm²/s for an effectively undrained test. This condition is usually satisfied with all clayey soils, unless the vane blade position coincided with a sand lens. Obviously, very much larger times to failure are required if a drained shear strength is to be measured. In this case, times to failure of several days may be required. This type of very slow vane test has been carried out by Williams (1980) who used

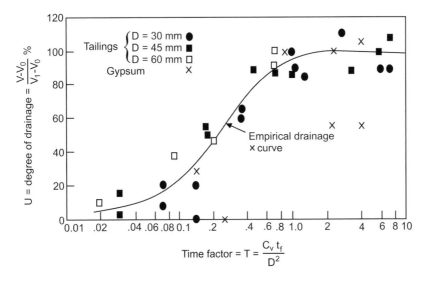

Figure 7.39 Empirical drainage curve for the vane shear test.

a motorized slow drive to rotate the vane at a constant rate over a period of days. In silty soils, it is possible to measure a drained shear strength within a reasonable shearing time. U will exceed 90% if $T = 0.8$. Hence, if $c_v = 300 \, m^2/y$, t_f must be at least 6 minutes.

Vane size and shape

The vane can be of any size, but usually a diameter D of 50 mm is used as a minimum. To minimize disturbance effects on insertion, the area ratio should not exceed 10 to 15%. The area ratio is defined as the ratio of the cross-sectional area of the cruciform (i.e., 2Dt with $t =$ blade thickness) to the circular plan area swept by the blades ($= \pi D^2/4$), or $8t/\pi D$.

As stated above, smaller vanes have sometimes been used in stiff clay. For example, Blight (1967a) reports on a residual fissured clay indurated by calcium and iron salts with an undisturbed vane shear strength which often exceeded 600 kPa. The vane apparatus had blades with H = 60 mm, D = 30 mm and 2 mm thick. This corresponds to an area ratio of 17%. As vane testing of very stiff soils does not appear to have been carried out previously, the effect of time to failure on vane shear strength was investigated. A series of tests was carried out at the same depth in a deposit of lime-indurated clay derived from the *in situ* weathering of a norite gabbro. The time to failure in these tests ranged from 5 seconds to 1 hour. The results, shown in Figure 7.40, indicated a very slight decrease in measured shear strength as the time to failure increased. This decrease, however, was less than the scatter in shear strength measurements at any one time to failure. As a result of this investigation a convenient time to failure of 1 minute has been adopted for tests on indurated clays.

In addition to the standard rectangular vane of ratio H/D = 2, rectangular vanes with other ratios and also triangular shaped (rhomboidal) vanes with various apex

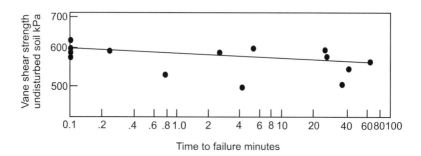

Figure 7.40 Effect of time to failure in vane shear tests on a stiff lime-indurated residual clay.

angles have been employed, mainly for studying in-situ shear strength anisotropy (Aas, 1965; Blight *et al.*, 1970; Richardson *et al.*, 1975; Silvestri & Aubertin, 1988; Silvestri *et al.*, 1993).

For calculating the shear strength ratio S_H/S_v, it is usually assumed that the shear strength is fully mobilized and uniformly distributed across the entire shear surface. The expression for the vane torque can then be written in the form:

for a rectangular vane: $T = \pi DH \cdot DS_v/2 + \pi D^2 \cdot DS_H/6$

$$= \pi D^2(HS_v/2 + DS_H/6) \tag{7.8a}$$

for a triangular vane: $T = 4S_\beta \cdot \pi L^3(\cos^2 \beta)/3 \tag{7.8b}$

where S_v and S_H are the undrained shear strengths acting on vertical and horizontal planes respectively and S_β is the shear strength on planes inclined at angles of $\pm\beta$ to the horizontal, and L is the length of the side. Figure 7.41a shows the results of a series of vane tests using vanes of various shapes to establish the strength anisotropy in a weathered mudstone (Blight *et al.*, 1970). In this example, S_H proved to be appreciably larger than S_v. This arose because the measurements were made in a sliding mass in which horizontal stresses parallel to the direction of sliding were considerably reduced by the slide. Figure 7.41b shows a similar set of measurements in a clayey sand residual from sandstone which showed no systematic difference between S_v and S_H.

If $S_v = S_H = S$, equation 7.8a reduces to:

$T = S\pi D^2(H/2 + D/6)$, or

$S = 2T/\{\pi D^2(H + D/3)\}$ \tag{7.8c}

The horizontal to vertical stress ratio K can also be estimated from shear strengths measured by means of vanes having different shapes. This can be done by writing

$S_v = KS_H,$

so that equation 7.7a becomes:

$T = \pi D^2(KHS_H/2 + DS_H/6)$, or

$K = 2T/\{\pi D^2 S_H(H + D/3)\}$ \tag{7.8d}

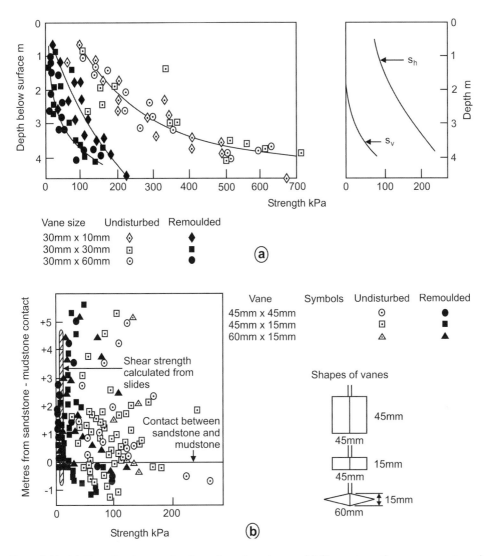

Figure 7.41 (a) Directional strength of weathered mudstone. (b) Shear strength measurements and shapes of vanes in a residual weathered sandstone.

Remoulded vane shear strength

Once the undisturbed or peak shear strength has been measured, the vane is rotated 20 to 25 times and the torque is remeasured. This represents the resistance offered by the remoulded shear strength. As mentioned earlier, and illustrated in Figures 7.5, 7.41b and 7.42, the remoulded strength represents the strength along an artificial fissure or joint in the soil. In stiff jointed or fissured saprolitic soils, the remoulded strength approximates to the strength of the soil in bulk.

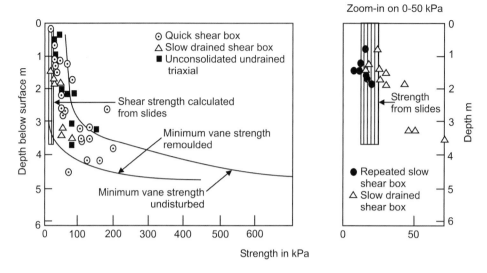

Figure 7.42 Laboratory strength measurements on weathered mudstone.

Comparison of vane shear strength of residual soils with other types of measurement

Figure 7.5 shows a comparison of vane shear strengths measured in a soil residual from weathered shale with unconsolidated undrained triaxial strengths (c_{uu}) measured on 76 mm diameter specimens and quick unconsolidated shear box tests, measured in a 76 mm diameter circular shear box. It will be seen that the small-scale laboratory tests correlate quite well with the remoulded vane shear strength. Figure 7.42 shows a similar comparison for a soil residual from a weathered mudstone (Blight *et al.*, 1970). Here again the remoulded vane strengths correlate reasonably well with similar small-scale laboratory shear tests. Note that in both Figures 7.5 and 7.42, the measured shear strengths are somewhat greater than the shear strength of the soil in mass, as back-calculated from large-scale soil shearing movements. Also, quick shear box test results do not differ very much from those of slow drained shear box tests. This is indicative of the low A value of the soil. (See section 7.1.1). The diagram to the right of Figure 7.42 shows that repeatedly sheared slow shear box strengths agree very well with the strength calculated for a slide, showing that the strength of the mudrock in mass is controlled by the strength of the fissures it contains, i.e., by its residual strength.

7.3.4 The pressuremeter test

The use of the pressuremeter test to determine soil moduli and predict settlements has been referred to in Chapter 6 (section 6.4.4). The limit pressure p_L measured in the test can be used as a measure of the shear strength of the soil. In the Menard-type test, the relationship that is used to assess the unconsolidated shear strength c_{uu} is:

$$c_{uu} = \frac{p_L - \sigma_{vo}}{N_c} \qquad (7.9a)$$

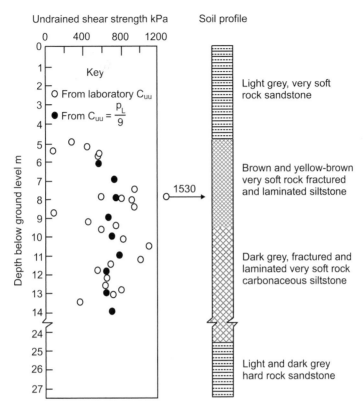

Figure 7.43 Comparison of strengths of weathered siltstone measured in triaxial and by Menard pressuremeter with $N_c = 9$.

If the pressuremeter is carried out sufficiently slowly for drained conditions to prevail, a drained strength c_D can be measured, and is given by

$$c_D = \frac{p_L - \sigma'_{vo}}{N'_c} \qquad (7.9b)$$

Pavlakis (1983) has shown that if N_c is taken as 9, a good correlation between pressuremeter results and unconsolidated undrained triaxial strengths can be obtained. His results, for a very soft rock residual weathered siltstone, are shown in Figure 7.43. Figure 7.44 shows the pressuremeter results of Figure 6.21 interpreted as shear strengths. As many of the tests did not reach a limit pressure p_L, arrows in the diagram indicate that strengths should be higher than shown. The comparison with the triaxial shear strength envelope of Figure 7.38 shows that equation 7.9 has some general validity, provided the correct value is selected for N_c.

7.3.5   Standard penetration test (SPT)

The appropriateness of the SPT for use in residual soils has been doubted. Blight (1985) stated that the test may at most give a rough index to soil strength or compressibility.

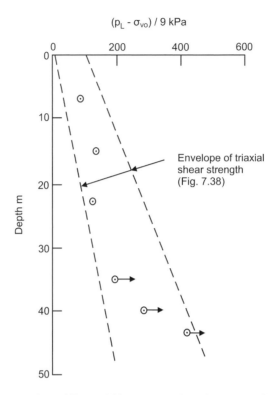

Figure 7.44 Pressuremeter data of Figure 6.21 interpreted as shear strength of a residual andesite profile.

This critique may have its justification because, even when used for transported soils, the test has a poor reproducibility and great variability. Serota & Lowther (1973) demonstrated that under laboratory controlled conditions N-values were reproducible within a standard deviation of about 15%. The poor reproducibility in field tests is, however, not only due to the variability of the soils and the testing principles, but also because of the variety of testing equipment in use, even in one country, and the lack of enforcement of equipment standards and testing procedures. The SPT will probably continue to be used as part of routine borehole investigation, regardless of its shortcomings, because it permits a rapid and economic evaluation of ground conditions in both difficult and easy situations.

Principles of the test

Present practice involves driving a standard split sample tube (or spoon) of heavy wall construction (see Figure 7.45) a distance of 450 mm into the undisturbed residual soil (or soft rock) at the bottom of a borehole. Driving is accomplished under the impact of a 63 kg hammer with a free fall of 750 mm. The blows required to drive the sample tube the first 150 mm are considered to be the seating drive, affected by disturbance at the bottom of the hole. The number of blows required to drive the sample tube the

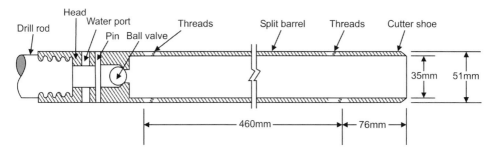

Figure 7.45 Standard Penetration Test (SPT) sampling tube.

next 300 mm is termed the SPT number or "N" value. After driving, the sample tube is withdrawn, dismantled, and the soil sample is used for identification and index tests.

Split spoon sample tube

The thick-walled split spoon has an external diameter of 51 mm and a length of 457 mm. There is a driving shoe at one end having the same diameter as the barrel, and a coupling at the other end (see Figure 7.45). The tube usually has a ball check valve to prevent sample loss. Sometimes the driving shoe contains a core retainer to prevent loss of sample.

Australian, British and South African practice also has an option to fit a solid 60° cone to replace the open shoe when probing in gravelly soil in order to minimize damage to the cutting edge of the drive shoe. This was originally proposed by Palmer & Stuart (1957). The N-values are of similar magnitude as with the shoe, or slightly higher. When applied in loose and medium dense sands having no significant gravel content, the cone may give significantly higher N-values.

The SPT was originally developed to explore the properties of cohesionless transported soils. As most residual soils are cohesive, adaptations to the interpretation of the SPT have had to be made. A useful correlation has been produced by Stroud (1974) from tests on various stiff clays and soft rocks in the United Kingdom, which relates the ratio c_u/N where c_u is the unconsolidated undrained shear strength, to the plasticity index. For plasticity indices between 35 and 65% the value of c_u/N lies between 4 and 5 kPa. The ratio appears to be essentially independent of depth and of discontinuity spacing. SPTs in clays may be relevant where stones prevent the extraction of undisturbed samples by means of thin-walled sampling tubes. Stroud (1974) also pointed out that in fissured clays the mass shear strength is only about one half of the shear strength of the intact material. The approach taken by Stroud has been found to be applicable to clayey residual soils and is quite widely used in South Africa. Stroud's correlation provides a first approximation to the mass shear strength of a residual soil and is taken as:

$$c_u = 5N \ [\text{kPa}] \tag{7.10}$$

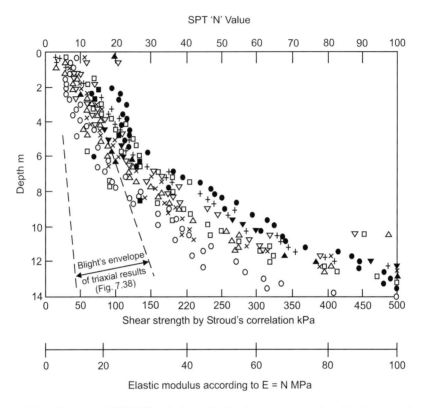

Figure 7.46a Variation of SPT "N" with depth for 8 holes in weathered residual andesite lava.

The SPT N value has also been correlated with the elastic modulus of clayey residual soils by the equation

$$E = 200c_u = 1000N \text{ [kPa]} \quad \text{or} \quad E = N \text{ [MPa]} \tag{7.11}$$

Figure 7.46a shows the variation of SPT N with depth for a number of sites on residual andesite lava. As indicated on Figure 7.46a, the correlation with triaxial strength is reasonably good at depths of up to 5 m. At greater depths than this, Stroud's expression appears to overestimate the shear strength. However, another set of data for the same weathered andesite (Pavlakis, 1985) in which the SPT tests were taken to much greater depths (50 m as opposed to 14 m) shows a much better correlation with depth (see Figure 7.46b). This is clearly because the depth of weathering on the sites represented in Figure 7.46b was greater than for the sites represented in Figure 7.46a. This illustrates one of the many pitfalls of making generalized assumptions about and applying empirical relationships to residual soil profiles.

7.3.6 Cone penetration test (CPT)

The quasi-static cone penetration test, or CPT has been applied in residual soils to a limited extent. Residual soils are often very stiff and dense and penetration may be limited

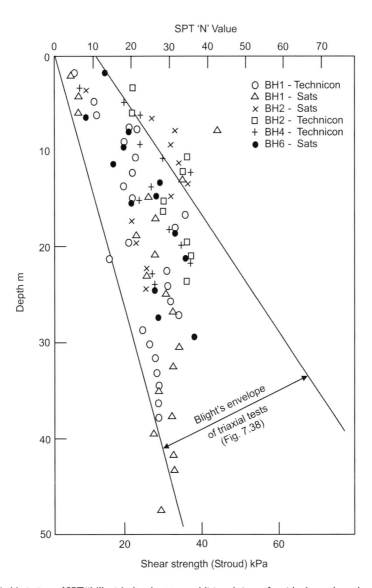

Figure 7.46b Variation of SPT "N" with depth at two additional sites of residual weathered andesite lava.

to the top few metres. Moreover, corestones and pedogenic inclusions such as lime nodules, which often occur in residual soil profiles cannot be penetrated or may deflect the cone. Still, the survey by Brand & Phillipson (1985) showed that the quasi-static CPT is fairly widely used in residual soils, mainly for shallow foundation and pile design.

Field penetrometer testing of residual soils

As the modified vane apparatus has an upper limit of shear strength measurement of 600 kPa, soils with an undisturbed vane strength exceeding this limit must be tested

Note that the friction sleeve cone extends in two increments to reach the extended position shown in b. The cone is advanced by (69-33.5)=34.5mm and the friction sleeve is then dragged down by 45mm to measure cone + sleeve resistance. System is only effective if cone resistance does not decrease by penetrating into a layer of different strength as the sleeve is extending.

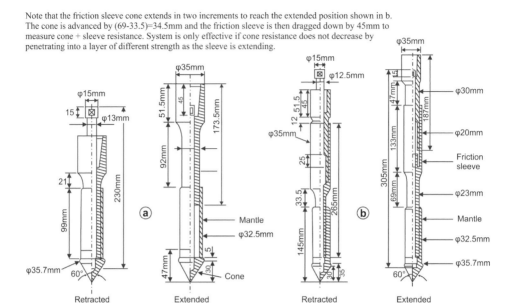

Figure 7.47 Two common types of mechanical cone: (a) Delft mantle cone. (b) Begemann friction sleeve cone.

by an alternative means. One instrument that has been adopted for this purpose is the cone penetration test. The most widely used mechanical penetrometer tips are the Delft mantle cone and the Begeman friction cone (Figure 7.47). Both have a mantle of slightly reduced diameter attached above the cone. A sliding mechanism allows a downward movement of the cone in relation to the push rods or casing. The force on the cone (penetration resistance) is then measured as the cone is pushed downward by applying a thrust on the inner rods. If the cone is equipped with a friction sleeve, a second measurement is taken when the flange engages in the friction sleeve and cone and friction sleeve are pushed down together a further depth increment.

In the electrical penetrometer tip the cone is fixed and the cone resistance and the force on the friction jacket are measured separately by means of force transducers built into the cone. Electric cables threaded through the push rods or other suitable means (e.g., solid state memory) transmit the transducer signal to a data recording system. The electrical cone penetrometer tip permits continuous recording of cone and sleeve resistance, as well as pore water pressure, measured over each push rod-length interval.

Electrical penetrometers are vastly superior to the mechanical type. They permit of measurements with higher precision and with optimum repeatability. As stated above, they allow the simultaneous recording of pore water pressure, cone resistance and sleeve friction. There is the option to install an inclinometer in the probe to check the verticality of the sounding. The string of rods may undergo considerable bending as it is pushed down, resulting in a sideways deflection of the penetrometer tip when exceeding a certain depth or when encountering an obstacle, such as a boulder or corestone.

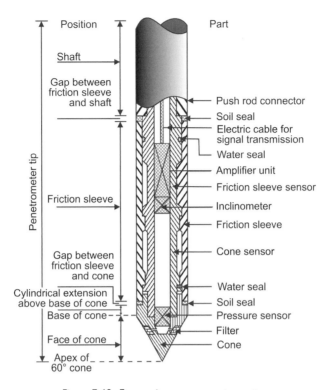

Figure 7.48 Fugro piezocone penetrometer.

The piezocone is a more recent development which incorporates a pore water pressure transducer within the tip of an electrical cone. Pore pressure transducers for this purpose should have a volume factor of less than 2.5 mm³/kPa (De Ruiter, 1982). Figure 7.48 illustrates the Fugro piezocone penetrometer. Examples of the specific use of the cone penetrometer in residual soils are given by Peuchen *et al.* (1996).

The relationship between the undrained shear strength c_u and the cone penetration resistance q_c is of the form:

$$c_u = \frac{q_c - \sigma_{vo}}{N_c} \tag{7.12}$$

where σ_{vo} is the total overburden stress at the depth of measurement, and

   N_c is a bearing capacity factor which must be evaluated empirically.

σ_{vo} is usually negligible in comparison with q and may be omitted from equation 7.12 with little error. Penetrometer measurements in stiff soils are usually made from the bottom of a 100 mm diameter augered hole. The mechanical probe is jacked in its closed position to a distance of 0.5 m below the bottom of the hole. The cone is then advanced separately until a steady penetration resistance is recorded. No study of the effects of the rate of penetration appears to have been made, but as a steady resistance is usually reached after a penetration of 25 mm a rate of penetration of

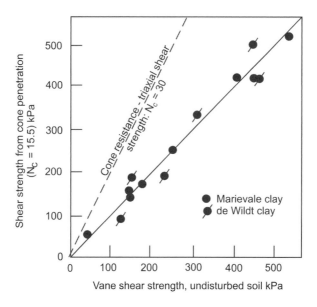

Figure 7.49 Correlation between cone penetrometer resistance and undisturbed vane shear strength for two indurated clays.

25 mm per minute has arbitrarily been used. This gives some similarity between times to failure in the penetrometer and vane tests. The value of N_c appears to be site- and soil-specific.

To evaluate N_c for stiff residual clays in South Africa, a series of comparative tests were made by Blight (1967a) at two sites with the field vane and the cone penetrometer. Strengths of up to 280 kPa were measured in a lacustrine deposit of lime-indurated clay, while strengths over 300 kPa were measured in a weathered residual norite clay. The results of the comparative tests are shown in Figure 7.49. When relating cone penetration resistance to undisturbed vane shear strength the best average value for N_c appears to be 15.5. In most stiff fissured clays the undisturbed vane shear strength is about twice the unconsolidated undrained triaxial strength. Hence in relating cone resistance to triaxial shear strength a value of $N_c = 30$ would be appropriate This agrees fairly well with extreme values for N_c of 28 to 35 reported by Thomas (1965) and Ward *et al.* (1965) for cone penetration tests on London clay.

Marsland & Quarterman (1982) and Powell & Quarterman (1988) correlated cone penetration data with shear strengths back-analyzed from plate load results obtained from various stiff clays and soft rock formations in the UK. They found a trend of N_c to increase with the plasticity index (Figure 7.50a), but there was also a distinct influence of the scale of the fabric, i.e., the spacing of the cracks and fissures in the clay in relation to the cone size (three classes of soil fabric, I, II and III, were distinguished, as shown in Figure 7.50b. Figure 7.50a shows that with large spacings of discontinuities, N_c values as high as 30 were obtained, whereas with closely spaced discontinuities, $N_c = 15$ would be appropriate.

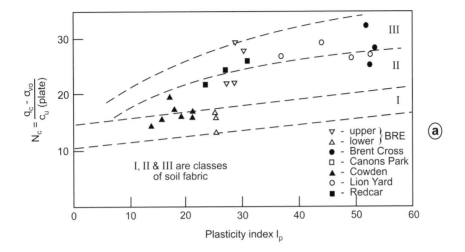

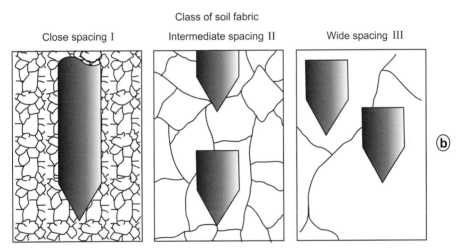

Figure 7.50 (a) Cone factors for stiff clays and soft rock based on plate load tests. (b) Three classes of fabric features with relation to cone size (Marsland & Quarterman, 1982).

For normally- and over-consolidated clays N_c has also been correlated with triaxial shear strengths via a pore pressure ratio, B_q, given by (Lunne *et al.*, 1997):

$$B_q = \frac{u_d - u_e}{q_c'} \tag{7.13}$$

where u_d = dynamic pore water pressure,
  u_e = equilibrium pore water pressure.

both measured by means of a piezocone.

The relationship between B_q and N_c is summarized by Figure 7.51. The original experimental points are widely scattered between and outside of the two limiting lines

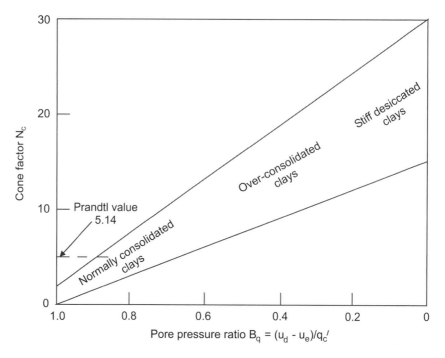

Figure 7.51 Summarized relationship between cone factor N_c and pore pressure ratio $B_q = (u_d - u_e)/q'_c$.

shown in Figure 7.51, but the N_c values cover a range from the theoretical Prandtl bearing capacity value (Prandtl, 1921) of

$$N_c = \pi + 2 = 5.14 \tag{7.14}$$

for normally consolidated clays to a maximum of about 30 for stiff desiccated clays which agrees with the $N_c = 30$ given in Figure 7.49.

REFERENCES

Aas, G. (1965) A study of the effect of vane shape and rate of strain on the measured values of *in situ* shear strength of clays. *6th Int. Conf. Soil Mech. & Found. Eng., Montreal, Canada.* Vol. 1, pp. 141–145.

Andresen, A. & Simons, N.E. (1960) Norwegian triaxial equipment and technique. *ASCE Res. Conf. Shear Strength of Cohesive Soils*, Boulder, USA. pp. 696–699.

Andresen, A. (1981) Exploration, sampling and *in situ* testing of soft clay. In: Brand, E.W. & Brenner, R.P. (eds) *Soft Clay Engineering*. Amsterdam, Elsevier. pp. 241–308.

Baldi, G., Hight, D.W. & Thomas, G.E. (1988) Re-evaluation of conventional triaxial test methods. In: *Advanced Triaxial Testing of Soil and Rock*, ASTM STP 977. Philadelphia, USA, ASTM. pp. 219–263.

Bishop, A.W. & Blight, G.E. (1963) Some aspects of effective stress in saturated and partly saturated soil. *Géotechnique*, 13 (3), 177–197.

Bishop, A.W. & Henkel, D.J. (1962) *The Measurement of Soil Properties in the Triaxial Test*, 2nd ed. London, Arnold.

Bishop, A.W. & Wesley, L.D. (1975) A hydraulic triaxial apparatus for controlled stress path testing. *Géotechnique*, 25 (4), 657–670.

Bishop, A.W., Alpan, I., Blight, G.E. & Donald, I.B. (1960) Factors controlling the strength of partly saturated cohesive soils. *ASCE Res. Conf. Shear Strength of Cohesive Soils, Boulder, USA*. pp. 503–532.

Bishop, A.W., Webb, D.L. & Lewin, P.I. (1965) Undisturbed samples of London clay from the Ashford Common shaft: Strength-effective stress relationships. *Géotechnique*, 15 (1), 6–20.

Bishop, A.W., Green, G.E., Garga, V.K., Anderson, A. & Brown, J.D. (1971) A new ring-shear apparatus and its application to the measurement of residual strength. *Géotechnique*, 21 (4), 273–328.

Black, D.K. & Lee, K.L. (1973) Saturating laboratory samples by back pressure. *J. Soil Mech. & Found. Div., ASCE*, 99 (SM1), 75–93.

Blight, G.E. (1963a) Bearing capacity and the unconsolidated undrained triaxial test. *S.A. Inst. Civ. Engrs'. Diamond Jubilee Conv.*, Johannesburg, South Africa. pp. 177–184.

Blight, G.E. (1963b) The effect of non-uniform pore pressures on laboratory measurements of the shear strength of soils. In: *Laboratory Shear Testing of Soils*, ASTM STP 361. Philadelphia, USA, ASTM.

Blight, G.E. (1963c) Effective stress properties of an undisturbed partly saturated, micaceous soil. *3rd African Reg. Conf. Soil Mech. & Found. Eng.*, Salisbury, Rhodesia. Vol. 1, 169–173.

Blight, G.E. (1967a) Observations on the shear testing of indurated fissured clays. *Geotech. Conf. Shear Strength Properties Natural Soils & Rocks*, Oslo, Norway. 1, 97–102.

Blight, G.E. (1967b) Effective stress evaluation for unsaturated soils. *J. Soil Mech. & Found. Div., ASCE*. 93 (SM2), 125–148.

Blight, G.E. (1968) A note on field vane testing of silty soils. *Canadian Geotech. J.*, 5 (3), 142–149.

Blight, G.E. (1969) Foundation failures of four rockfill slopes. *J. Soil Mech. & Found. Eng. Div., ASCE*, 95 (SM3), 743–767.

Blight, G.E. (1984) Uplift forces measured in piles in expansive clay. *5th Int. Conf. Expansive Soils*, Adelaide, Australia. pp. 363–367.

Blight, G.E. (1985) Residual soils in South Africa. In: Brand, E.W. & Phillipson, H.B. (eds) *Sampling and Testing of Residual Soils, a Review of International Practice*. Hong Kong, Int. Soc. for Soil Mech. & Found. Eng., Scorpion. pp. 159–168.

Blight, G.E. (1996) Properties of a soil residual from andesite lava. *4th Int. Conf. on Tropical Soils*, Kuala Lumpur, Malaysia. Vol. 1, pp. 575–580.

Blight, G.E., Brackley, I.J. & van Heerden, A. (1970) Landslides at Amsterdamhoek and Bethlehem – an examination of the mechanics of stiff fissured clays. *Civ. Eng. South Africa*, June, pp. 129–140.

Brand, E.W. (1988) Some aspects of field measurements for slopes in residual soils. *2nd Int. Symp. Field Meas. in Geomech., Kobe, Japan*. Vol. 1, pp. 531–545.

Brand, E.W. & Phillipson, H.B. (1985) *Sampling and Testing of Residual Soils*. Hong Kong, South East Asian Geotech. Soc., Scorpion.

Brand, E.W., Phillipson, H.B., Borrie, G.W. & Clover, A.W. (1983) *In situ* shear tests on Hong Kong residual soil. *Int. Symp. Soil & Rock Investigations In-situ Testing, Paris*. Vol. 2, pp. 13–17.

Brenner, R.P., Nutalaya, P. & Bergado, D.T. (1978) Weathering effects on some engineering properties of granite residual soil in northern Thailand. *3rd Cong. Int. Assoc. Eng. Geo.*, Madrid, Spain. Section 2, Vol. 1, pp. 23–36.

Burland, J.B., Butler, F.G.B. & Dunican, P. (1966) The behaviour and design of large diameter bored piles in stiff clay. *Symp. Large Bored Piles, Instn. Civ. Engrs.*, London, UK. pp. 51–71.

Casagrande, A. & Wilson, S.D. (1951) Effect of rate of loading on the strength of clays and shales at constant water content. *Géotechnique*, 2, 251–264.

Chandler, R.J. (1988) The in-situ measurement of the undrained shear strength of clays using the field vane. In: *Vane Shear Strength Testing of Soils, Field and Laboratory Studies*. ASTM STP 1014, Philadelphia, USA, ASTM. pp. 13–44.

Chang, M.F. & Goh, A.T.C. (1988) Laterally loaded bored piles in residual soils and weathered rocks. *2nd Int. Conf. Geomech. in Tropical Soils*, Singapore. Vol. 1, 303–310.

Cheung, C.K., Greenway, D.R. & Massey, J.B. (1988) Direct shear testing of a completely decomposed granite. *2nd Int. Conf. Geomech. in Tropical Soils*, Singapore. Vol. 1, 109–117.

Chu, B.L., Hsu, T.W. & Lai, (1988) *In situ* direct shear tests for lateritic gravels in Taiwan. *2nd Int. Conf. on Geomech. in Tropical Soils*, Singapore. 1, 119–126.

Cowland, J.W. and Carbray, A.M., (1988) Three cut slope failures on relict discontinuities in saprolitic soils. *3rd Int. Conf. Geomech. in Tropical Soils*, Singapore. 1, 253–258.

De Ruiter, J. (1982) The static cone penetration test. *2nd Euro. Symp. on Penetration Testing*, Amsterdam, Netherlands. 2, 389–405.

Florkiewicz, A. & Mroz, Z. (1989) Limit analysis for cracked and layered soils. *12th Int. Conf. on Soil Mech. & Found. Eng.*, Rio de Janeiro, Brazil. 1, 515–518.

Fredlund, D.G., Morgenstern, N.R. & Widger, R.A. (1978) The shear strength of unsaturated soils. *Canadian Geotech. J.*, 15 (3), 313–321.

Garga, V.K. (1988) Effect of sample size on shear strength of basaltic residual soils. *Canadian Geotech. J.*, 25, 478–487

Gibson, R.E. & Henkel, D.J. (1954) Influence of duration of tests at constant rate of strain on measured 'drained' strength. *Géotechnique*, 4 (1), 6–15.

Head, K.H. (1982) *Manual of Soils Laboratory Testing, Vol. 2: Permeability, Shear Strength and Compressibility Tests*. London, Pentech.

Ho, D.Y.F. & Fredlund, D.G. (1982a) Increase in strength due to suction for two Hong Kong soils. *ASCE Conf. Eng. and Constr. in Tropical and Residual Soils, Honolulu*. ASCE. pp. 263–295.

Ho, D.Y.F. and Fredlund, D.G. (1982b) A multistage triaxial test for unsaturated soils. *ASCE Geotech. Test. J.*, 5 (1), 18–25.

Howatt, M.D. (1988) The *in situ* strength of saturated decomposed granite. *2nd Int. Conf. Geomech. in Tropical Soils*, Singapore. Vol. 1, pp. 311–316.

Howatt, M.D. (1988) Written discussion. *2nd Int. Conf. Geomech. in Tropical Soils*, Singapore. Vol. 2, p. 603.

Howatt, M.D. & Cater, R.W. (1982) Passive strength of completely weathered granite. *1st Int. Conf. Geomech. in Tropical, Lateritic and Saprolitic Soils*, Brasilia, Brazil. Vol. 2, pp. 371–379.

Irfan, T.Y. & Woods, N.W. (1988) The influence of relict discontinuities in saprolitic soils. *2nd Int. Conf. Geomech. in Tropical Soils*, Singapore. Vol. 1, pp. 267–276.

James, P.M. (1969) *In situ* shear test at Muda Dam. *Conf. in situ Investigations in Soils & Rock*, London, UK. pp. 75–81.

Khalili, N. & Khabbaz, M.H. (1998) A unique relationship for the determination of the shear strength of unsaturated soils. *Géotechnique*, 48 (5), 681–687.

Lo, K.Y. (1970). The operational strength of fissured clays. *Geotechnique*, Vol. 20, No. 1, pp. 57–74.

Lo, K.W., Leung, C.F., Hayata, K. & Lee, S.L. (1988) Stability of excavated slopes in the weathered Jurong. *2nd Int. Conf. Geomech. in Tropical Soils*, Singapore. Vol. 1, pp. 277–285.

Lowe, J. & Johnson, T.C. (1960) Use of back pressure to increase degree of saturation of triaxial test specimens. *ASCE Res. Conf. Shear Strength of Cohesive Soils*, Boulder, USA. pp. 819–836.

Lumb, P. (1964) Multi-stage triaxial tests on undisturbed soils. *Civ. Eng. & Pub. Works Rev.*, 59, 591–595.

Lundgren, R., Mitchell, J.K. & Wilson, J.H. (1968) Effects of loading method on triaxial test results. *J. Soil Mech. & Found. Div., ASCE*, 94 (SM2), 407–419.

Lunne, T., Robertson, P.K. & Powell, J.J.M. (1997) *Cone Penetration Testing in Geotechnical Practice*. London, Chapman & Hall.

Marsland, A. (1972). The shear strength of fissured clay. In: Stress-strain Behaviour of Soils. Roscoe Memorial Symp. Henley- on Thames, UK.

Marsland, A. & Quarterman, R.S.T. (1982) Factors affecting the measurement and interpretation of quasi-static penetration testing in clays. *2nd Euro. Symp. Penetration Testing*, Amsterdam, Netherlands. Vol. 2, pp. 697–702.

Mirata, T. (1974). The *in situ* wedge shear test – a new technique in soil testing. *Géotechnique*, 24, 311–332.

Palmer, D.J. & Stuart, J.G. (1957) Some observations on the standard penetration test and the correlation of the test *in situ* with a new penetrometer. *4th Int. Conf. Soil. Mech. & Found. Eng.*, London. Vol. 1, pp. 231–236.

Pavlakis, M. (1983) Prediction of Foundation Behaviour in Residual Soils From Pressuremeter Tests. PhD. Thesis, Witwatersrand University, Johannesburg.

Peuchen, J., Plasman, S.J. & van Steveninck, R. (1996) *In situ* testing of tropical residual soils. *12th East Asian Geotech. Conf.*, Kuala Lumpur, Malaysia. Vol. 1, pp. 581–588.

Powell, J.J.M. & Quarterman, R.S.T. (1988) The interpretation of cone penetration tests in clays, with particular reference to rate effects. *1st Int. Symp. on Penetration Testing*, Orlando, USA. Vol. 2, pp. 903–909.

Prandtl, L. (1921) Uber die Eindringungs festigkeit (Harte) plastischer Baustoffe und die Festigkeit von Schneiden. *Zeitschrift von angewese Mathematik en Mechanik*. Vol. 18, pp. 83–85.

Premchitz, J., Gray, I. & Massey, J.B. (1988) Skin friction on driven precast concrete piles founded in weathered granite. *2nd Int. Conf. Geomech. in Tropical Soils*, Singapore. Vol. 1, pp. 317–324.

Richardson, A.M., Brand, E.W. & Memon, A. (1975) *In situ* determination of anisotropy of a soft clay. *ASCE, Spec. Conf. In situ Measurement of Soil Properties*, Raleigh, USA. Vol. 1, pp. 336–349.

Rodriguez, T.T. (2005) Colluvium Classification: A Geotechnical Approach. PhD. Thesis in Portuguese. Quoted by Futai, M.M., Almeida, M.S.S. & Lacerda, W.A. (2006). The laboratory behaviour of a residual tropical soil. In: Tan Phoon, K., Hight, D. & Leroueil, S. (eds) Characterization and Engineering Properties of Natural Soils. Singapore, Taylor & Francis. pp. 2477–2505.

Serota, S. & Lowther, G. (1973) SPT practice meets critical review. *Ground Eng.*, 6 (1), 20–22.

Silvestri, V. & Aubertin, M. (1988) Anisotropy and *in situ* vane tests. In: *Vane Shear Testing of Soils: Field and Laboratory Studies*. ASTM STP 1014. Philadelphia, USA, ASTM. pp. 88–103.

Silvestri, V., Aubertin, M. & Chapuis, R.P. (1993) A study of undrained shear strength using various vanes. *Geotech. Test. J.*, 16 (2), 228–237.

Sowers, G.F. (1985) Residual soils in the United States. In: Brand, E.W. & Phillipson, H.B. (eds) Sampling and Testing of Residual Soils, a Review of International Practice. Hong Kong, Scorpion. pp. 183–191.

Stroud, M.A. (1974) The Standard Penetration Test in insensitive clays and soft rocks. *Proc. Euro. Symp. on Penetration Testing*, Stockholm, Sweden. Vol. 2, pp. 367–375.

Thomas, D. (1965) Static penetration tests in London clay. *Géotechnique*, 15 (2), 177–187.

Vargas, M. (1974) Engineering properties of residual soils from south-central region of Brazil. *2nd Int. Congr., Int. Assoc. Eng. Geol.*, Sao Paulo, Brazil. Vol. 1, pp. 5.1–5.26.

Vaughan, P.R. (1988) Characterising the mechanical properties of *in situ* residual soil. *2nd Int. Conf. Geomech. in Tropical Soils, Singapore*. Vol. 2, pp. 469–487.

Ward, W.H., Marsland, A. & Samuels, S.G. (1965) Properties of the London clay at the Ashford Common shaft : *In situ* and undrained strength tests. *Géotechnique*, 15 (4), 342–350.

Warlam, A.A. (1960) Recent progress in triaxial apparatus design. *Res. Conf Shear Srength Cohesive Soil.*, Boulder, USA, p. 865. Pub. American Society of Civil Engineers.

Williams, A.A.B. (1980) Shear testing of some fissured clays. *7th Reg. Conf. for Africa Soil Mech. and Found. Eng., Accra, Ghana*. Vol. 1, pp. 133–139.

Wroth, C.P. (1984) The interpretation of *in situ* soil tests. *Géotechnique*, 34 (4), 449–489.

Chapter 8

Case histories involving volume change and shear strength of residual soils

G.E. Blight

Chapters 1 to 7 have described the origin and formation of residual soils and how their behaviour and properties differ from those of transported soils. The index properties of residual soils and the difficulties and pitfalls attendant on their measurement have also been set out, in addition to methods of classifying these materials. This was followed by a description of site exploration, profile description and sampling methods suitable for residual soils.

Because residual soils very often, even most often, occur in semi-arid to arid climates where the wet and dry seasons are clearly demarcated, many residual soils exist in a seasonally or permanent state of unsaturation and even desiccation. For this reason, the concepts and methods of unsaturated soil mechanics were very early introduced and integrated into the descriptions given in each chapter.

Two more technical aspects of residual soil mechanics followed, namely compaction and the mechanics of compacted residual soils, and permeability and fluid flow through saturated and unsaturated soils. These led to the two most extensive chapters, on compressibility, settlement and heave and also shear strength behaviour and the measurement of shear strength.

Because Geotechnical Engineering is a science practised both in the laboratory and the field, most of the impetus for laboratory or field research, as well as for the development of new engineering methods of construction and materials has come from the need to solve practical problems in design, construction and remedial work. It is therefore appropriate to end the book by describing a series of case histories that illustrate applications of the techniques described earlier to the solution of practical geotechnical problems. In order of appearance, the case histories will describe the following:

- settlement of two tower blocks on residual andesite lava,
- settlement of an earth dam embankment constructed of residual soil,
- settlement of an apartment block on loess,
- pre- and post-heaving of expansive clay sites as a preventive or remedial measure for damage by differential heave,
- prediction of the rate of heave of a house and an asphalt-surfaced road built over expansive clays,
- design investigation for piles supporting a power station on expansive residual siltstone and alluvial clay,

- failure of embankment dams by piping along interfaces between compacted soil and concrete,
- failure of natural and cut slopes of residual soil.

8.1 SETTLEMENT OF TWO TOWER BLOCKS ON RESIDUAL ANDESITE LAVA

Brand & Phillipson's (1985) survey shows that rafts and spread foundations have been used to transfer load to residual soil in a number of countries, including Brazil, Hong Kong, India, Nigeria, Singapore, South Africa and Sri Lanka. These types of foundation are also widely used in the United States (e.g., Barksdale *et al.*, 1982).

Developments in the field of foundation design of the mid 1960s to the 1980s generally overshadow more recent advances, so that the then state-of-the-art usually remains as relevant today as it was 30 years ago.

Two classic settlement studies on adjacent structures founded on weathered andesite lava in Johannesburg, South Africa were made in 1978 and 1983 by Jaros (1978) and Pavlakis (1983). Both buildings were founded within the residual andesite profile described in Figure 1.13. Jaros (1978) used a pseudo-elastic finite element method to predict the settlement of the two multi-storey structures (see Figure 8.1)

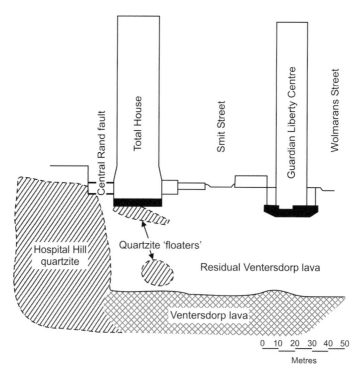

Figure 8.1 Section through the Total House and Guardian Liberty Centre buildings, Johannesburg (Jaros, 1978).

with reasonable success. Recognizing that the material behaved as if overconsolidated, he used rebound curves from oedometer tests to derive his pseudo-elastic constants. He did not however, attempt a full time-settlement analysis. The time settlement records for the two buildings, Total House and G. L. Centre which are both founded on rafts at depths of 18 m and 15 m respectively, are given in Figures 8.2 and 8.3. Both of these records appear to show that most of the settlement occurred during construction with post-construction settlement amounting to a small proportion of total settlement. (Although settlement measurements on the G. L. Centre building ceased at the end of construction.) If the settlement measurements on Total House had been commenced after construction, very little movement would have been recorded. As far as the amount of settlement is concerned, both analyses proved reasonably accurate. In the case of Total House, Jaros overpredicted the actual settlement by 26%, while for Guardian Liberty centre, his calculation underpredicted by 21%. Jaros ascribed the discrepancy in the case of Total House to the presence of large quartzite inclusions or floaters (see Figure 8.1) in the lava, the effect of which could not adequately be considered in the analysis, as their location and extent were not known with certainty.

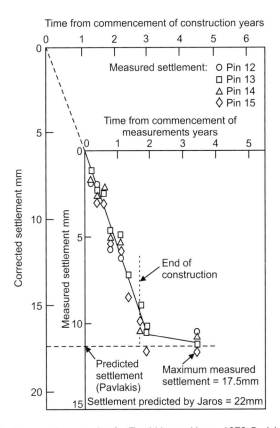

Figure 8.2 Time-settlement plot for Total House (Jaros, 1978; Pavlakis, 1983).

Pavlakis (1983) re-analyzed the settlement records for the two buildings. Using a conventional analysis based on pressuremeter tests, and correcting for the presence of the quartzite floaters, he was able to predict the measured settlements very closely, as recorded on Figures 8.2 and 8.3. However, Pavlakis was working towards a known result, whereas Jaros' prediction was made before construction, 5 years earlier. His prediction must therefore be given greater credence as an unbiased prediction than Pavlakis'. Pavlakis' pressuremeter prediction was therefore no more accurate than Jaros' prediction based on oedometer tests. The difference in the predictions arises from Pavlakis' adjustment of the time zero for the two time-settlement curves. Nevertheless, these two independent analyses illustrate that the settlement of structures on residual soils can be predicted with adequate accuracy for most engineering purposes.

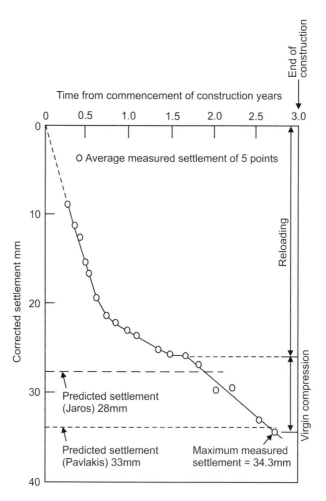

Figure 8.3 Time-settlement curve for Guardian Liberty Centre (Jaros, 1978; Pavlakis, 1983).

8.2   SETTLEMENT OF AN EARTH DAM EMBANKMENT CONSTRUCTED OF RESIDUAL SOIL

As stated in section 4.3, a compacted soil will always contain entrapped air in its pores and will therefore be partly saturated after compaction. As a result, the time-settlement curve for a compacted soil subjected to loading will feature an immediate compression or settlement that will occur as the load is applied (AB in Figure 8.4), followed by time-dependent settlement (BC in Figure 8.4) during which pore fluid is being expelled from the pores. The fluid will consist mostly of air, with water only being expelled if the applied stress is large enough to saturate the soil by compression and change the soil water suction to a positive pore pressure. However, as soon as the pressure in any expelled pore water reduces, air will start coming out of solution, and the air will be expelled in preference to the water because of the greater permeability of soil to air flow rather than water flow and the greater pressure in the air. (See sections 4.15, 5.4 and 5.8.)

If the instantaneous compression is separated from the total compression, as shown in Figure 8.5, it will usually be found that the instantaneous compression forms the major part of the total compression. This is very similar to the time-settlement behaviour observed for Total House (Figure 8.2) and Guardian Liberty Centre Figure 8.3), both structures being built on unsaturated residual soil. Concerning the settlement of an earth dam, the time-dependent and especially the post-construction settlement will usually be more important than the instantaneous settlement, which occurs during construction and therefore is "not seen".

The data for Figures 8.4 and 8.5 are for the Mita Hills dam, a rolled earth fill dam, constructed in 1957/58 in Zambia (14.7°S, 28.7°E). This dam also supplied

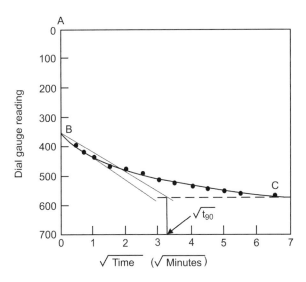

Figure 8.4 Typical oedometer time-settlement curve for soil from Mita Hills dam embankment, measured during design of dam.

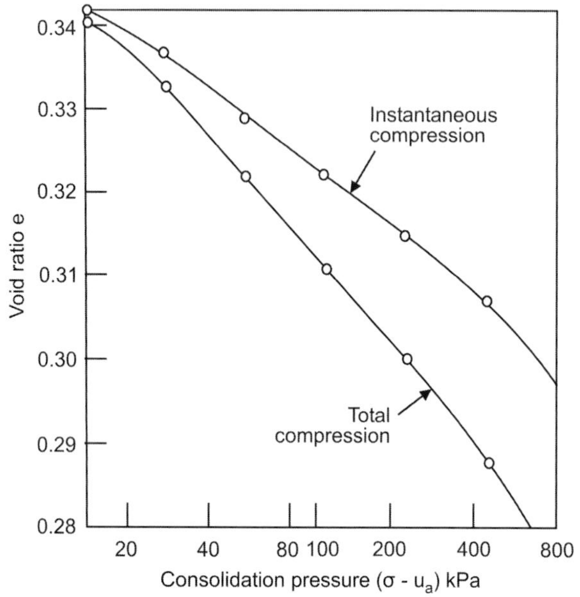

Figure 8.5 Relationship between void ratio and total pressure for soil from Mita Hills dam embankment, measured during design of dam.

the data given in Figures 4.9 and 4.17. Post-construction settlements were monitored and later calculated by Annandale (1979), from the accumulated test data used to design the dam and monitor its construction, as well as information given by Legge (1970). Figure 8.6 shows details of the dam embankment, both in plan and section. The settlement calculation considered the post-construction settlement for the crest settlement beacon located at the section (AA), having maximum height above the rock foundation.

The assumption was made that settlement would result from compression of the air-filled voids in the compacted soil and subsequent dissipation of the pore air pressure, as shown in Figures 8.4 and 8.5. To do this, both the pore air pressure built up and dissipated during construction and the post-construction dissipation of pore air pressure was considered by means of the equation:

$$\frac{\delta u_a}{\delta t} = c_v \left(\frac{\delta^2 u_a}{\delta x^2} + \frac{\delta^2 u_a}{\delta y^2} \right) + B_a \cdot \gamma \cdot \frac{\delta h}{\delta t} \qquad (8.1)$$

In equation 8.1 the increase of pore air pressure as the overburden at any depth increased was represented by:

$$\frac{\delta u_a}{\delta t} = B_a \cdot \gamma \cdot \frac{\delta h}{\delta t} \qquad (8.1a)$$

where $B_a = u_a/\gamma h$ and B_a is less than 1.0.

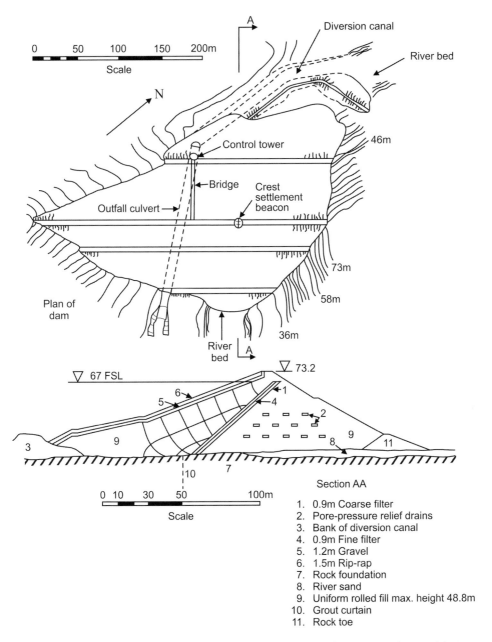

Figure 8.6 Plan and typical section of Mita Hills embankment showing dimensions and internal drainage and indicating steady-seepage flow net.

The remaining term represented the two-dimensional dissipation of the above-atmospheric pore air pressure. The coefficient of consolidation c_v was found by conventional means, as indicated in Figure 8.4.

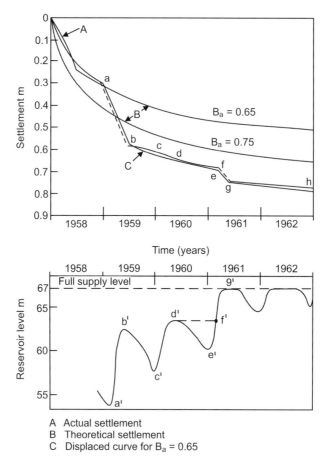

Figure 8.7 Post-construction time-settlement record for the highest section of Mita Hills dam as well as record of reservoir water levels. (See Figure 8.6.).

The upper diagram in Figure 8.7 shows the results of the post-construction set-tlement calculation (lines B for $B_a = 0.65$ and 0.75) compared with the measured settlement (line Oabcdefgh) and the fitted calculated line C.

The lower diagram in Figure 8.7 shows the variation in reservoir level from 1959 to 1962. Comparing this diagram with the measured settlement line C in the upper diagram, it is clear that much of the post-construction settlement occurred as an instan-taneous response to the increasing water load. Thus, portion ab on the settlement diagram corresponds to the increase a'b' in the reservoir level. Level d' on the reservoir level diagram is only slightly above level b' and hence d on the settlement line is only slightly below b. Level f' is on the same level as d' and hence very little settlement occurred between points d and f, but level g' is well above f' and, correspondingly, significant settlement occurred along fg. Point g' represents the full supply level, and once this had been reached, relatively little further settlement occurred.

The post construction time-settlement curve could not be predicted as a time-settlement relationship because it was dependent on the water-load applied to the embankment and there was no way of predicting the time or time-sequence in which this would happen.

The post-construction settlement of the crest had been calculated in 1956/1957 (Blight *et al.*, 1980a) in order to over-build and allow for time-dependent settlement of the final crest profile of the embankment. This was calculated for the self-weight of the embankment, ignoring the effect of water loads, with the following results:

Total settlement	2.17 m
Instantaneous settlement	1.43 m
Post-construction settlement	0.74 m

The actual post-construction settlement was 0.79 m. Hence the prediction was only 9% less than the measured value, at the end of 1962.

8.3 SETTLEMENT OF AN APARTMENT BLOCK BUILT ON LOESS IN BELGRADE

A 13-storey apartment building was constructed between May 1971 and August 1972, 1.5 km distant from the right bank of the Danube river (Popescu, 1998). It was founded on strip footings with an average contact pressure of 130 kPa. Site investigations showed that the thickness of the loess layer below the ground surface was about 15 m.

Settlement analyses predicted a maximum settlement of 110 mm and differential settlements up to 30 mm. Loess samples had been taken by auger and were mechanically disturbed. Accordingly, deformation properties obtained from laboratory tests and used in the settlement analysis were not representative of the *in situ* soil. Recorded settlement values were several times larger than the calculated ones. Subsequent investigations showed that the additional settlement was caused by post-construction wetting of the loess beneath the foundation resulting in collapse settlement. Figure 8.8 shows the measured settlement versus time relationships from the beginning of 1972 until the end of 1973. It is quite clear from this diagram that the settlement was negligible until approximately half of the load had been applied. From July 1972 onwards, rapid settlement occurred. This had virtually ceased by the end of 1972, but further rapid settlement occurred in May–June 1973. No further settlement is mentioned after the end of 1973, but it is quite likely that further settlement could have occurred, for example, during an exceptionally wet year.

8.4 PREHEAVING OF EXPANSIVE CLAY SOILS BY FLOODING

A large-scale field experiment, reported by Blight & de Wet (1965), demonstrated the essential requirements for effective preheaving by flooding, as well as illustrating the mechanism of preheaving. It was recognized that the only simple way to accelerate the penetration of water into an impervious soil is to reduce the length of the maximum flow path. In the experiment, this was achieved by drilling a grid of vertical

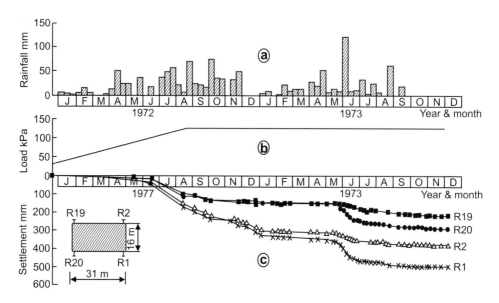

Figure 8.8 Settlement record for a 13-storey apartment house in Belgrade.

holes to facilitate the entry of water into the soil by horizontal radial flow. By this means, it was possible to induce almost full heave of a 7.5 m deep expansive clay profile within 3 months.

Preheaving by flooding has two potential practical drawbacks. These are:

- The time necessary to achieve substantially full heave of a profile is of the order of two to three months, even if a grid of holes is used to accelerate entry of water into the soil. Careful planning is required to enable a period as long as this to be included at the start of the construction schedule.
- At the end of flooding the soil surface is soft, muddy and untraffickable. To provide immediate access to the site a pioneer layer consisting of a geofabric separation layer covered by a layer of crushed rock, clinker or similar free-draining, highly frictional material must be provided. Alternatively, a working platform of lime stabilized soil can be constructed. Possibly because of these disadvantages, flooding has not become a popular method of dealing with expansive soils.

Figure 8.9 illustrates the changes in effective stress that occur during and after the preheaving process.

In the virgin desiccated profile the pore water pressure at depth z is u_0 (point a in Figure 8.9a). u_0 is in dynamic equilibrium with the water table at depth h below the surface. The corresponding point on an effective stress versus movement relationship (Figure 8.9b) would be A. At the end of flooding a temporary perched water table has been established at the ground surface and the pore water pressure is u_f (point d in Figure 8.9a). The surface has heaved and the effective stress has been reduced. Point B in Figure 8.9b describes the state of the soil in terms of effective stress and heave.

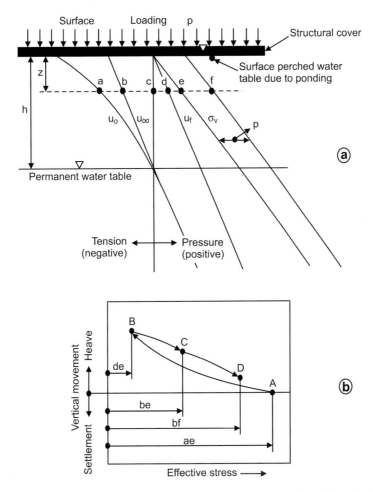

Figure 8.9 (a) Changes of pore pressure and total stress during flooding of a profile, and after flooding. (b) Change of effective stress and corresponding vertical movement.

When the soil surface is covered by the structure, the profile will contain an excess of moisture which will gradually drain away until static equilibrium is established with the permanent water table. The pore water pressure will then be u_∞ (point b in Figure 8.9a). The effective stress will have increased somewhat, and the surface will have settled, bring the soil to point C in Figure 8.9b. The effects of the surface loading imposed by the structure (ef in Figure 8.9a) will be to increase the effective stress further and cause further settlement to take place to point D in Figure 8.9b.

Figure 8.10 shows the time-heave-settlement curve observed in the (Blight & de Wet, 1965) field experiment: Flooding was maintained for 96 days, after which period the surface of the test area was covered and sealed by casting a concrete slab. Heave was observed to continue for a further year, after which a slow settlement commenced. After 7.5 years the settlement had virtually ceased. In this particular case,

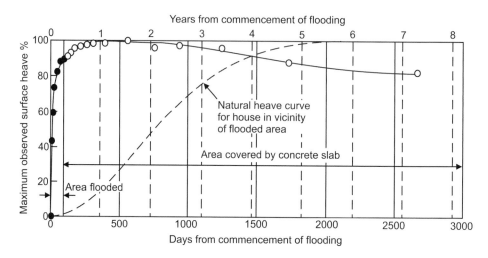

Figure 8.10 Heave during flooding and subsequent settlement of test area.

post-flooding settlement amounted to about 20% of maximum heave. It is likely that post-flooding settlement will be more in cases where the permanent water table is deeper or the superimposed loading is greater. Particular care is necessary not to plant any deep-rooted vegetation with a high evapo-transpiration requirement (e.g., euca-lyptus or poplar trees or fast-growing ornamental creepers) near to the preheaved area. By re-desiccating the soil and causing localized settlement, such plantings could be and have been disastrous for light structures founded at the surface of a pre-heaved area. The broken curve superimposed on Figure 8.10 shows the normalized time-heave curve for a house constructed on the same heaving profile. Here, heave occurred by gradual seasonal accumulation of moisture under the house. A comparison of this curve with the experimental time-heave curve shows the marked acceleration of the heave process that can be brought about by flooding.

The pre-heaving technique has been successfully applied in constructing a large shopping complex on alluvial and residual expansive clays in the town of Vereeniging, South Africa (Blight *et al.*, 1980b). The soil profile consisted of a thin surface layer of fill underlain by 1.5 m to 2 m of stiff, slickensided sandy clay alluvium. The alluvium in turn, rests on 9 m to 10 m of residual shale, weathered to a stiff, slickensided clayey silt. This overlies less weathered shale. The water table before flooding was at a depth of 11 m.

The main structure was to be supported on piles, and the pre-heaving was intended to stabilize the soil underlying the slab-on-grade floors. Prewetting holes of 300 mm diameter and 6 m deep were drilled on a 3 m grid with a wider spacing of 3.6 m adjacent to main column grid-lines. The holes were filled with crushed rock to prevent them from collapsing when filled with water.

Expansion of the soil was monitored by means of surface pegs and 4 multi-depth extensometers. The extensometers enabled the progression of heave with time at

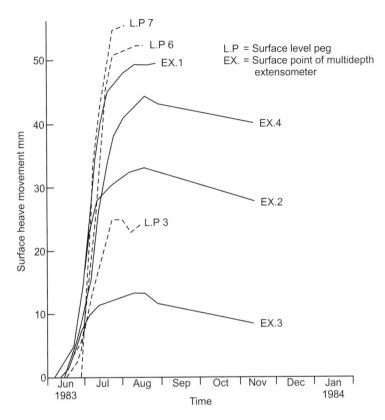

Figure 8.11 Time-surface heave relationships observed during flooding of the site for a shopping complex in Vereeniging, South Africa.

various depths in the profile to be monitored. Flooding was maintained for a period of 2½ months. Figure 8.11 shows available records of surface heave versus time, while Figure 8.12 shows profiles of heave recorded at the four extensometers shortly before flooding was terminated.

The maximum recorded surface heave varied from 12 mm to 55 mm over the 11 000 m² area of the site. The time-heave curves given in Figure 8.11 show very similar features to those of the time-heave curve shown in Figure 8.10. Unfortunately, the surface pegs were destroyed once construction was started and the extensometers were destroyed shortly thereafter. By this time, the extensometer curves were showing a steady settlement as the excess water drained out of the soil, but it was not possible to follow the continuing course of the settlement with time.

Figure 8.12 shows the variation of heave with depth recorded by the four extensometers. These curves show that most of the heave took place within the upper 5 m of the profile where desiccation should have been greatest. No post-construction monitoring has been carried out on the structure. However, no problems of excessive settlement have been experienced and that the owners of the complex are satisfied with the result.

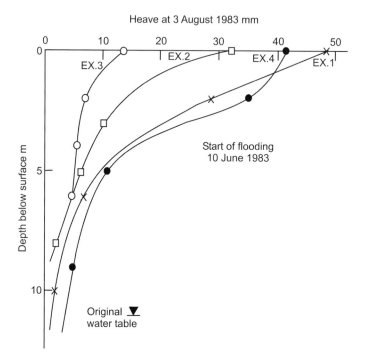

Figure 8.12 Heave-depth relationships after flooding of the site for a shopping complex in Vereeniging, South Africa.

Blight (1984) described a large scale experiment on a group of seven anchored tension piles. The piles were all 1050 mm diameter and 33 m long. They were installed in a profile of residual siltstone that has weathered to a stiff fissured clayey silt. The water table before flooding was at a depth of 14 m. (Also see section 8.6.)

The object of the experiment was to measure the uplift tensions induced in the piles as the profile heaved. Flooding was used to accelerate the heave. The piles were installed on a 2.63 m grid and the flooding holes bisected the grid. The flooding holes were 75 mm in diameter and 25 m deep. Each flooding hole contained a perforated hose pipe over its full depth. Heave of the soil was monitored by a single multipoint extensometer.

The pile group was first flooded for 10 days. After a 60-day interval to study the effects of this initial wetting, flooding was resumed, and maintained for a further 50 days.

Figure 8.13 shows a record of movement with time, at the surface and at 7 m and 14 m below surface. The diagram also shows the volume of water injected via the watering system. There is a very clear correlation between the volume of water injected and the heave curves. Unfortunately, the volume of soil absorbing the injected water is not known with any precision. Hence a meaningful relationship or ratio between the volume of water injected and the volume of heave induced cannot be established. Approximate calculations show that up to half of the flooding water must have been lost from the test area by lateral seepage into the surrounding soil.

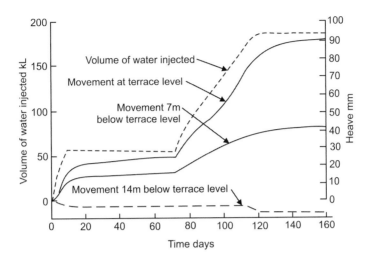

Figure 8.13 Relationships between time and heave and time and water injected for flooding of experimental pile site.

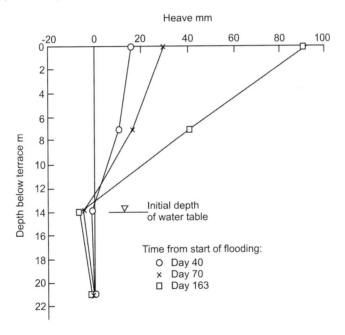

Figure 8.14 Heave-depth relationships during flooding of experimental pile site.

Figure 8.14 shows relationships between depth and heave at various times after the start of flooding. The data are rather similar in form to those shown in Figures 8.11 and 8.12.

In this case, therefore, the technique of flooding proved very successful as a means of accelerating the development of uplift tensions in the piles.

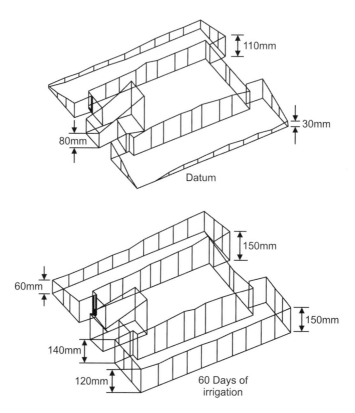

110mm

30mm

80mm

Datum

150mm

60mm

150mm

140mm

120mm

60 Days of
irrigation

Figure 8.15 Effect of surface irrigation on distortion of apartment block damaged by heave.

Much of the heave at a particular site will take place in the upper 2 m to 3 m of the soil profile, where desiccation effects are likely to be most severe. This indicates that surface flooding or irrigation should also be effective for preheaving, provided sufficient time is allowed for the water to penetrate the soil. The time required will probably be at least 2 to 3 months. Williams (1980) describes the use of surface irrigation as a remedial measure for an apartment building damaged by heave. A three storey load-bearing brick apartment block was erected on shallow strip footings on a soil profile that consists of 1 m of calcareous windblown sand overlying a great depth of desiccated residual shale. No water table was found within 23 m of the surface. The apartment block consisted of two buildings on either side of a courtyard, the two being linked by walkways on either side to form a hollow square. The courtyard was surfaced with asphalt concrete. No thought appears to have been given to the possible occurrence of heave.

Cracking of the buildings was noticed before they were occupied and occupation was delayed because of this. Within three years the structures were so severely distorted and cracked that it was feared they would have to be demolished. Examination of the complex, and level measurements showed that the two apartment buildings had heaved

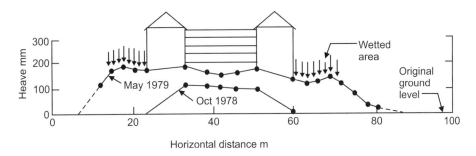

Figure 8.16 Effect of surface irrigation on ground profile under apartment block.

and tilted outwards, moisture having accumulated under the buildings and the paved courtyard.

It was decided to attempt to reduce the distortion by wetting the whole area using spray irrigation. This was undertaken and within 60 days the distortion had been much reduced. Figure 8.15 shows the progressive changes of level of the perimeter of the complex, while Figure 8.16 shows typical transverse sections of the site before and after irrigation. However, it will be noted from Figure 8.15 that some of the differential movements were maintained, e.g., 80 mm across the front side of the building in Figure 8.15 and 90 mm between the right and left hand corners at the back. To maintain the site in a heaved condition, the area around the apartments was paved with open-jointed concrete blocks bedded in a sand layer. This type of surfacing allows rainfall to penetrate into the soil, but minimizes water losses by evaporation from the surface.

Ten years later the remedial measures continued to be successful, although there had been a slow settlement of the area since irrigation ceased.

The case histories have shown that preheaving of expansive sites by flooding can be successfully carried out. Preheaving is a useful technique if done rationally, with careful monitoring of the effects. There are certain disadvantages, the principal being the time taken for the water to penetrate the soil. However, careful planning and scheduling can overcome this shortcoming.

8.5   HEAVE ANALYSIS FOR A PROFILE OF DESICCATED EXPANSIVE CLAY AT AN EXPERIMENTAL SITE

The observed time-heave record for a house at a test site at Vereeniging, South Africa, has been shown in Figure 6.43 which has been repeated here as Figure 8.17a. This record covers the first seven years after building (2550 days). The primary movement of the house appears to have ended after the first six years. Since then only seasonal movements and movements due to the effects of faulty drainage and gardening activities have occurred. The movement of a 6.1 m wide black-top road adjacent to the house is shown in Figure 8.17b. The heave movement at these two structures will be analyzed in what follows.

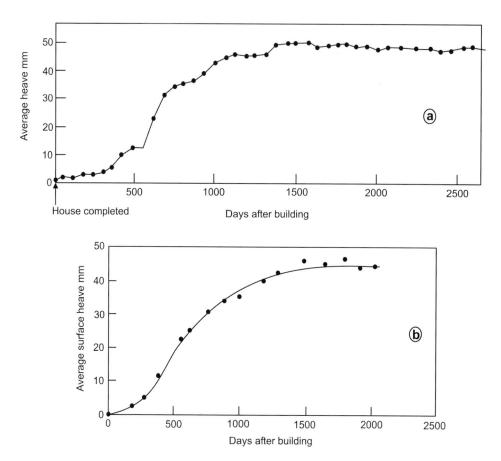

Figure 8.17 Heave vs time curves for (a) House measuring 13.2 m by 7.1 m. (b) Asphalt surfaced 6.1 m wide road adjacent to the house.

8.5.1 Similarities between heave and settlement analyses

Settlement of a saturated clay occurs when applied loading has produced pore water pressures that are out of static equilibrium with the ground-water regime. The settlement proceeds by the expulsion of pore water until the distribution of pore water pressure beneath the settling structure has regained static equilibrium with the water table. The relationship between changes of pore water pressure and settlement in a saturated soil can be written as:

$$\frac{\Delta h}{h} = C\,\Delta(\sigma_v - u_w) \tag{8.2}$$

where $\Delta h/h$ is the settlement strain,

C is the compressibility of the soil in a vertical direction,

σ_v is the vertical total stress in the soil due to overburden and superimposed loading,

u_w is the pore water pressure.

In most cases of settlement the vertical stress σ_v remains virtually constant after application of the superimposed loading, and equation 8.2 can be simplified to:

$$\frac{\Delta h}{h} = -C\Delta u_w \text{ with } \sigma_v \text{ constant.} \tag{8.2a}$$

In this equation, for a given value of C, the settlement of a saturated soil is approximately proportional to the change in pore water pressure. As the excess pore water pressure decreases, the settlement strain increases.

The rate at which settlement occurs depends on the coefficient of consolidation of the soil, c_v, on the drainage conditions, i.e., on the dimensions of the structure and the location of free surfaces and permeable strata in which the pore pressure is constant.

In order to perform a settlement analysis and produce a prediction of the time-settlement curve, it is therefore necessary:

- by means of laboratory tests on undisturbed samples, to measure the compressibility and coefficient of consolidation of the clay,
- to estimate the profile of excess pore pressure set up by the superimposed load, and
- to assess the drainage conditions for the problem.

In a partly saturated soil profile, under vegetation, the pore water pressures or suctions are in dynamic equilibrium between the water table and the weather conditions prevailing on the surface. In the area referred to in this case history, there is a net moisture deficiency in the profile and hence suctions exceed those required for static equilibrium with the water table. (Also see Figure 8.9.)

When the surface of the soil profile is developed and is covered by a structure, the moisture deficiency is gradually recovered and the pore pressure profile beneath the structure approaches static equilibrium with the water table. As moisture accumulates under the structure, the soil swells and heave of the surface results. The difference between the suction profiles for conditions of dynamic and static equilibrium represents the excess negative pore pressure for the heave process.

The relationship between heave and changes in pore water pressure in a partly saturated soil can be written:

$$\frac{\Delta h}{h} = C\{\Delta\sigma_v - \Delta[\chi u_w]\} \tag{8.3}$$

In this equation:

$\Delta h/h$ is the heave strain, and

χ is the Bishop effective stress parameter (less than unity) which depends mainly on the pore water pressure u_w.

Once again, σ_v will remain virtually constant after its application, and equation 8.3 can be rewritten as:

$$\frac{\Delta h}{h} = -C\Delta[\chi u_w] \tag{8.3a}$$

or, if $-u_w = p''$, the suction of the pore water, then:

$$\frac{\Delta h}{h} = C\Delta[\chi p''] \tag{8.3b}$$

As the factor χ varies with p'', there is no approximate direct proportionality between heave and changes in suction.

The rate at which heave occurs depends on the coefficient of swell of the soil, c_s, and on the drainage conditions. For an unsaturated soil the coefficient of swell depends considerably on the suction. The drainage conditions for the heave process depend on the dimensions of the structure and the sources of water available for the swelling process.

The heave and settlement processes are, therefore, essentially similar. The main differences arise from differences in behaviour between saturated and partly saturated soils.

8.5.2　The profile of excess pore water pressure for heave

In the case being considered, in order to measure the coefficient of swell for the expansive clay and also to estimate the suction profile under vegetative cover, undisturbed block samples were taken from a 760 mm diameter test hole between the experimental house and an adjacent asphalt surfaced road. Undisturbed triaxial samples 38 mm in diameter were subsequently trimmed from the block samples, subjected to overburden pressure in a triaxial cell, and the suction at various depths was measured using the axis translation technique. The results of these measurements are shown in Figure 8.18.

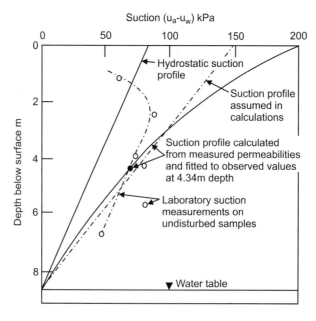

Figure 8.18　Measured, calculated and assumed suction profiles at experimental site.

This Figure also shows a suction profile calculated from measured soil permeabilities using the equation:

$$\frac{\delta}{\delta x}\left\{k\left(\frac{\delta u_w}{\delta x}\right)\right\} = 0 \tag{8.4}$$

where k is the permeability of the soil to unsaturated diffusional flow, and
u_w is the pore water pressure at a distance x above the water table.

The calculated profile has been fitted to the 9 m deep expansive profile at a depth of 4.3 m. It is thought that rainfall a few days before sampling was responsible for the low suction values in the upper part of the profile.

In the absence of better data it was assumed that the suction profile before the commencement of heave lay somewhere between the calculated and observed profiles. The assumed suction profile is shown as a chain-dotted line on Figure 8.18.

The excess suction for heave is represented by the difference between the assumed suction profile and the line of hydrostatic suction.

8.5.3   Measurement of the coefficient of swell, c_s, for diffusional flow

A clear distinction must be made between diffusional flow and rain water penetration. In the former process the supply of water is not sufficient to fill the fissures and large voids in the soil and flow takes place by diffusion through the fine pores of the soil. In rain water penetration, an abundance of water is present and flow takes place through fissures in the soil. Penetrational flow consequently takes place at a considerably more rapid rate than diffusional flow. In this case all time effects are considered to be due to diffusional flow. Rain water penetration is assumed to take place in a negligible time.

The coefficient of swell for diffusional flow through the expansive clay (c_s) was measured on samples from depths of 2.4 and 5.5 m. The samples were subjected to over-burden pressure in a triaxial cell apparatus and the initial suction was measured. A controlled pore water pressure was then applied at the base of the sample to give a decrement of suction and the rate of water intake was measured by observing the rate of movement of a mercury thread 2 mm^2 in cross-section. In this way an accurate water intake volume versus time curve was observed for each suction decrement. The mean value of the coefficient of swell for each stage of swell was then calculated by fitting the observed intake-time curve to the appropriate curve for the classical consolidation theory at a fractional intake of 50%.

Figure 8.19 shows the measured values of the mean coefficient of swell (c_s) at various suctions.

8.5.4   Drainage conditions for the heave process

The most difficult problem is to decide on the source of the moisture that is necessary for the heave movement to take place. Discounting accidental sources such as broken drains, there are three possible sources of moisture:

- upward diffusional flow of moisture from the water-table over a vertical distance of 9 m,

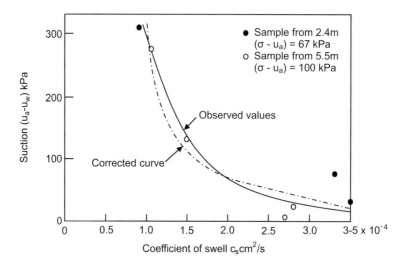

Figure 8.19 Relationships between suction and coefficient of swell for partly saturated expansive clay.

- penetration of rain-water down vertical or near vertical fissures in the soil profile, along the perimeter of the structure, followed by lateral diffusional flow into the soil beneath the structure. The house measured 13.2 m by 7.1 m, hence the minimum wetting path length would be 3.5 m, and
- lateral penetration of rain-water under the structure in the layer of sandy surface soil, followed by downward diffusional flow towards the water-table. (A drainage path length of 9 m.)

It appears likely that the heave movement takes place through the supply of moisture from all three of these sources in combination. However, it is just as likely that the effects of one or two of the sources of supply will have an overriding effect.

The relative importance of each of the above mechanisms was investigated by comparing the observed behaviour of the two structures with that predicted on the assumption of different drainage conditions.

Figure 8.20 shows a smoothed curve representing the average heave of the internal and external pegs in the experimental house replotted in dimensionless form. The curve shows the variation of the percentage average heave with a dimensionless time factor:

$$T = \frac{c_s t}{D^2} \tag{8.5}$$

where c_s is the coefficient of swell,
$\qquad$ t is the time after completion of building, and
$\qquad$ D is the length of the wetting path.

c_s has been taken as the measured value at a suction of 65 kPa which is about the average suction in the soil profile during the heave process.

For a constant isotropic c_s, at any time T would be inversely proportional to D^2 and hence lateral diffusional flow ($D^2 = 12.25$) would be 6.6 times more effective than either form of vertical flow ($D^2 = 81$).

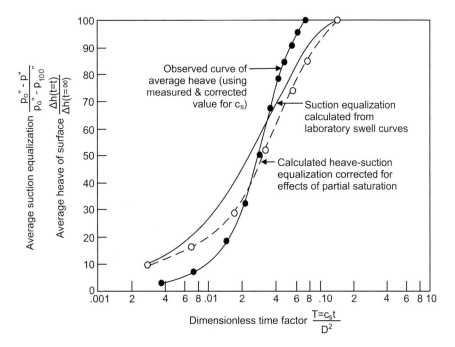

Figure 8.20 Comparison between observed dimensionless time-heave curve for the house and predicted dimensionless curves for suction equalization and heave.

For the 6.1 m wide road, the minimum wetting path is 3 m and hence lateral diffusion would also be the most likely mechanism for heave.

Upward diffusional flow from the water-table

The curve, if it is assumed that heave takes place by upward flow of moisture from the water table only, is shown to the right of Figure 8.21. It will be noted that the time factor corresponding to 50% completion of heave is 15 times greater than the observed time factor. Hence upward flow from the water-table was not an important factor in the heave process for the house (or for the road).

Vertical rainfall penetration followed by lateral diffusional flow

An examination of the movement records for the depth points at the experimental house showed that seasonal variations in level occur at all depths down to 24 feet. It is therefore probable that rainfall penetrates to the water-table at this site.

Let it be assumed that the rainfall of the first wet season after building reduced the suction in the profile around the perimeter of the structure to hydrostatic values and that thereafter this condition was maintained. The dimensionless heave curve calculated on this assumption is shown in Figure 8.21 together with the calculated curve, if heave by combined horizontal inflow and upward flow from the water table is assumed. There is little difference between these two curves and at 50% heave the observed and calculated time factors differ by a factor of only 1.2.

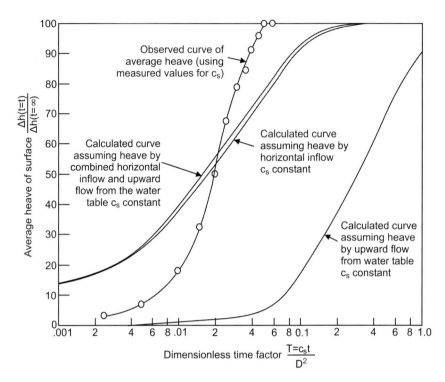

Figure 8.21 Comparison between observed dimensionless time-heave curve for the house and calculated dimensionless heave curves for different assumed drainage conditions.

If the measured values of c_s are accepted, the postulated most likely mechanism of heave fits the observed facts surprisingly well.

Lateral rainfall penetration followed by downward diffusional flow

This mechanism of moisture accumulation can probably also be discounted in the case of the experimental house, where the strip footings would act as cut-offs through the pervious surface stratum. In the case of the road, however, lateral rainfall penetration is more likely (but with a much longer drainage path).

It can thus be concluded that of the three mechanisms of moisture accumulation suggested, that described by the second bullet on page 306 appears to be the most important, although the influence of the other two will also have an effect on the resultant time-heave relation.

It must be stressed that this conclusion applies only to the two structures under consideration. With different structural dimensions, soil conditions or water-table depths a different conclusion may apply.

8.5.5   The relationship between heave and changes in suction

It was seen earlier that no direct proportionality exists between changes in suction in a partly saturated clay and corresponding changes in volume. The relationship between these two properties was given by equation 8.3b.

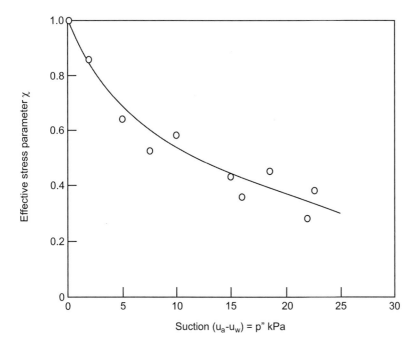

Figure 8.22 Relation between the effective stress parameter and suction for swelling of the partly saturated expansive clay.

The relationship between the suction p'' and the effective stress parameter χ for swelling of the clay is shown in Figure 8.22. It will be noted that as the suction in the soil decreases, i.e., as the heave progresses, the parameter χ increases. As a consequence, when the average equalization of suction beneath the structure has reached 50 per cent, less than 50 per cent of the heave movement will have taken place.

It can be shown from equation 8.3b that if at any time U_h is the percentage of the total surface heave which has taken place, and p'' indicates suction $(u_a - u_w)$ then

$$U_h = \frac{(\chi p'')_0 - \chi p''}{(\chi p'')_0 - (\chi p'')_{100}} \tag{8.6}$$

where $(\chi p'')_0$ is the average value of $\chi p''$ in the profile before the start of heave, and
$(\chi p'')_{100}$ is the average value of $\chi p''$ in the profile at the completion of heave.

Equation 8.6 shows that the value of c_s appropriate to changes in volume will be less than that for suction equalled by a factor of 1.3 (in this case).

The steps in the calculation of a predicted time-heave curve are as follows:

- The observed suction equalization curve or a finite difference method is used to calculate the curve of pore water pressure equalization for the structure. As the shapes of the experimental equalization curves do not vary very much with suction, it was sufficiently accurate, in this case, to use a mean curve,
- The resulting curve is corrected, using equation 8.6 to obtain the time-heave curve.

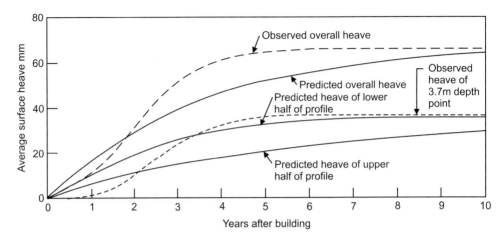

Figure 8.23 Comparison between observed and predicted time-heave curves for the experimental house.

The predicted heave curve shown in Figure 8.21 was calculated using average values for the whole depth of expansive material. In order to test the validity of the assumptions with regard to drainage conditions and in an attempt to improve the accuracy of the prediction, the calculation for the experimental house was repeated, using average soil properties for the upper and lower halves of the profile. The results of the recalculation are shown in Figure 8.23, in comparison with the observed time-heave curve for the 3.7 m depth point and the overall movement of the house.

8.5.6 Accuracy of the time-heave prediction

The predicted time-heave curves agreed fairly well with the observed time-heave curves for the house and the road. In both cases the predicted movements at first lead the observed movements and later lag behind them, but on the average the agreement is fairly close.

The initial lead of the predicted movements is probably caused by a delay in the establishment of the assumed boundary conditions. The lag of the prediction in the latter stages of the heave may possibly be accounted for by the presence of water from extraneous sources such as a faulty drainage system. (No irrigation of gardens was allowed around the experimental house.)

The accuracy of the time-heave predictions is summed up in Figure 8.24a, which shows the error in the time prediction at various proportions of the total heave. Using average properties for the whole soil profile, virtually the same accuracy is achieved for both the house and the road even though the structures are dissimilar in type and dimension but have similar water ingress conditions.

Using a two-layer calculation for the house does not alter the accuracy of prediction materially and the accuracy of prediction for the lower half of the profile is not very different from that for the complete profile.

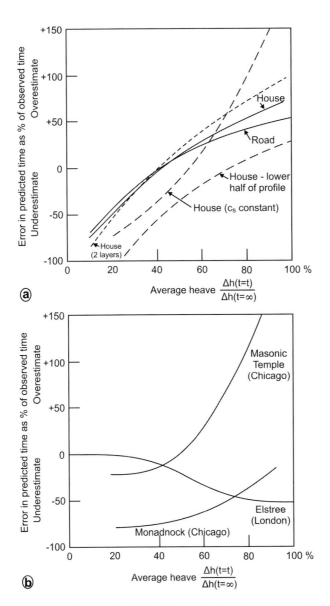

Figure 8.24 (a) Error diagrams for predicted time-heave relation of experimental house and road. (b) Error diagrams for predicted time-settlement relations of three structures on saturated clay (after Skempton & Bjerrum, 1957).

To put the error lines of Figure 8.24a into correct perspective, consider Figure 8.24b which shows similar error graphs for the benchmark settlement analyses of three structures founded on fairly uniform deposits of saturated transported clays (Skempton & Bjerrum, 1957). The errors incurred in these settlement analyses are of the same order as those illustrated in Figure 8.24a.

8.6   THE PERFORMANCE OF TENSION PILES SUBJECTED TO UPLIFT BY EXPANSIVE CLAYS

Lethabo thermal power station in South Africa is founded on a deep profile of desiccated clayey silt residual from weathered siltstone. The water table at the site is at a depth of about 20 m. Studies of the potential heave of the site had indicated that a surface heave of up to 120 mm could be expected to occur. (Although surface heave in the first 8 years after construction of up to 260 mm was observed – see Figure 6 45.)

The operation of power station plant and other installations such as cooling towers and cooling water ducts is very sensitive to differential movement. For this reason, an early design decision was to found all structures on bored cast-*in-situ* reinforced concrete piles. The piles were to penetrate the residual siltstone and be founded in a less weathered dimensionally stable carbonaceous shale that underlies the expansive weathered siltstone. All structures would be supported on piles with a 300 mm void between the soil and the underside of the structure. The design of piles for heavy loads would not be problematic, as the carbonaceous shale provides a high bearing capacity. Lightly loaded piles would have to be anchored in the carbonaceous shale and resist the uplift forces imposed on them by the surrounding expansive soil. This would also apply to piles that ultimately would be heavily loaded, but which would be in the ground for long enough to be affected by heave and upward tension before the loads were applied.

There were three basic problems associated with the pile design.

Collins (1953) had proposed a procedure for designing piles subject to uplift by expansive clays. This method had been widely used in the intervening 30 years, without ever having been checked against the performance of an actual pile installation. Donaldson (1967) had carried out measurements of the load distribution in an instrumented pile of 230 mm diameter 10 m in length. These measurements confirmed the form of Collins' expression for pile tension which is:

$$P = \pi D \int_0^L (c' + K\sigma'_v \tan \varphi')dz \qquad (8.7)$$

In which P is the tension in the pile,
D is the pile shaft diameter,
L is the length of the pile subjected to uplift, and therefore requiring anchorage,
c' and φ' have their usual meaning.

However there was uncertainty as to whether the expression would apply to the very large piles that would be needed at Lethabo (35 m long and up to 2000 mm in diameter):

- There was uncertainty as to the values of c' and φ' to use.
- What value of K should be used?

Donaldson's study had not attempted to compare measured pile tensions with tensions predicted from measured shear strengths.

A simple paper exercise showed that with so many piles to be constructed, even small economies on each pile would make a full-scale field testing exercise a financially viable proposition. It was therefore decided to embark on such an exercise.

8.6.1 Shear strength

It was known from published work (see Chapter 7) that the appropriate strength to use in designing the piles probably corresponded to the lower limit of laboratory strength measurements. But, should an adhesion factor be applied and what was the value of K?

To answer these questions a series of plug pulling tests was arranged. The plugs were of 1050 mm diameter and 2 m length and were constructed as shown in Figure 8.25. The plugs were installed at a series of depths and the surrounding soil was soaked by filling the hole with water. After a period of 4 weeks the water was allowed to drain away until there was no free water in the standpipe/suction-breaker (see Figure 8.25). This ensured that the effective stress in the soil surrounding the plugs was close to the value of the total stress as the pore pressure must have been close to zero. The plugs were then pulled by jacking up a beam around which the cables from the plug were fastened.

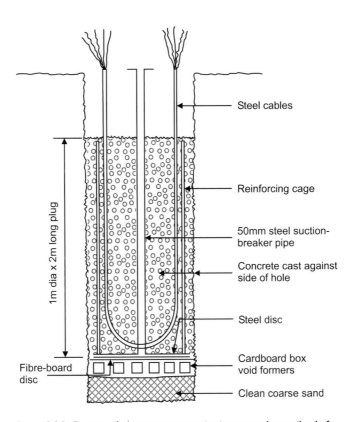

Figure 8.25 Design of plugs to measure *in situ* strength on pile shafts.

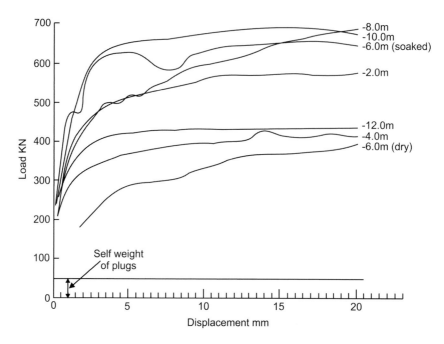

Figure 8.26 Load-displacement curves for plug-pulling tests in weathered siltstone.

Figure 8.26 shows the resultant load displacement curves. Note the difference between the two plugs centred at a depth of 6 m, one of which was pulled dry, the other after soaking. Also note the small displacement of 5–7 mm required to mobilize full resistance.

Figure 8.27 summarizes all the available results of laboratory drained shear strength tests on the siltstone, as well as showing the results of the plug-pulling tests. With the exception of one test, the strengths from the plugs coincided quite closely with the lower limit to the laboratory strength measurements. This result confirmed expectations, in the light of previously published work. However, the way in which the test had been performed and the fact that at a given depth in Figure 8.27, the laboratory shear tests correspond to the same effective stress as the plug tests, eliminated the need to know values for either the adhesion factor or the lateral stress ratio K. In fact, the indication was that both of these factors equalled unity.

Design of piles

The piles could be designed using Collins' expression (equation 8.7) in terms of the changing effective stress and the progressive heave predicted for the site. Whereas the laboratory shear tests had shown a pronounced peak in the stress displacement curve, with one exception, the field curves showed no peak. As shown by Figure 8.28, using strengths derived from the plug-pulling tests rather than the mean laboratory curves had a considerable effect on the maximum design tension.

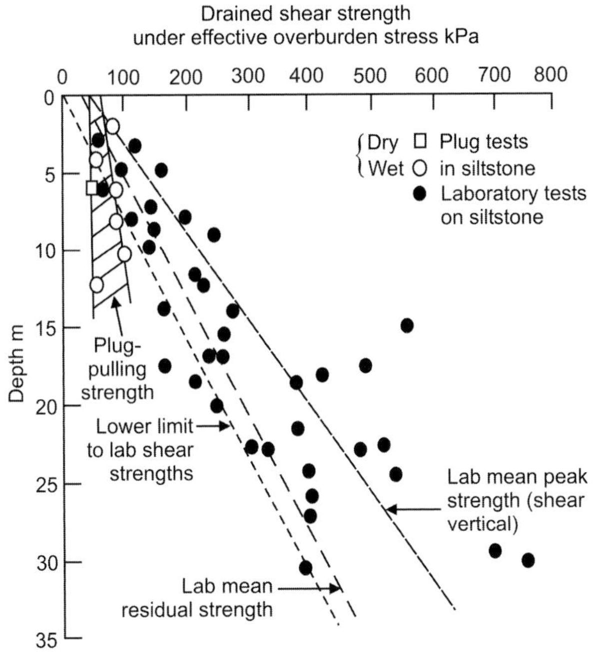

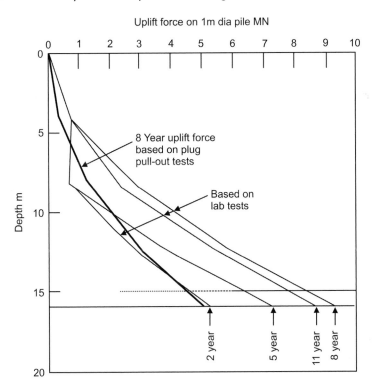

Drained shear strength
under effective overburden stress kPa
0
100
200
300
400
500
600
700
800
Depth m
0
5
10
15
20
25
30
35
Dry □ Plug tests
Wet ○ in siltstone
● Laboratory tests
on siltstone
Plug-
pulling
strength
Lower limit
to lab shear
strengths
Lab mean peak
strength (shear
vertical)
Lab mean
residual strength

Figure 8.27 Summary of laboratory and *in situ* strength measurements made on siltstone.

Uplift force on 1m dia pile MN
0
1
2
3
4
5
6
7
8
9
10
Depth m
0
5
10
15
20
8 Year uplift force
based on plug
pull-out tests
Based on
lab tests
2 year
5 year
11 year
8 year

Figure 8.28 Pile design curves based on Collins' formula. Curves based on lab tests take account of progression of heave with time and effect of peak in stress-displacement curve.

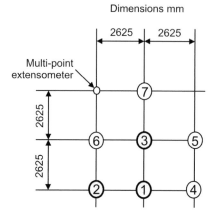

Figure 8.29   Layout of piles in test pile group.

8.6.2   Field test on an instrumented pile group

The one question remaining was whether the Collins expression correctly represented the distribution of tension in the shaft of a pile subjected to uplift and whether the uplift forces could correctly be predicted from the measured shear strengths. To answer the question, it was decided to install a group of instrumented test piles that could be subjected to accelerated swell. The test group consisted of 7 test piles arranged on a 2.625 × 2.625 m grid, as shown in Figure 8.29. All piles were straight-shafted, 1050 mm in diameter and 33 m long. Referring to Figure 8.29, the positions of the three instrumented piles were selected so that they represented a side pile (1), a corner pile (2), and an interior pile (3) in a typical pile group. (For details of the instrumentation, the reader is referred to Blight (1984). Note that two types of gauge – electric resistance and vibrating wire – were used as a check on each type).

The soil surrounding the test pile group was flooded by means of a grid of boreholes and the strain at various depths in each instrumented pile was recorded as the soil took up water and swelled. (See Figures 8.13 and 8.14.)

Figure 8.30 shows the variation at various times of the recorded tension in pile 2 (see position in Figure 8.29). Tensions were calculated from the recorded strains on the assumption that the concrete forming the pile had cracked and the steel reinforcing was carrying the entire load. Strains at depths of up to 5 m and more than 22 m were not sufficient to crack the concrete and the tensions were therefore underestimated at either end of the pile.

The design tension curve calculated from the results of the plug-pulling tests via Collins' equation (Equation 8.7) has been superimposed on Figure 8.30 and shows reasonable agreement with the measured pile tensions, especially considering the known underestimation at depths less than 5 m.

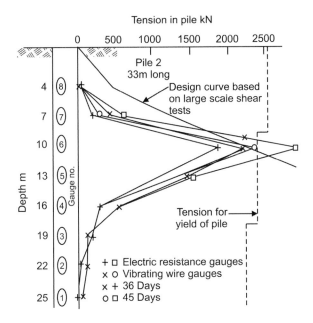

Figure 8.30 Development of tension in test pile 2 with time and comparison of measured and predicted depth-tension relationships.

Figure 8.31 shows the variation of shaft shear stresses down the three test piles. The diagram shows how uplift shears developed above a depth of about 10 m were counteracted by downward anchorage shears on the lower part of the pile shaft. This diagram also provides a check on the validity of the tension measurements, as the area under the uplift shear curve must equal that under the anchorage shear curve if vertical equilibrium is to be preserved. For piles 1 and 2, each shear area (upward and downward) equaled just more than 2000 kN, while for pile 3, each shear area equaled 1500 kN. The difference between the edge piles 1 and 2 and interior pile 3 is obviously because of their position in the group. Another check was provided by comparing the measured shaft extension to the integration of the measured axial strains. These differed by 15%, with the strain integration, as it should have been, giving the lower figure. The strength versus depth relationship established from the plug-pulling tests has been superimposed on the upper part of Figure 8.31. This agrees reasonably with the shear stress-depth curves calculated for the piles, if one bears in mind that the pile shaft shear stresses were underestimated at depths of less than 5 m.

8.6.3 Effect of loading a pile previously subjected to uplift

As many of the piles would be installed two to three years prior to being loaded, it was decided to study the effect of this delayed loading by applying a compression load to two of the instrumented piles after full tension had developed in them. A 1000 kN compressive load was applied by installing anchors into the carbonaceous shale and jacking off the anchors. As had been expected, the applied load was resisted by shears

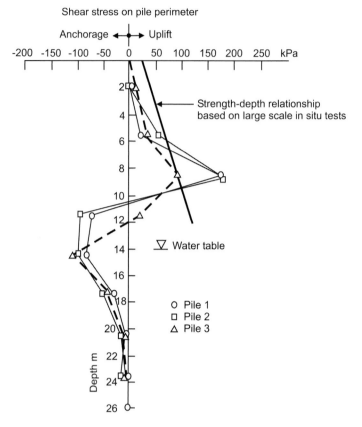

Figure 8.31 Shear stresses developed down length of piles and comparison with measured shear strength vs depth relationship.

on the pile shafts and was completely carried by the upper 6m of pile. Tensions in the shafts below this depth were unaffected.

8.6.4 Conclusions

The programme of field testing and monitoring achieved all its objectives:

(i) The Collins expression for pile shaft tension was validated for the large diameter piles used at Lethabo.
(ii) Appropriate design parameters for the piles were established.

In the past 30 years, the piles at Lethabo appear to have performed as expected. Problems were experienced when one of the cooling water ducts started to lift. When the supporting piles were opened up, it was found that the void between the underside of the duct and the soil had carelessly been allowed to fill up with soil washed in by a thunderstorm before the sides of the duct were backfilled. The underlying soil had

heaved and lifted the duct off the piles, but the piles had not moved. This was rectified by removing the washed-in soil.

Another problem arose with one of the cooling tower ponds which also started to lift. Here, it was found that the cardboard void formers, that should have created a void beneath the ponds, had carelessly been omitted. The remedy was to undermine the pond, by hand, creating a 1 m void. Plate C26 shows the void created on the underside of one of the ponds during the undermining.

8.7   THE MECHANISMS OF PIPING FAILURE ALONG CONCRETE OUTLET CONDUITS

Section 1.9 briefly dealt with problems caused by dispersive soils and gave an example, illustrated by Plate C18, of a catastrophic failure of an earth dam by piping along the side of the outlet conduit. This section also mentioned that while failures of this type are probably more likely to occur when the earthfill is dispersive, poor compaction of fill against soil to concrete interfaces and soil arching over conduits usually also plays a role in the mechanism of failure. Indeed, the reduction of lateral total stresses by soil arching can by itself result in hydraulic fracture of the fill and failure by piping.

8.7.1   Gennaiyama and Goi dams – failure along outlet conduit

This case history (Ngambi et al., 1999) will briefly summarize the results of pressure measurements made on the top and sides of two reinforced concrete conduits passing through earthfill dams, the Gennaiyama and Goi dams in Japan.

Both dams are relatively low (13 m and 20 m respectively). The Gennaiyama dam has a box-shaped conduit as shown in Figure 8.32. Pressure cells and settlement cross-arms were installed as shown. The impervious fill (permeability $k = 10^{-7}$ cm/s $= 0.3$ m/year) consisted of a mixture of river deposits (75% by volume) and volcanic ash (25% by volume), compacted in 0.3 m thick layers to a bulk unit weight of 18 kN/m^3. The Goi dam has a horseshoe-shaped conduit with pressure cells mounted as shown in Figure 8.33. In this case the impervious fill consists of a mixture of three equal parts by volume of soil residual from sandstone and mudstone, weathered mudstone and crushed unweathered mudstone, compacted to a bulk unit weight of 16.7 kN/m^3. The permeability of the mixed fill was 10^{-6} m/s ($=3$ m/year).

Figure 8.34 shows the settlement at Gennaiyama adjacent to the pressure cells, measured by means of the cross-arms. Note that at level 493 m (4 m above the top of the conduit) the reduced settlement shows the effects of soil arching over the rigid structure, even though it had been given a nominally negative projection by locating it in a trench 2 m below the level of the surrounding fill.

Figure 8.35 shows records of overburden pressure and pressures measured on the various pressure cells at the two dams as the trench was backfilled and the embankments of the two dams were constructed. Considering Figure 8.35a, for Gennaiyama dam, the final overburden pressure above the conduit was 110 kPa, whereas the vertical pressures recorded on the horizontal top of the box-conduit, 2 years after embanking started, were 280 (E − 4), 290 (E − 5) and 390 kPa (E − 6). These pressures, up to nearly four times overburden, illustrate the effect of soil arching over the rigid conduit

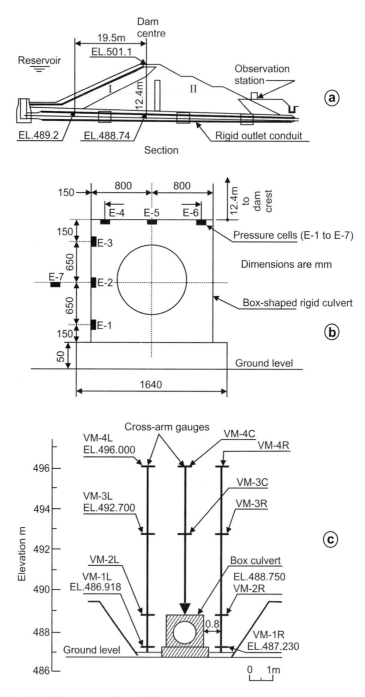

Figure 8.32 Details of Gennaiyama dam: (a) Section along outlet conduit. (b) Cross-section of outlet conduit showing positions of pressure cells. (c) Positions of cross-arms for measuring settlement.

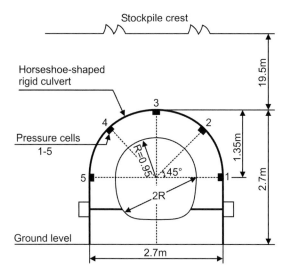

Figure 8.33 Cross-section of outlet conduit for Goi dam showing positions of pressure cells.

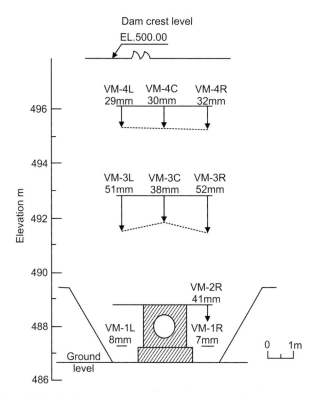

Figure 8.34 Settlement measured on cross-arms at Gennaiyama dam.

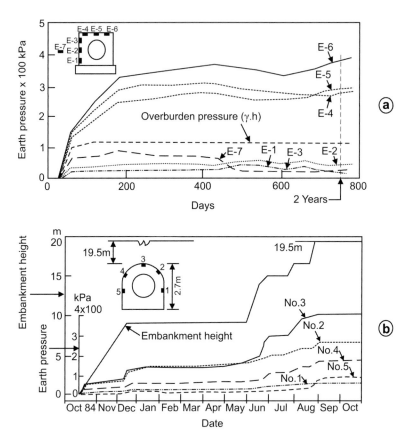

Figure 8.35 Pressures measured on top and sides of outlet conduits at (a) Gennaiyama dam (b) Goi dam.

which behaved as if it had a positive projection. Horizontal pressures measured at E − 1, E − 2 and E − 3, on the vertical side of the box were only of the order of 30 kPa and were very similar to the vertical pressure reading of pressure cell E − 7 "floating" in the backfill away from the side of the box. Hence the soil near the sides of the box was subject to a near–isotropic total stress of only 30 kPa. (The axial horizontal stress was not measured.) Thus, a pore pressure in this region of only 3 to 4 m of water head (30 to 40 kPa) would have reduced the effective stress to zero. In other words, if

$$\frac{\sigma_h}{\gamma_w h} < 1 \tag{8.8}$$

a danger of hydraulic fracture would exist, where h is the hydraulic head. In the case of Gennaiyama, both horizontal and vertical stresses were very, very low in proximity to the conduit and a distinct danger of hydraulic fracture on first filling, leading to failure by piping, must have existed.

For Goi (Figure 8.35b), the stresses measured 10 months after the start of embanking for the crown of the conduit were as follows: No. 3: 390 kPa, No. 2: 270 kPa,

No. 4: 190 kPa. For the vertical sides of the conduit, the measured stresses after 10 months were: No. 1: 60 kPa, No. 5: 90 kPa. Vertical stresses in the soil near the sides of the horse-shoe were not measured, but it is very likely that they were also in the range of 60–90 kPa. (Strangely, the authors do not say in the paper that the two dams failed as a result of piping along their outlet conduits. It must be assumed, however, that both dams did fail and that the facts set out in the paper formed parts of the post-failure investigations.)

8.7.2 Cut-off trench, Lesapi dam, Zimbabwe

Clay-filled cut-off trenches for water storage dams may also be vulnerable to piping failure after hydraulic fracture. Figure 8.36 (Blight, 1973) shows a cross-section of the clay cut-off trench for the Lesapi dam in Zimbabwe. During construction there was concern that hydraulic fracturing, followed by piping might occur in the narrow lower

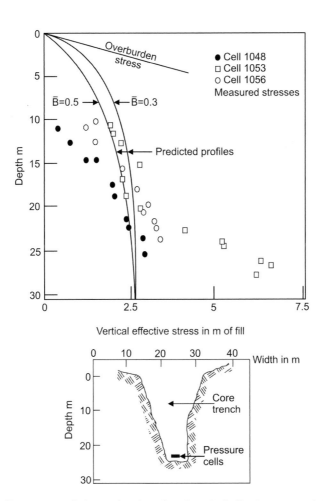

Figure 8.36 Comparison of observed and predicted vertical effective stresses in Lesapi dam.

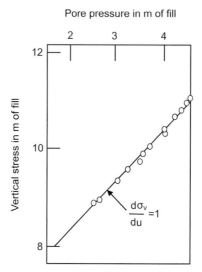

Figure 8.37 Measured relationship between pore pressure and vertical stress in the base of the core trench at Lesapi dam.

section of the trench which was only 5 m wide within hard rock walls. Three pairs of pressure cells and piezometers were installed in this zone at various plan positions along the length of the cut-off trench, as indicated in Figure 8.36. The soil backfill was a compacted sandy clay residual from granite.

The three pressure cells were all set horizontally at the same depth in the trench, and each pressure cell was accompanied by a double tube piezometer. The predicted vertical pressure profiles were calculated by silo theory for a value of K (horizontal to vertical pressure ratio) of 2, which value had been estimated by means of *in situ* shear vane tests in the adjacent compacted backfill. (See section 7.3.3., equation 7.8d.)

Figure 8.37 is a plot of measured pore water pressure versus the corresponding vertical stress recorded by the pressure cells and piezometers after filling of the reservoir. This shows that each increment of pore water pressure was accompanied by an equal increment of vertical stress and presumably an equal increment of horizontal stress. As the vertical stresses, although reduced by silo action within the core trench, and much less than overburden stress, were relatively large (see Figure 8.36), and the horizontal stresses, as a result of the heavy compaction, were about twice the vertical stresses, it would not have been possible to reach a condition of zero effective stress in the core and hydraulic fracture could also not occur.

In the 40 years since the dam was first filled there have been no problems attributable to excessive seepage through the cutoff trench.

8.7.3 Concrete spillway, Acton Valley dam, South Africa, soil to concrete interfaces

The Acton Valley dam is a relatively small farm water storage dam with a height of 14 m. It is constructed of compacted soil residual from shales and sandstones, having

a clay core of residual shale and flanks of residual sandstone. The dam has a central concrete spillway with an embankment crest level of 4.6 m above the spillway crest. It was designed by international consultants famous for their work on the Sydney opera house. Figure 8.38 shows details of its layout and construction.

The first signs of the leak that developed into a piping failure, became apparent during first filling on 10 January 1987, shortly after the water in the reservoir had covered the concrete slab on the crest of the spillway, but before the water spilled over the crest. The sound of running water could then be heard at the training wall on the right flank of the spillway, near the crest. Hence it appears that the leak must have originated not far below the spillway crest under a head of a few hundred millimeters of water at the most.

Figure 8.38 shows a reconstruction of the flow paths taken by the water. Section AA shows the main flow path and plan BB shows the two entry points that either formed simultaneously or sequentially.

The first point of entry was under the key beam at the upstream edge of the spillway crest slab about 12 m from the training wall, and took its course along the underside of this slab and towards the training wall. Plate 8.1 shows two men standing at this entry point. The second point of entry was under the edge of the crest slab close to the training wall. A whirlpool appeared in this area in an early stage of events leading to the failure. (For clarity of the drawing, Figure 8.38 omits the baffles shown in Plate 8.1.)

The two flow channels appear to have joined below the vertical training wall, as shown in plan B-B in Figure 8.38 and eventually gouged out a large erosion cavity behind and against the spillway training wall. The exit of the flow path or rathole adjacent to the training wall is shown in Figure 8.38 and Plate 8.2 which shows the erosion cavity downstream of the rathole as well as the vertical side of the training wall and the underside of the inclined splash slab. The entry point shows as a slit of daylight at the far end of the splash slab. The fact that the leak originated under such a low head indicates that the water found paths along pre-existing voids or zones of loose uncompacted soil.

There are several likely positions for such voids to have formed or to have pre-existed the failure.

Figure 8.38 shows that the spillway slab bridges over the compacted clay core of the dam [marked (CL) for clay of low plasticity]. The downstream edge of the slab rests on the top of the filter drain which is backed by the downstream clayey sand shell [marked (SM)]. The clay core would be expected to settle more than the shells that flank it, because clayey soils tend to be more compressible than sandy soils. There is no detail at the junction of the spillway crest slab and the spillway apron to accommodate this differential settlement, hence a void could be expected to form under the spillway crest slab as the clay core settled away from it. This would have provided the path of flow from point of entry 1 in Figure 8.38.

Apart from the visual evidence of Plate 8.2, remedial grouting under the spillway slab in 1988 and 1989 confirmed that an extensive void had formed under the spillway crest slab as a result of unaccommodated differential settlement.

The L drain shown on the drawings for the dam was probably provided to intercept any small leak, such as the initially small flow that eventually resulted in failure. However, the drain did not fulfil its function. This was probably because the void that

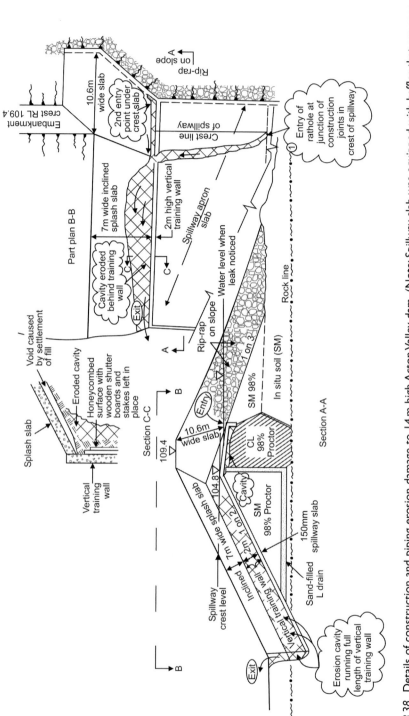

Figure 8.38 Details of construction and piping erosion damage to 14 m high Acton Valley dam. (Note: Spillway slab was equipped with baffles that are seen in Plate 8.1, but omitted from Figure 8.38).

Plate 8.1 Entry points of erosion pipes in Acton Valley dam.

initiated the leak allowed the flow to pass over the top of the drain and/or because entry of water to the drain had been prevented by carelessly allowing it to be blocked by grout during concreting the spillway slab.

In the area behind the training walls the L drain had no chance to fulfil its function, as the flow paths bypassed the drain, cutting in to behind the training walls downstream of the drain.

The height of the dam along the centerline of the embankment changes abruptly by 4.6 m where the spillway crosses the dam. As a result, differential settlement has occurred between the spillway crest and the embankments on either side. At the end of the spillway where the failure occurred, a settlement differential of 50 mm had developed between the end of the spillway crest and a point 12.25 m distant (i.e., in line with point of entry 1) by September, 1987. This differential movement, which had hogged the crest of the spillway, would have opened up the transverse joints in the spillway crest slab, and possible facilitated the formation of entry point 1.

In addition to the dragging down of the ends of the spillway, the rock on which the lower ends of the training walls are founded would have settled less than the fill below

Plate 8.2 (View from right). One of the erosion cavities under spillway of Acton Valley dam.

the training walls. There is no provision in the detail of the junction of the splash slabs and the training wall to take up this differential movement. Therefore a void would have formed under the splash slabs and close to the training walls. The type of void is illustrated in section C-C of Figure 8.38.

After inspecting the dam in September 1989, and noting evidence of continuing differential settlement between sections of the training walls on both sides of the spillway, the first author telephoned the owner of the dam to warn him that voids, as yet undiscovered, would probably also exist under the splash slabs on the other side of the spillway. On investigation, this was found to be the case. Not only had settlement resulted in a void under the splash slabs, but severe hidden erosion damage had already occurred on the other side of the spillway as well.

Section CC depicts the differential settlement void discovered between the underside of the splash slab and the backfill, on the left of the spillway. It is an exactly similar void that is thought to have occurred under the splash slabs to the right of the spillway where the main piping failure occurred.

The erosion void behind the training wall also revealed that the back of the wall was honeycombed in places and it was found that, wooden shutter boards and stakes had been left in place behind the wall, and backfill placed against them. It is very likely in view of this carelessness, and the obstructions against the wall that the backfill against the wall was not well compacted and was certainly not in contact with the concrete surface. Hence the distance between the initial rathole and the void under the splash slab was probably riddled with voids, honeycombing in the concrete and poorly compacted fill against the back of the training wall.

Soil to concrete contacts are notoriously prone to the development of leaks and piping erosion in the abutting soil, yet the importance of careful preparation of the back of the training wall to receive the fill and the importance of good compaction against the wall was not noted on the drawings. There was also no specific warning on this point in the specification.

The fact that similar erosion damage occurred on both sides of the spillway is an indication that the original failure was not just an unfortunate accident, but arose from faulty design and specification. The designers failed to recognize that differential settlement would occur at various points in the dam, and failed to recognize the consequences of such settlement. As a result the details of the design were not such as to accommodate differential settlement. The designers also appear to have failed to recognize the importance of meticulous preparation of concrete surfaces against which fill would be compacted and the importance of good compaction against such surfaces. Neither preparation for receiving backfill, nor the importance of good compaction against concrete surfaces is emphasized on the drawings or in the technical specification.

This case history had a sad ending. Unable to obtain compensation for the damage to the dam and the crops lost because of the loss of irrigation water, the farmer battled on under a huge debt burden, but was, in 2006, eventually forced to sell his farm, which his family had owned and developed from virgin land and worked since 1866.

8.8 THE STABILITY OF SLOPES OF RESIDUAL SOIL

8.8.1 General comments and observations on the stability of slopes in residual soil

A natural slope in residual soil will usually have formed by a process of erosion that has, over millennia, removed weathered material from the upper reaches of the slope and deposited it on the lower slopes. If undercutting of less weathered zones or corestones has occurred, minor slumps will accompany the process of slope degradation. Figure 8.39 shows measurements of soil transport on a natural slope formed on a very ancient, 4200 million year-old micaceous schist (or greenstone from its greenish-hued talc content). The slope is covered with grass, bushes and small trees. As the measurements show, the surface of the slope is not stable, but is gradually changing profile, with the upper slopes becoming steeper and the slopes at the toe, formed of soil transported from above, gradually flattening. The upper part of the slope is eroding at rates approaching 50 T/ha/y at the top, reducing to zero at about half the slope length (1 T = 10 kN). The lower part is accreting at much the same maximum rate. The area under the erosion rate curve (in $Tm^2/ha/y$) is slightly less than that under the accretion rate curve, but erosion and accretion are close to being in balance. The measured difference of 150 $Tm^2/ha/y$ in 1300 $Tm^2/ha/y$ (11.5%) probably arose because erosion occurring from the slope above the boulders was not recorded, but contributed to the accretion. Eventually, the upper slope will flatten by sliding, establishing a drastically different profile, and the erosional transportation process, steepening the newly established upper slope, will continue.

The point is that natural slopes in residual soil are inherently unstable and, if disturbed by more than a small amount, will slide. Plate C27 (which has featured on the

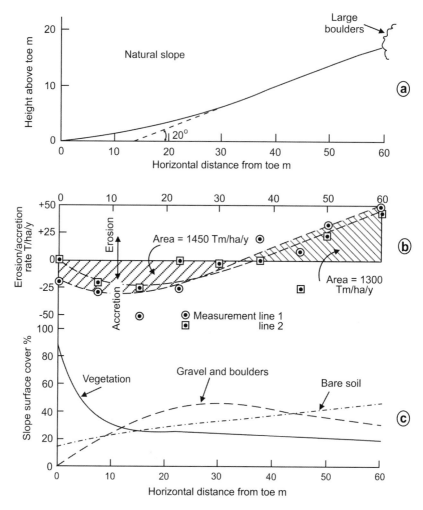

Figure 8.39 Erosion of a natural residual soil slope in South Africa. (a) profile of slope, (b) distribution of erosion/accretion, (c) surface cover of slope.

cover of editions 1 and 2 of this book) shows a slide in a slope of residual shales. A small cut-to-fill platform had been constructed to build the timber house shown in the picture. The disturbance was sufficient to cause a slide that half-demolished the house during a short wet period. After building a series of retaining walls to locally stabilize the site, the house was completed, and 16 years later stands, apparently stable. However, it is quite possible that the next extremely wet period may re-activate the slide.

Plate C28, taken in 1963, shows one of a series of slides that occurred in a natural slope of residual mudstone (see Figures 8.48 and 8.49). Many of the houses built on the slope which has a sea view, were 60 to 70 years old at the time, and the evidence of cracked and repeatedly repaired steps, garden and house walls, etc. showed that the

slope had been in limiting equilibrium with slight seasonal movement, all that time. 1964 was the wettest year in the past 50 years, with the result shown in Plate C28. 45 years later, 2009 was another very wet year and the first author received a call to say that most of the houses demolished or damaged in 1964 had, shortly thereafter, been repaired or rebuilt and in 2009 were still owned or occupied by the children or adult grandchildren of the owners in 1963. Much the same damage had recurred with history repeating itself two generations later. Once again, the houses have been repaired or rebuilt.

Cuts in slopes of residual soil may apparently be stable for many years, even though they must always be steeper than the original slope. As shown by Plate C29, a cut in residual soil removes the near-surface, highly weathered soil and exposes less weathered soil beneath. This may ensure temporary stability, but as the strengthening effect of the interparticle bonds and possibly also negative water pressures caused by dilation of the soil, are gradually destroyed as weathering of the soil continues (see Figure 7.11), the stability decreases. This ensures that most slopes cut in residual soils, without flattening the whole slope, will eventually fail, also as shown by Plate C29.

Another example of long-term instability can be seen in Plate C30 which shows a failure in a steep (1:1) railway cutting in weathered andesite lava. The cutting had been made in the late 1920s and had stood for well over 70 years when two adjacent failures developed, one on each side of the railway. The failures were deep-seated and heaved up the rail-bed, completely blocking the single line. Because repair was urgent, an attempt was made to effect a repair by constructing a reinforced soil retaining wall, with the even more disastrous effects shown in Plate C31. A simple surveyed cross-section made after this disaster, clearly showed that the mass of soil retained by the hastily constructed reinforced "remedial" wall had added to the disturbing weight and was not the stabilizing buttress it was thought to be when building began.

8.8.2 The largest ever failures of slopes in residual soils

In terms of volume of soil mobilized, the largest ever failure in slopes of residual soil took place at Vajont in northern Italy, in 1963. A 276 m high concrete arch dam had been constructed across a narrow gorge through which the Vajont (often spelt Vaiont) river flowed. It was completed in 1960. The estimated capacity of the reservoir was 169 million m^3. In 1959 concerns were raised about the stability of the hillside slopes forming the sides of the reservoir, and detailed analyses of their stability were undertaken. The various analysts agreed that a stability problem existed, but disagreed as to the seriousness and extent of the problem (Kiersch, 1964; Muller, 1964, 1987). As a result, a system of monitoring instruments was installed on the slopes in 1960 and it was decided to carefully control filling of the reservoir in conjunction with measurements of surface movement of the slopes. This was done, but the problem was that the limiting amounts and rates of movement that would be dangerous could not be identified.

On 9 October, 1963, the southern hillside suddenly slid over a length of 2 km and, moving at a speed estimated at about 25 m/s (90 km/h) an estimated volume of soil and weathered rock of 275 million m^3 moved into and over the reservoir to come to rest 140 m above the full supply level on the opposite hillside. As the slide debris entered the reservoir, it displaced the water and slopped a wave of water (estimated at 250 m high)

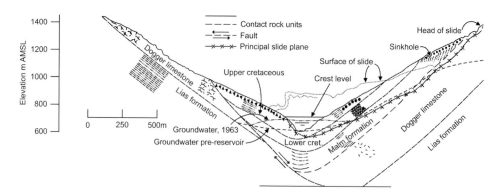

Figure 8.40 Vajont slide – section across reservoir showing principal slide surface and extent of slide.

over the crest of the dam. The dam itself was undamaged, but the flood wave destroyed five villages downstream, killing 2040 people. Figure 8.40 shows a section across the reservoir, indicating the main slide surface and the top surface of the slide debris.

The next biggest slope failure in residual soil and weathered rock was the slide that was triggered by the cutting of the Culebra cut during and after the construction of the Panama canal (McCullough, 1977). The Culebra cut takes the Panama canal through the highest point of the Isthmus of Panama, at a water surface elevation of 25.9 m AMSL. Excavation of the cut, firstly by the French Compagnie Universelle du Canal Interoceanique, fronted by Ferdinand De Lesseps, and later by the US Corps of Engineers gave problems from first to last. These problems continue a century later as the canal is being widened. The cause of the problem is a formation of weathered clay shale, the Cucaracha shale, which collapsed into the cut with terrifying frequency and debilitating cost. Plate 8.3, photographed from a newspaper of the time shows a steam shovel almost buried in slide debris. Figure 8.41 shows (above) the planned profile of the cut, the extent to which the French managed to excavate, starting in January 1882, before the Canal company went bankrupt in February 1889, as well as the volume of excavation by the US before the canal was opened to traffic in 1914.

The lower part of Figure 8.41 shows profiles of the cut as the side slopes successively sloughed away. The diagram shows the extent to which the walls of the cut had flattened 33 years later, in 1947. Sporadic slides still occurred until recently. An American who worked on the cut during construction of the canal, recorded the following reminiscences (McCullough, 1977).

"The whole top of the hill, is covered with boiling springs", he recalled. "It is composed of a clay that is utterly impossible for a man to throw off his shovel once he gets it on. He has to have a little scraper to shove it off." Nothing they had tried had kept the hill from sliding. "It won't stay there . . . ."

Why? he was asked.

"The rainy season will saturate the earth and it will slough off."

"Did it do so while you were there?"

"Yes, we had a cut right alongside of where the canal was going to be built and it sloughed off, not only over the top of our track, but we found it was going to be so expensive to move it that I cut the track away there and laid another one. And a

Plate 8.3 Steam shovel half-buried by slide of Cucaracha shale, Panama canal, 1912.

year or so afterwards the same thing took place and I laid another track, and where the present track is, there are two underneath."

"... when I was there at Culebra that week, my house was up on the hill about four hundred to five hundred feet from the canal and I got up one morning and come out and the land had gone off and left a crack there two to three feet wide, and I did not say anything, but I knew what it was .... The whole side of that mountain is going down into the canal. Every rainy season, whenever it rains a little, the earth becomes saturated and it slides right off on this strata of blue clay."

"It slides on the blue clay?"

"It slides on the blue [Cucaracha] clay."

8.8.3 Specific comments and observations on various causes of instability

The effects of unusually heavy and prolonged rainfall

The effects of unusually wet weather on the stability of slopes in residual soils have been examined by many researchers, including Lumb (1975), Brand (1982) and

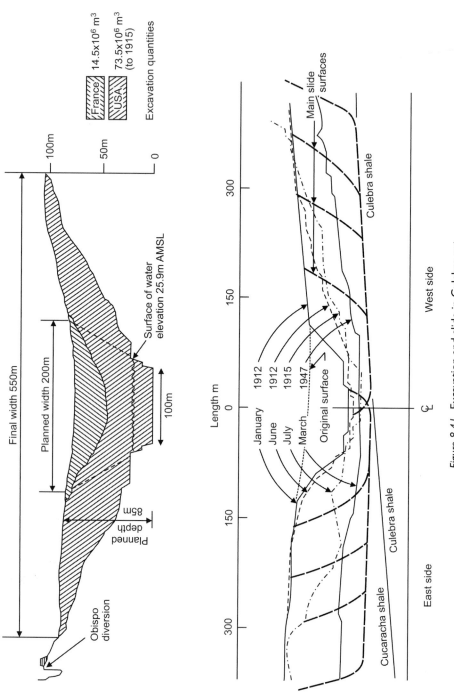

Figure 8.41 Excavation and slide in Culebra cut.

Lim *et al.* (1996). Many natural slopes of residual soil exist in an unsaturated condition and their margin of safety against sliding depends on the capillary tensions that exist in the pore water and enhance the strength of the soil. In a natural slope of weathered mudstone (Blight *et al.*, 1970), the capillary tension was found to vary, after light rain, from zero at the surface to 1000 kPa at a depth of 1 m. Brand (1982) and Lim *et al.* (1996) have observed suctions in residual soil slopes in Hong Kong and Malaysia. At shallow depths, suctions have been observed to decrease to zero during prolonged rain. Infiltration during prolonged rainfall can reduce capillary tensions to a point where the slope becomes unstable.

Lumb (1975) suggested that the limiting rate of infiltration of rain into a homogeneous soil in the absence of surface ponding is numerically equal to the saturated permeability of the soil. The water advances into the soil as a wetting front which travels at an approximate velocity of $v = k/(1 - S)n$ where k is the permeability, S the initial degree of saturation and n the porosity. Using typical values for soil permeability and rainfall intensity, Lumb showed that such a wetting front could reach a potentially critical depth in a slope (such as the contact between soil and rock) within a few hours. Open cracks and fissures in a soil have the effect of accelerating the advance of such a wetting front.

One of the most extensive series of storm-associated slides on record was that which occurred in the Serra das Araras district of Brazil in 1967 (Da Costa Nunez, 1969). During and following a single night in which the rainfall intensity reached 70 mm/hour, an area 24 km long and 7.5 km wide was devastated by a series of landslides that killed an estimated 1000 people and caused untold damage to property.

Van Schalkwyk & Thomas (1991) record a similar set of circumstances in the KwaZulu Natal province of South Africa during 1987. Rainfalls totalling as much as 800 mm fell during a period of 4 days resulting in a loss of life of 380 and damage to property and infrastructure equivalent to 500 million (1991) U.S. dollars. During this period, 211 slope failures occurred that damaged housing, roads and railways. In almost every case, the failures were associated with slopes in residual soils that had been subject to man's interference by cutting, terracing or constructing cut-to-fill road or railway alignments.

Effects of seismic events

Yamanouchi & Murata (1973) describe a number of failures in slopes of shirasu (a residual volcanic soil) that occurred during the 1968 Ebino earthquake. As shirasu is a relatively rigid, brittle material, the earthquake loading produced multitudinous shear cracks in the soil which caused slabs of material to slough off. Although natural slopes appear to have been affected, cut slopes suffered more severe damage.

Human interference

There are many types of human interference that may affect the stability of natural slopes in residual soil. Of these the following are probably the most common:

- Removal of toe support by cutting or by erosion may precipitate a failure. The introduction of a cut at the toe of a long natural slope may precipitate a slide. The slide that occurred at Bethlehem, South Africa (Blight *et al.*, 1970) is a typical

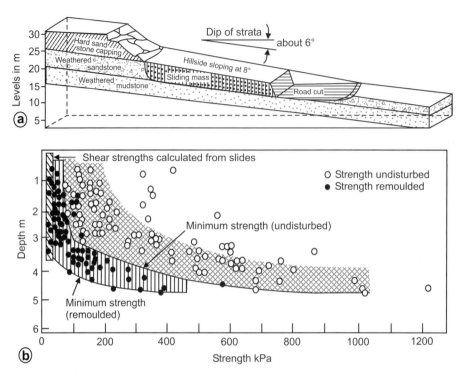

Figure 8.42 (a) Sectional block diagram showing the geology of the slide at Bethlehem. (b) Vane strength measurements in weathered mudstone at Amsterdamhoek.

example of this type of failure (see Figure 8.42a). The slide occurred when a shallow road cut was made in a hillside which slopes at a gentle 8°. On investigation it was found that the slide involved a block of weathered sandstone that was sliding on its contact with an underlying stratum of weathered mudstone. Conditions for failure were exacerbated by a high water table and the existence at the sandstone-mudstone contact of a concentration of illite and montmorillonite clays leached out of the sandstone layer.

Material removed from the toe of a slope by erosion can also cause a previously stable slope to become unstable (e.g., Yamanouchi & Murata, 1973). (But see the converse effect caused by erosion at the head of the slope, described by Figure 8.39.)

- Changes in the soil water regime of a slope may cause instability:

    If the soil water regime of a slope is drastically altered by irrigation, the removal of vegetation or partial inundation by impounded water, instability may ensue (Richards, 1985). The most spectacular and tragic recorded example of instability by raised water levels in a natural residual soil slope is the failure in the Vajont valley described in section 8.8.2.

- The effect of deforestation may also affect the stability of natural slopes: The following appear to be the more important factors related to deforestation: If deep-rooted vegetation is removed, capillary tensions will be reduced and the phreatic

surface within a slope will probably rise (e.g., Blight 1987). Roots mechanically reinforce the soil. The stability of a slope will decrease as the root systems decay after removal of surface vegetation. The relative importance of the above factors is dependent on the climatic environment. In a humid climate rotting of the root system may be rapid and decisive in producing instability. Under semi-arid conditions, however, the reduction in evapotranspiration rates caused by deforestation may outweigh all other effects.

Failures of cut slopes in residual soils

Many of the agencies that promote failure in natural slopes (usually heavy rainfall, seismic activity, toe erosion, changes in soil water regime) may also affect the stability of cut slopes especially in the long term. However, the stability of cut slopes appears to be dominated by the structural features of the soil.

Residual soils that do not contain pronounced structural features such as relict jointing, bedding or slickensiding may sometimes be cut to remarkably steep stable slopes. For instance, Wesley (1973) cites several stable slopes cut by river erosion into a latosol clay. These slopes have heights of about 10 m and inclinations of up to 70°. (see Figure 8.43). He also mentions a 45° cut in similar material which is 20 m high and has been stable for over 40 years. Cut slopes in loess also show remarkable stability, stable vertical cuts of 5 to 6 m in height being quite common (Mitchell, 1976).

Woo et al. (1982) list cut and natural slopes in cemented gravel deposits in north west Taiwan. These gravels are ancient alluvial deposits that have been lateritized

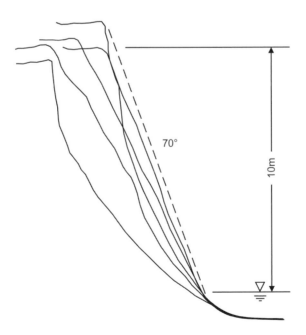

Figure 8.43 Cross sections along stable red clay bank (after Wesley, 1973).

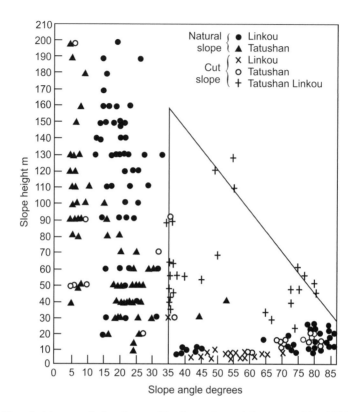

Figure 8.44 Slope height vs angle for slopes of ferricreted gravel deposits in North Western Taiwan.

in situ. Woo *et al.*'s observations are summarized in Figure 8.44. As the gravels probably have angles of shearing resistance exceeding 35° and have been cemented by the lateritization, it is only the cases that fall within the triangle marked in Figure 8.44 that may be unexpectedly steep, and most of these are relatively low slopes.

The stability of the slopes illustrated in Figures 8.43 and 8.44 can be explained by invoking the effects of pore water capillary tensions. Blight (1980) showed that a vertical slope with the water table at its toe can theoretically stand stably to any height if the capillary stresses within the slope are in static equilibrium with the water table. This applies whether or not the soil is cohesive in effective stress terms and is clearly how many such slopes remain stable over long periods of time.

Slides caused by the presence of hidden joint surfaces, slickensides, etc., are very difficult to predict or design against as these features are generally not visible in borehole specimens and may be difficult to locate, measure or map even in test pits or trenches. Even if potentially dangerous features are identified during site or road route exploration, it is usually practically impossible to assess their extent, frequency, variation in dip and other factors that would be needed to design a cut slope (e.g., Sandroni, 1985).

St. John *et al.* (1969), mention a number of failures in cut slopes that were attributable to the presence of structural weaknesses. In one series of incidents a 5^1/$_2$ km

stretch of road traversing a mountainous volcanic tuff area of Puerto Rico suffered 40 separate slides in cuts ranging from 6 to 25 m in depth. The slides were of two basic types which are illustrated in Figure 8.45. In the first type (Figure 8.45a) sliding took place along slickensided surfaces that were flatter than the cut slope. In the second type (Figure 8.45b), sliding occurred on intersecting slickensided surfaces that were steeper than the cut slope. In all cases, stress relief caused by opening the cuts, allowed fissures and joint planes in the soil to open. All the slides occurred during rain when these open clefts filled with water. As the plans and sections sketched in Figure 8.45 show, the failures were three-dimensional, the sliding mass generally being bounded by intersecting structural features. Three-dimensional wedge-type failures in which the sliding mass is bounded by intersecting structural features are typical of cuts in saprolitic residual soils.

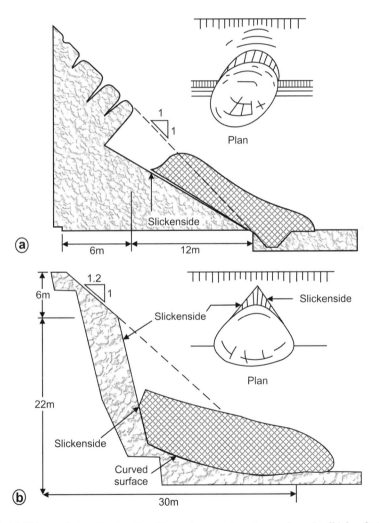

Figure 8.45 (a) Slide on slickensided surface flatter than cut slope in weathered tuff (after St. John *et al.*, 1969). (b) Slide on intersecting slickensided surface steeper than cut slope in weathered tuff (after St. John *et al.*, 1969).

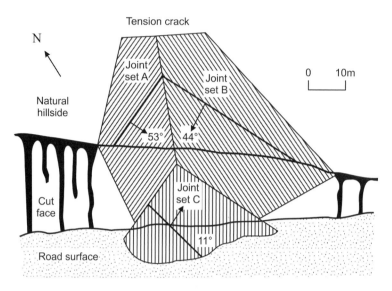

Figure 8.46 Plan showing three dimensional nature of slip in weathered diabase (after Pells & Maurenbrecher, 1974).

Another example has been described by Pells & Maurenbrecher (1974) (see Figure 8.46) in which a wedge of material slid out of an 18¹/₂ m high cut in weathered diabase. The failure occurred over a period of weeks at the height of the dry season and the pre-existing sliding surfaces were found to be covered by a hard polished layer of a hornblende-like secondary mineral. The completely three-dimensional nature of this failure is clearly shown by Figure 8.46. Plate C32 shows a similar failure with two almost intact wedges of soil that moved out of the slide in the same way as shown in Figure 8.46.

8.8.4 Design of slopes and analysis of slides

Because of the unknown and potentially treacherous nature and effects of the structural discontinuities in a residual soil, design is difficult to perform on a completely rational basis. For this reason, many building excavations in residual soil are completely supported, the support system being designed for a nominal coefficient of earth pressure (e.g., Flintoff & Cowland, 1982). In many cases these supported slopes would stand unsupported, but the economic and safety implications of an unexpected slide may be vastly more serious than the cost of providing the support. In less confined situations such as road or rail cuts, slopes are often cut to a nominal angle and if slides occur, ad hoc remedial measures are taken (not always successfully, as shown by Plate C31). As the surfaces exposed by a slide in residual soil are often stable even though steeper than the original cut face (e.g., Figure 8.45) remedial measures may consist simply of clearing away the fallen debris.

In other cases drainage measures of varying complexity may be required or the provision of toe support in the form of retaining walls of various types. Da Costa Nunes (1969) describes the use of gravity, counterfort, cantilever and crib walls.

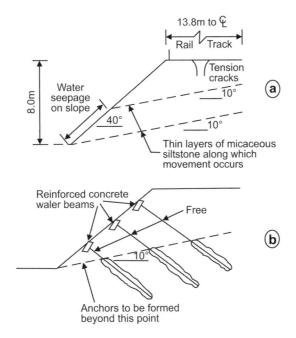

Figure 8.47   (a) Slip in a cut in residual micaceous siltstone. (b) Stabilization by means of stressed soil anchors (after Wagener & Neely, 1975).

Gabion or reinforced earth walls are other possibilities. In certain cases it may be necessary to tie back the slope using stressed soil anchors (e.g., Wagener & Neely, 1975). A section through this potential slide showing the preventive measures taken appears in Figure 8.47.

  If rational design and analysis methods are to be applied to slopes in residual soils, a decision has to be made as to what shear strength the soil has in bulk, so that this value may be used in the design. Lumb (1975) suggested a design procedure for cuts in which it is assumed that either:

(a)   stability is controlled by the cumulative peak strength over the whole potential failure surface,

        or

(b)   stability is controlled by the least peak strength anywhere along the potential failure surface. This procedure provides bounds within which the actual soil behaviour should lie. Because of the effects of seasonal wet weather, peak strengths are established in terms of effective stresses for saturated conditions.

  There is, however, considerable evidence (e.g., Burland *et al.*, 1966; Blight, 1969) that the shear strength of a stiff fissured material in bulk corresponds to the lower limit of strengths measured in small-scale laboratory or *in situ* tests (see, e.g., Figures 8.27, 8.41b and 7.41b). This was also shown by Blight (1969) and Blight *et al.* (1970) by back analysing failures of four waste rock dumps founded on residual soil profiles, and in relation to series of slides that took place in a mantle of weathered cretaceous

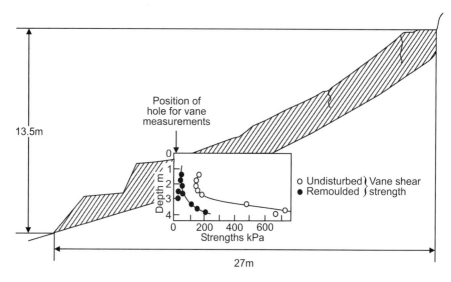

Figure 8.48 Detailed section through slide at Amsterdamhoek showing typical vane strength profile.

mudstone at Amsterdamhoek, South Africa (Figures 8.48 and 8.49). A section through the slide and a typical vane strength profile is shown in Figure 8.48. Figure 8.49 shows the shear strengths, back-calculated from several slides at Amsterdamhoek, plotted as a strength versus effective stress diagram. These values agree reasonably well with reversed shear box tests done in the laboratory. A similar principle can be shown to apply to three residual soils in Hong Kong by re-analysing data published by Malone & Shelton (1962). The data are shown in the form of strength diagrams in Figure 8.50. It will be noted that in each case the lower limit to the strength measured in the laboratory (triaxial shear) corresponds to the lower limit to strengths back-figured from the slides. The analyses assumed zero pore pressure in the slopes ($r_u = 0$) whereas there was probably a small suction present. Hence the mean effective stress σ' in Figure 8.50 has probably been under-estimated. If the lower limits to the strength measured in the laboratory had been used to analyze the various slopes a conservative but realistic result would have been obtained.

8.8.5 Types of failure of natural slopes

Many attempts have been made to classify slides in natural slopes of residual soil. Classifications depend on two main characteristics – the geometry or shape of the sliding soil mass and the velocity of its motion.

Geometry of slides

There are two main types of slide geometry. If the residual soil mantle is shallow in comparison with the length of the slope, a planar slide may result. Commonly,

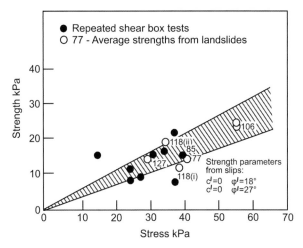

Residual strength characteristics of cretaceous mudrock

Average shear strengths calculated for five
landslides at Amsterdamhoek

Location of slide	Average shear strength kPa
House 77	15
House 85	16
House 106 (i)	24
House 106 (ii)	21
House 118 (i)	13
House 118 (ii)	19
House 127	15

Mean: 18 kPa

Figure 8.49 Shear strengths back-calculated for a series of slides in a weathered cretaceous mudrock.

the soil mantle slides over the underlying rock surface. The thickness of the sliding mass is usually roughly constant and the failure surface may be plane, slightly convex or slightly concave. The Canaleira slide described by Vargas & Pichler (1975) (see Figure 8.51) and the slide at Vajont (Figure 8.40) could be taken as typical planar slides. Other typical planar or near-planar slides in natural residual soil slopes have been those at Bethlehem (Figure 8.42) and the potential slide shown in Figure 8.47.

The second typical slide geometry occurs when the residual soil mantle is deep and a rotational or gouging slip or a block gliding occurs. The slide at Estrada do Jequia described by Morgenstern & de Matos (1975) typifies the rotational slide (Figure 8.52), while the Danube slide of a river bank in loess described by Kezdi (1969) typifies the block gliding failure (Figure 8.53).

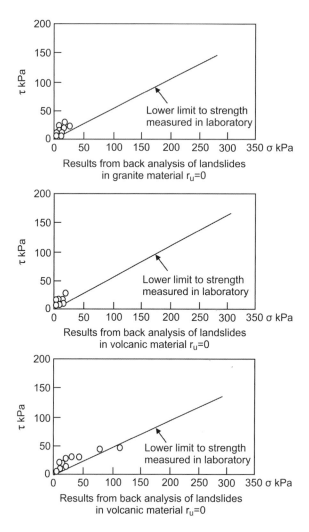

Figure 8.50 Comparison of lower limit to strength measured in laboratory with strengths back-figured from slope failures in three materials in Hong Kong (after Malone and Shelton, 1982).

Slide velocity

The velocity of a land-slide may vary from barely perceptible creep velocities to speeds of several kilometres per hour. Gray (1974) has recorded steady creep velocities in afforested slopes in Oregon U.S.A., averaging as little as 1mm/year, while a slope on the southern California coast quoted by Yen (1969) has been found to creep at 100 mm/year, and one in the Caucasus at 3 m/year (Ter-Stepanian, 1963). Creep velocities are by no means constant with time, but vary seasonally and may be accelerated by wet weather and retarded by drought. An acceleration of the creep rate may lead to

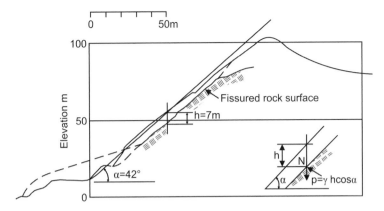

Figure 8.51 Section through a typical planar slide in residual soil at Caneleira, Brazil (after Vargas and Pichler, 1957).

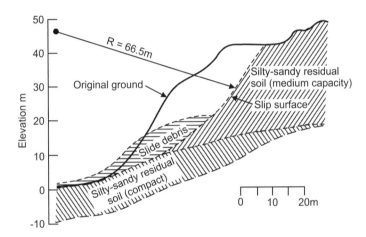

Figure 8.52 Section through a typical rotational slide in residual soil at Estrada do Jeqia, Brazil (after Morgenstern and de Matos, 1975).

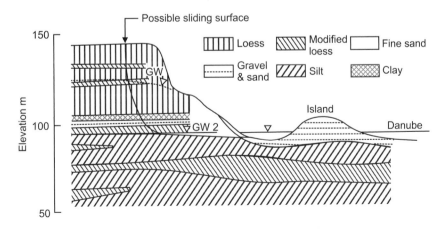

Figure 8.53 Section through unstable bank of river Danube (after Kezdi, 1969).

failure at a more rapid speed, for example 10 km/h as has happened in certain instances in Hong Kong (Lumb, 1975) and the enormous estimated speed of 90km/h of the slide at Vaiont (Muller 1964). Flow slides in residual soil that occur during heavy rain may travel considerable distances – a travel distance of over 5 km has been recorded by Da Costa Nunes (1969).

REFERENCES

Annandale, G.W. (1979) *Settlement of Embankments Constructed of Unsaturated Compacted Soil.* MSc(Eng) Dissertation, University of the Witwatersrand, Johannesburg, South Africa.

Barksdale, R.D., Bachus, R.C. & Calnan, M.B. (1982) Settlements of a tower on residual soil. *Eng. & Const. Tropical & Residual Soils, ASCE, Geotech. Divn. Spec. Conf.*, Honolulu, USA. pp. 647–664.

Blight, G.E. (1969) Foundation failures in four rockfill slopes. *J. Soil Mech. & Found. Eng. Div., ASCE*, 95 (SM3), 743–767.

Blight, G.E. (1973) Stresses in narrow cores and core trenches of dams. *11th ICOLD Int. Congr. Large Dams*, Madrid, Spain. pp. 63–79.

Blight, G.E. (1980) Partial saturation can assist the soil engineer. *6th S.E. Asian Conf. Soil Eng.*, Taipei, Taiwan. Vol. 1, pp. 15–29.

Blight, G.E. (1984) Uplift forces measured in piles in expansive clay. *5th Int. Conf. Expansive Soils*, Adelaide, Australia. Vol. 1, pp. 240–244.

Blight, G.E. (1987) Lowering of the groundwater by deep rooted vegetation. *9th Euro. Conf. Soil Mech. & Found. Eng.*, Dublin, Ireland. Vol. 1, pp. 285–288.

Blight, G.E. & de Wet, J.A. (1965) The acceleration of heave by flooding. *In: Moisture Equilibria and Moisture Changes in Soils Beneath Covered Areas.* Sydney, Australia, Butterworth. pp. 89–92.

Blight, G.E., Brackley, I.J. & van Heerden, A. (1970) Landslides at Amsterdamhoek and Bethlehem – an examination of the mechanics of stiff fissured clays. *Civ. Eng. South Africa*, June, 129–140.

Blight, G.E., Legge, G.H.H. & Annandale, G.W. (1980a) Settlement of a compacted unsaturated earth embankment. *Civ. Eng. South Africa*, 22 (2), 25–29.

Blight, G.E., Schwartz, K., Weber, H. & Wiid, B.L. (1980b) Preheaving of expansive soils by flooding – failures and successes. *7th Int. Conf. Expansive Soils*, Dallas, USA. pp. 131–136.

Brand, E.W. (1982) Analysis and design in residual soils. *Eng. & Const. Tropical & Residual Soils, ASCE, Geotech. Div. Spec. Conf.*, Honolulu, USA. pp. 89–143.

Brand, E.W. & Phillipson, H.B. (1985) *Sampling and Testing of Residual Soils.* Hong Kong, Scorpion.

Burland, J.B. Butler, F.G.B. & Dunican, P. (1966). The behaviour and design of large diameter bored piles in stiff clay. Symp. Large Bored Piles. Instn. Civ. Engrs., London, pp. 51–71.

Collins, L.E. (1953) A preliminary theory for the design of under-reamed piles in expansive clay. *South African Inst. Civ. Engrs.*, 3 (11), 17–23.

Da Costa Nunes, A.J. (1969) Landslides in soils of decomposed rock due to intense rainstorms. *7th Int. Conf. on Soil Mech. & Found. Eng.*, Mexico City, Mexico. Vol. 2, pp. 547–554.

Donaldson, G.W. (1967) The measurement of stresses in anchor piles. *4th Reg. Conf. Africa Soil Mech. & Found. Eng.*, Cape Town, South Africa. pp. 253–256.

Flintoff, W.T. & Cowland, J.W., (1982) Excavation design in residual soil slopes. *Engineering & Construction in Tropical & Residual Soils, ASCE Geotech. Div. Spec. Conf.* Honolulu, USA. pp. 539–556.

Gray, D.H. (1974) Reinforcement and stabilization of soil by vegetation. J. Geotech. Div., ASCE., 100 (GT6), 695–699.

Jaros, M.B. (1978) The settlement of two multistorey buildings on residual Ventersdorp lava. *S.A. Inst. Civ. Engrs. Symp. on Soil-Structure Interaction*, Durban, South Africa. pp. 20–25.

Kezdi, A. (1969) Landslide in loess along the bank of the Danube. *7th Int. Conf. on Soil Mech. & Found. Eng.*, Mexico City, Mexico. 2, 617–626.

Kiersch, G.A. (1964) Vajont reservoir disaster. *ASCE Civil Engineering*. pp. 32–40.

Legge, G.H.H. (1970) Mulungushi and Mita Hills dams operation and maintenance. *10th Int. Conf. on Large Dams*, Montreal, Canada. pp. 269–290.

Lim, T.T., Rahardjo, H. & Change, M.F. (1996) Climatic effects on negative pore-water pressures in a residual soil slope. *4th Int. Conf. on Tropical Soils*, Kuala Lumpur, Malaysia. Vol. 1, pp. 568–574.

Lumb, P. (1975) Slope failures in Hong Kong. *Quart. Jour. Eng. Geol.*, 8, 31–65.

Malone, A.W. & Shelton, J.C. (1982) Landslides in Hong Kong 1978–1980. *Engineering & Construction in Tropical & Residual Soils, ASCE, Geotech. Div. Spec. Conf.*, Honolulu, USA. pp. 424–442.

McCullough, D. (1977) *The Path Between the Seas, the Creation of the Panama Canal, 1870–1914*. New York, Simon & Schuster.

Mitchell, J.K. (1976) *Fundamentals of Soil Behavior*. New York, Wiley. pp. 49–57.

Morgenstern, N.R. & de Matos, M. (1975) Stability of slopes in residual soils. *5th Pan Amer. Conf. Soil Mech. & Found. Eng.*, Buenos Aires, Argentina. Vol. 3, pp. 369–384.

Muller, L. (1964) The rockslide in the Vaiont Valley. *Rock Mech. & Eng. Geol.*, 2 (3–4), 148–212.

Muller, L. (1987) The Vajont catastrophe – a personal review. *Eng. Geol.*, 24, 423–444.

Ng'ambi, S.C., Shimizu, H., Nishimura, S & Nakano, R. (1999) Case studies on the failure mechanism of embankment dams along the outlet conduit due to massive leakage. In: Blight, G.E. & Fourie, A.B. (eds), *Geotechnics for Developing Africa*. Rotterdam, Wardle, Balkema. pp. 647–652.

Pavlakis, M. (1983) *Predictions of Foundation Behaviour in Residual Soils from Pressuremeter Tests*. PhD Thesis, University of the Witwatersrand, Johannesburg, South Africa.

Pells, P.J.N. & Maurenbrecher, P.M. (1974) Cutting failures near Waterval Boven. *Civ. Eng. South Africa*, 16 (5), 180–181.

Popescu, M. (1998) Engineering peculiarities of loessial collapsible soils along the river Danube lower course. *UNSAT '98*, Beijing, China. pp. 272–279.

Richards, B.G. (1985) Geotechnical aspects of residual soils in Australia. In: Brand, E.W. & Phillipson, H.B. (eds) *Sampling and Testing of Residual Soils*. Hong Kong, Scorpion. pp. 23–30.

Sandroni, S.S. (1985) Sampling and testing of residual soils in Brazil. In: Brand, E.W. & Phillipson, H.B. (eds) *Sampling and Testing of Residual Soils*. Hong Kong, Scorpion. pp. 31–50.

Skempton, A.W. & Bjerrum, L. (1957) A contribution to the settlement analysis of foundations on clay. *Géotechnique*, 7, 168–178.

St. John, B.J., Sowers, G.F. & Weaver, C.H.F. (1969) Slickensides in residual soils and their engineering significance. *7th Int. Conf. on Soil Mech. & Found. Eng.*, Mexico City, Mexico. Vol. 2, pp. 126–130.

Ter-Stepanian, G. (1963) *On the Long Term Stability of Slopes*. Norwegian Geotech. Inst., Pub. No. 52, Oslo, pp. 1–13.

van Schalkwyk, A. & Thomas, M.A. (1991) Slope failures associated with the floods of September 1987 and February 1988 in Natal and KwaZulu, South Africa. In: Blight, G.E., Wardle, J. & Fourie, A.B. (eds), *Geotech. in the African Environment*. Rotterdam, Balkema. pp. 57–64.

Vargas, M. & Pichler, E. (1975) Residual soil and rock slides in Santos, Brazil. *4th Int. Conf. on Soil Mech. & Found. Eng.*, London. Vol. 2, pp. 394–398.

Wagener, F. von M. & Neely, W.J. (1975) Stability of a railway cutting in micaceous silt stones. *6th Reg. Conf. for Africa Soil Mech. & Found. Eng.*, Durban, South Africa. Vol. 1, pp. 213–218.

Wesley, L.D. (1973) Some basic engineering properties of halloysite and allophone clays in Java, Indonesia. *Géotechnique*, 23 (4), 471–494.

Williams, A.A.B. (1980) Severe heaving of a block of flats near Kimberley. *7th Reg. Conf. for Africa on Soil Mech. & Found. Eng.*, Accra, Ghana. Vol. 1, pp. 301–310.

Woo, S.M., Guo, W.S., Yu, K. & Moh, Z.C. (1982) Engineering problems of gravel deposits in Taiwan. *Engineering & Construction in Tropical & Residual Soils, ASCE, Geotech. Div. Spec. Conf. Honolulu*, USA. pp. 500–518.

Yamanouchi, T. & Murata, H. (1973) Brittle failure of a volcanic ash soil – shirasu. *8th Int. Conf. Soil Mech. & Found. Eng.*, Moscow, Russia. Vol. 1, pp. 495–500.

Yen, B.C. (1969) Stability of slopes undergoing creep deformation. *J. Soil Mech. & Found. Div.*, ASCE, 95 (SM4), 1075–1096.

Subject index

Colour plates

C.1 (View from right) Near-surface profile of a weathered andesite lava.

C.2 (View from right) Deeper continuation of weathered andesite lava profile.

C.3 Residual profile of a diabase dyke.

C.4 Residual weathered andesite lava flow containing quartzite floaters.

C.5   Residual andesite lava profile.

C.6   Profile of residual volcanic ash showing saprolitic features.

C.7 Lateritic gravel formed in a profile of residual weathered shale.

C.8 Layer of calcrete formed in weathered cretaceous mudstone.

C.9 Profile of dolerite sill fragmented into blocks by weathering.

C.10 Profile of granite deeply weathered to a clayey sand.

C.11   Relatively shallow weathering of a mudstone profile.

C.12   Shale profile weathered to a stiff fissured residual clay.

C.13 Colour variability in a clay residual from shale.

C.14 River course that has cut through ancient weathered volcanic ash underlying a village buried in ash by a recent eruption.

C.15 Near-surface rock pinnacles formed by differential weathering of dolomite.

C.16 Damaged building constructed over differentially weathered dolomite.

C.17 "Bad-land" topography in Turkish Cappadocia.

C.18 Piping failure on first filling of rolled earthfill dam.

C.19 Extensive slickenside exposed in residual shale overburden of iron ore mine.

C.20 Slickenside forming part of failure surface in road cutting.

C.21 Rapid weathering of lava flow on Hawaii island.

C.22 Roman aqueduct in Toledo, Spain showing little sign of weathering after 2000 years of exposure.

C.23 "Salt road" in Namibian coastal town of Swakopmund.

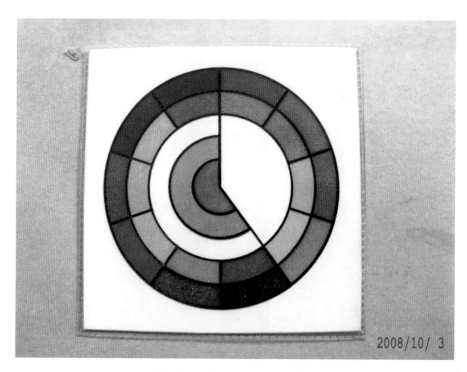

2008/10/ 3

C.24 Burland's pocket colour chart.

C.25 Example of variability in soil depth and composition of residual norite profile.

C.26 Post-construction undermining of cooling tower pond floor to relieve uplift by heave of weathered mudstone.

C.27 Shallow slide in residual shale slope cut into to found house.

C.28 One of a series of recurrent slides in slope of residual mudstone.

C.29 Failure of slope cut in residual granite in Malaysia.

C.30 Slide in 65 year old cut in residual andesite lava.

C.31 Abortive attempt to remediate failure of cut in residual andesite.

C.32 Three-dimensional wedge failure in residual diabase.